Springer Series in Reliability Engineering

Series editor

Hoang Pham, Piscataway, USA

For further volumes:
http://www.springer.com/series/6917

D. N. Prabhakar Murthy
Nat Jack

Extended Warranties, Maintenance Service and Lease Contracts

Modeling and Analysis for Decision-Making

Springer

D. N. Prabhakar Murthy
School of Mechanical and Mining Engineering
University of Queensland
Brisbane
Australia

Nat Jack
Dundee Business School
University of Abertay Dundee
Dundee
UK

ISSN 1614-7839 ISSN 2196-999X (electronic)
ISBN 978-1-4471-6955-0 ISBN 978-1-4471-6440-1 (eBook)
DOI 10.1007/978-1-4471-6440-1
Springer London Heidelberg New York Dordrecht

Printed on acid-free paper

Springer is part of Springer Science+Business Media (www.springer.com)

Dedicated to

Wallace R. Blishcke, in Memorium

D. N. Prabhakar Murthy

Rhona for her patience, understanding and support

Nat Jack

Preface

Engineered objects (products, plants, infrastructures) are built to meet the needs of individuals, businesses and societies. These objects are getting more complex, and also more expensive, to meet the ever-increasing demands.

The performance over time (reliability) of an engineered object is of great interest to its owners. Every object is unreliable in the sense that it degrades with age and/or usage and ultimately fails (when it is no longer capable of performing as expected). Maintenance actions are actions to control the degradation process and to restore a failed object to normal operation. These are termed preventive and corrective maintenance actions respectively.

Maintenance actions can be done by the owner (in-house maintenance) or by an external agent (outsourcing of maintenance). There is a growing trend towards outsourcing. In the case of products, a warranty (or more precisely a base warranty) requires the manufacturer to rectify any failure occurring within the warranty period as long as the owner operates as per the terms of the warranty. There is no cost to the customer as the warranty is integral to the sale and the manufacturer has factored the maintenance cost into the sale price. Customers can buy extended warranties either at the time of purchase or just before the base warranty expires by paying an additional amount. For plants and infrastructures, the owners can outsource some or all of the maintenance to an external service agent through a maintenance service contract. The contract specifies the tasks to be carried out by the service agent and the payments made by the owner to the service agent.

Maintenance outsourcing raises new challenges as it involves two (or more) parties each with several players and the objectives (or goals defined through outcomes) of each player are different. Each player has more than one choice and the decision of each player affects the outcomes of the others. If one assumes that the players are acting in a rational manner, they need to take into account these effects. A proper framework is needed to arrive at the optimal decisions for the parties involved.

Game theory provides the most appropriate framework for determining the optimal decisions (strategies) for the different players. An issue that plays a crucial role in obtaining the optimal decisions is the information available to the different players. This includes the usage profile of the object, the competence of service provider, assessing the condition of the object and so on. When there is asymmetry in information (different players having different information) this can lead to adverse selection (wrong choice of the service agent) or moral hazard (cheating by the owner of the object or the service agent providing the maintenance). These issues need to be addressed in determining the optimal strategies for the different players.

Over the last few decades, there has been a growing trend towards leasing rather than owning where the lessee (the user or operator) leases an object from a lessor (owner of the object) under a lease contract. In this case, the maintenance of the object can be the responsibility of either the lessee or the lessor depending on the terms of the contract. Here again we have two parties (with several players in each party) with different objectives or goals. Again, game theory provides a framework to determine the optimal decisions with players acting rationally.

This book deals with three topics—extended warranties, maintenance outsourcing and leasing. For each, we first give an overview of the issues involved and then review the different game-theoretic models that have been proposed to assist in the decision-making process of the different players involved.

The book is aimed at three groups of people.

1. People from industry to get a better understanding on how decisions should be made.
2. Students in Master's and Doctoral programmes to get an appreciation of extended warranties, maintenance service contracts and lease contracts.
3. Researchers working in extended warranties, maintenance service contracts and lease contracts as there is a need for a lot more new research—theoretical as well as applied (to bridge the gap between theory and practice).

The first author is grateful to his ex-students—Dr. Ezzatollah Ashgharizadeh whose thesis dealt with maintenance outsourcing and Jarumon Jaturonnatee (nee Pongpech) whose thesis dealt with maintenance of leased equipment. A special thanks to Mr. Eric Arnum, Editor of Warranty Week, for giving us the permission to use material from several issues of Warranty Week. Professor Wallace Blischke provided useful comments on the detailed outline of the book proposal and was to write an introduction to the book. Unfortunately, he passed away a few weeks before the final manuscript was completed.

We are grateful to the staff at Springer Verlag for their support. We especially want to thank Anthony Doyle for his early interest and encouragement, and Garrett Ziolek for his valuable guidance during the preparation of the final manuscript. Finally, we would like to thank Ms. Gayathri Umashankar and Mr. V. Ramasubramaniyan for their efforts which transformed the manuscript into a book.

Brisbane, Australia D. N. Prabhakar Murthy
Cupar, Scotland Nat Jack

Contents

Part I Background Material

Part IV Management Issues

Acronyms

1-D	One-dimensional
2-D	Two-dimensional
ABM	Age-based maintenance
AFT	Accelerated failure time
AHP	Analytical hierarchy process
AT	Agency theory
BOO	Build own, operate
BOOT	Build, own, operate, transfer
BOT	Build, operate transfer
BR	British rail
BW	Base warranty
CAPEX	Capital expenditure
CBM	Condition-based maintenance
CCA	Contingent claim analysis
CDF	Cumulative distribution function
CLW	Cost limit warranty
CM	Corrective maintenance
CS	Cost subsidisation
CTMC	Continuous-time Markov chain
DB	Design and build
DBFO	Design, build, finance and operate
DCF	Discounted cash flow
DM	Decision maker
DOM	Design-out maintenance
DTMC	Discrete-time Markov chain
ELA	Equipment Leasing Association
EU	European Union
EW	Extended warranty
FM	Facilities management
FRW	Free replacement warranty
FTA	Fault tree analysis
GT	Game theory
HPP	Homogeneous poisson process
ICD	Individual cost deductible

LC	Lease contract
LCC	Life cycle cost
LCS	Lump sum cost sharing
LDO	Lease, develop, operate
LIC	Limit on individual costs
LTC	Limit on total cost
MAPI	Machine and Applied Products Institute
MCF	Mean cumulative function
MLS	Labour (or material) cost sharing
MRO	Maintenance, repair, overhaul
MSC	Maintenance service contract
NE	Nash equilibrium
NHPP	Non-homogeneous poisson process
NPV	Net present value
O&M	Operation and maintenance
OBM	Opportunity-based maintenance
OPEX	Operating expenditure
OPRAF	Office of passenger rail franchising
PBE	Perfect Bayesian equilibrium
PFI	Public financing initiative
PH	Proportional hazards
PLC	Product life cycle
PM	Preventive maintenance
PPP	Public–private partnership
PRS	Private sector
PRW	Pro-rata warranty
PUS	Public sector
RBD	Reliability block diagram
RCF	Rolling contact fatigue
RIW	Reliability improvement warranty
ROCOF	Rate of occurrence of failures
ROSCO	Rolling stock leasing companies
SA	Service agent
SLA	Service level agreements
SPE	Specified parts excluded
TOC	Train operating company
TRAC	Terminal rental adjustment clause
UBM	Usage-based maintenance
UCC	Uniform commercial code
UP	Uptime bonus
UTB	Uptime target and bonus

Chapter 1
Introduction

1.1 Introduction

Developed societies are complex social structures consisting of several individuals, businesses and government agencies who all depend on various products to carry out their daily operations. Plants are collections of many different products which businesses need to produce goods.[1] Finally, several types of infrastructures are needed for the smooth functioning of a society. All of these (products, plants/service facilities and infrastructures) can be viewed as systems (collections of several interconnected elements).

Every system degrades with age and/or usage, and its performance deteriorates. A system is deemed to have failed when its performance falls short of a specified (desired) level. Maintenance is the set of actions that is used to (1) control the rate of degradation and (2) restore the performance of the system when it fails.

Customers (individuals, businesses and government agencies) need assurance regarding performance when they buy a product or have a plant or infrastructure constructed. Product manufacturers and builders of plants and infrastructures provide this assurance through a warranty.[2] During the warranty period subsequent to the customer acquiring the product (or having a plant or infrastructure built), the manufacturer (builder) is responsible for all the maintenance actions. Once the warranty expires, it is the responsibility of the owner of a system to carry out the maintenance actions needed to ensure satisfactory performance. There are several options available to the owners:

1. Perform the maintenance themselves (also called in-house maintenance).
2. Buy an extended warranty (EW) from the manufacturer, retailer or some third party in the case of products.

---

[1] Service facilities (such as hospitals, retail outlets, schools, banks) use a range of products similar to plants to deliver services. We will use the term "plant" to cover both the production of goods and services.

[2] A more appropriate term is Base Warranty (BW).

D. N. P. Murthy and N. Jack, *Extended Warranties, Maintenance Service and Lease Contracts*, Springer Series in Reliability Engineering,
DOI: 10.1007/978-1-4471-6440-1_1, © Springer-Verlag London 2014

3. Outsource some or all of the maintenance of a product, plant or infrastructure to an external service agent through a maintenance service contract (MSC).[3]

In the case of EWs and MSCs, the owner of a system has to pay an extra amount, depending on the terms and conditions of the EW and MSC.[4]

There is a growing trend for customers to lease systems rather than purchase them. In this case, a customer (referred to as the lessee) leases a system from an owner (referred to as the lessor) under a lease contract (LC). There are many types of LCs. The maintenance can be the responsibility of either the lessor or the lessee, depending on the type of LC.

In all the three cases—EWs, MSCs and LCs—there are two (or more) parties involved and the maintenance is an important issue as it impacts on the performance of the system.[5] Each party has a different objective (goal), and the decision-making process needs to take into account the interaction between the parties involved. This book deals with decision-making in the context of EWs, MSCs and LCs with a special emphasis on maintenance-related decisions.

This chapter discusses some basic concepts to set the background to define the scope and focus of the book and give an outline of the book structure and a brief description of the chapters. Section 1.2 deals with system classification and decomposition, and Sect. 1.3 looks at system performance and degradation. Section 1.4 looks at maintenance and the outsourcing-related issues. Section 1.5 deals with EWs/MSCs, and Sect. 1.6 looks at LCs. In Sect. 1.7, we list some of the decision problems faced by the different parties involved in EWs, MSCs and LCs. Section 1.8 discusses the framework needed to solve the decision problems of these parties. These sections set the background for defining the focus and scope of the book, which is discussed in Sect. 1.9. In Sect. 1.10, we indicate the structure and outline of the book.

1.2 System Classification and Decomposition

As mentioned earlier, a system can refer to a product, plant/service facility or infrastructure. We first look at the classification of systems and then the decomposition of a system.

1.2.1 Products

Products are physical objects designed and built for a specific purpose. Products can be fairly simple (e.g. an electric kettle) or very complex (e.g. an aircraft). They can be classified in different ways as indicated below:

---

[3] There is a lot of confusion regarding the terms EW and MSC. This is discussed in Chap. 5.

[4] This is in contrast to the BW which is integral to the sale (or construction contract) and included in the sale price (price of construction).

[5] There can be many other parties involved as will be discussed later in the book.

- Standard (off-the-shelf) and custom-built
- Consumer, industrial and commercial and defence.

Consumer Products These are standard products (e.g. television sets, appliances, automobiles, PCs) which are consumed by society at large.[6] They are characterised by a large number of consumers (customers) for the products. The complexity of the products can vary considerably, and the typical small consumer is often not sufficiently well informed to evaluate product performance, especially in the case of complex products.

Industrial and Commercial Products These can be either standard or custom-built products (e.g. large-scale computers, CNC machines, pumps, X-ray machines, commercial aircraft, hydraulic presses) and are characterised by a relatively small number of consumers and manufacturers. The technical complexity of such products and the mode of usage can vary considerably. The products can be either complete units, such as cars, trucks, pumps and so forth, or product components needed by manufacturers, such as batteries, drill bits, electronic modules and turbines blades.

Defence Products These are specialised products (e.g. military aircraft, ships, rockets) which are characterised by the presence of a single consumer and a relatively small number of manufacturers. The products are usually complex and expensive and involve "state-of-the-art" technology with considerable research and development effort required by the manufacturers. They are usually designed and built to consumer specifications.

1.2.2 Plants

A plant is a collection of products used to produce different types of goods. They can be classified into several categories:

Mining Plants These extract raw materials (ore, fuels, etc.) from the ground.

Processing Plants These convert the raw material into commodities used by businesses—e.g. ore to produce metals (mineral processing) and different kinds of fuel from crude oil (chemical processing).

Manufacturing Plants These convert the outputs of processing plants to produce goods—e.g. the production of cars involves the use of many different kinds of metals and plastics.

Power Plants These produce energy (electrical, heat, mechanical) using different types of fuels—e.g. thermal power plants producing electricity and heat using coal; diesel power plants—for propelling large ships or producing electricity producing diesel oil; and nuclear power plants producing electricity using nuclear fuel.

---

[6] These products are also consumed by businesses and government agencies.

1.2.3 *Infrastructures*

Infrastructures are either large physical structures or networks with nodes and arcs, where the nodes are discrete units and the arcs are links between nodes with a spatial dimension. The arcs can be physical entities (such as cables in an electricity network; pipes in the case of gas, water and sewerage networks; roads in a road network; and rails in a rail network) or non-physical entities (such as in airline, shipping and satellite communication networks). The nodes can be complex objects (equipment such as power plants, buildings such as airports or shipping terminals and transmitters).

Infrastructures are inherently complex and are almost always custom-designed. They can be classified into several types, depending on their function as indicated below:

- *Transport Infrastructures* These can be further subdivided into rail, road, air and marine.
- *Utility Infrastructures* These can be further subdivided into water, gas, electricity, oil, sewerage and communications.
- *Large Physical Structures* These include dams, bridges, public and commercial buildings.

1.2.4 *System Decomposition*

Any system can be decomposed into a hierarchy of different levels, with the system at the top level down to parts at the lowest level. The following is an example of a seven-level decomposition:

Level	Characterisation
1	System (product)
2	Subsystem
3	Assembly
4	Subassembly
5	Module
6	Submodule
7	Component

The decomposition of a system (product, plant or infrastructure) is important in the context of maintenance as maintenance needs to be done at different levels. For example, in the case of a product, the failure of a module might involve replacing the whole module, a submodule or only the failed component.

1.2.5 Some Examples

Example 1.1 Air conditioners (Product/Plant) The function of an air conditioner (often simply called an AC unit) is to dehumidify and extract heat from an enclosed space. The cooling is based on the refrigeration cycle—evaporation (condensation) occurs when heat is absorbed (released). Air conditioners use a compressor to cause pressure changes between two compartments, and a refrigerant is pumped into the evaporator coil located in the compartment to be cooled. The low pressure causes the refrigerant to evaporate into a vapour and in the process taking heat with it. At the opposite side of the cycle is the condenser located outside the cooled compartment where the refrigerant vapour is compressed and forced through another heat exchange coil, condensing the refrigerant into a liquid, thus rejecting the heat absorbed from the cooled space.

Small ACs used in households are products, whereas those used in large buildings are plants. The usage pattern can vary significantly. In large public and commercial buildings, the air conditioners are used in continuous mode, whereas in small houses and apartments, they tend to be used intermittently. □

Example 1.2 Automobile (Product) The automobile is a self-propelled passenger vehicle designed to operate on ordinary roads. Automobiles can be classified into several types based on (1) structure and usage—passenger cars (PC), light trucks (LT), heavy trucks, vans, buses, etc. and (2) the primary energy source—petrol, diesel, electric, hybrid (combinations of petrol and electric) and others such as hydrogen and solar, which are still in the experimental stages of development.

Individuals normally buy one automobile at a time, whereas a business might buy a fleet either for use by its staff or for renting out. □

Example 1.3 Lift in a Building (Product) A passenger lift is a vertical transport mechanism to move people between floors of a building, vessel or other structures. Lifts are powered by electric motors that drive either traction cables or counterweight systems (similar to a hoist) or pump hydraulic fluid to raise a cylindrical piston (similar to a jack).

A goods lift (freight elevator) is designed to carry goods, rather than passengers, and often has manually operated doors and a rugged interior finish. □

Example 1.4 Power Plant (Plant) A power plant is used for the generation of electric power. At the centre is a generator, a rotating machine that converts mechanical power into electrical power by creating relative motion between a magnetic field and a conductor. The energy source used to turn the generator varies widely, and the process can involve (1) burning fossil fuels such as coal, oil and natural gas, (2) creating fission in a nuclear reactor and (3) using cleaner renewable sources such as solar, wind and hydroelectric.

A thermal power station is a power plant in which the prime mover is steam driven. Water is heated, turns into steam and spins a steam turbine which drives an electrical generator. After it passes through the turbine, the steam is condensed in a condenser and recycled to where it was heated. Some thermal power plants also deliver heat energy for industrial purposes, for district heating or for desalination of water as well as delivering electrical power. □

Example 1.5 Rail Transport (Infrastructure) Rail transport involves wheeled vehicles running on rail tracks. Track usually consists of steel rails installed on sleepers and ballast on which the rolling stock (wagons and carriages) fitted with metal wheels moves. Rolling stock in railway transport systems has a lower frictional resistance than vehicles on highways and roads and can be coupled together to form longer trains. In some countries, the rail transport is public owned (by the government), and in others, it is private owned (by private businesses) or jointly private and public owned. The two major subsystems are (1) infrastructure and (2) rolling stock. The infrastructure is managed by the track operator (often publicly owned company or agency), and the rolling stock is managed by rolling stock operators (can be either public or private). Together, they provide transport between train stations (for passenger and freight transport) and between two terminals (for freight)–such as a mine or manufacturing/processing plant and a port. Power is provided by locomotives which either draw electrical power from an electrical network or produce their own power (usually using diesel engines). Most tracks are accompanied by a signalling system to ensure smooth and safe operation of trains. □

1.3 System Performance

The performance of a system (product, plant or infrastructure) is a complex entity involving many dimensions, and it depends on the performance of its elements. A system is designed and manufactured (built) to some desired performance, and this is discussed further later in the section.

1.3.1 Performance Degradation and Failure

Elements of a system degrade with age and/or usage, and this in turn lowers the performance of the system. The rate of degradation depends on several factors. Decisions made during the design stage (e.g. material selection) and during the production stage (e.g. heat treatment, quality of welding) are factors that have an impact on the rate of degradation and are under the control of the manufacturer (builder). Similarly, the usage mode, usage intensity and operating environment are factors that also affect the rate of degradation, and these are under the control of the customer.

When the system performance falls below the desired specified level, then it is deemed to have *failed*. Failures occur in an uncertain manner and are influenced not only by the factors discussed above but also by human factors which are also important and can lead to system failure.

1.3.2 Consequences of Failures

Customer's Point of View When a failure occurs, no matter how benign, its impact is felt. For customers, the consequences of a failure may range from being only a mere nuisance value (e.g. the failure of air conditioner) to serious economic loss (e.g. the failure of a freezer) to something resulting in serious damage to the environment and/or loss of life (e.g. brake failure in an automobile).

When the customer is a business enterprise, failures lead to downtimes and this affects the production of services and goods. These delays in turn affect the goodwill of the clients as well as resulting in a financial loss to the business.

Manufacturer's Point of View Failures result in warranty costs (arising from having to service claims under the BW). The annual warranty costs for large manufacturers (such as GM, Ford, Toyota and HP) can amount to billions of dollars.[7]

1.3.3 Performance Measures

Performance is best characterised through a vector of variables, where each variable is a measurable quantity of the system. For products and plants, the performance measures also depend on the particular perspective—manufacturer or customer. For infrastructures, the measures can involve many other parties (such as the public and regulators). There might be measures common to more than one party, and others might be of interest to only one. The variables contained in the measures can be divided broadly into two categories—reliability related and non-reliability related.

1.3.3.1 Reliability-Related Performance Measures

The reliability of a system conveys the concept of dependability or the absence of failure (the inability of the system to perform as expected).[8] Some of the reliability measures used in system design are as follows:

- Interval reliability: The probability of no failure over a specified interval.
- Interval availability: The fraction of time that the system is in an operational (non-failed) state over a specified interval.

---

[7] System failures impact the manufacturer in many another ways. One of these is the impact on sales due to the negative word-of-mouth effect resulting from customer dissatisfaction with failures. This in turn affects the market share and the manufacturer's reputation.

[8] Reliability theory deals with various issues such as the scientific understanding of the failure mechanisms and the engineering insights needed into the design of reliable assets.

- Asymptotic availability: The fraction of time that the system is in an operational (non-failed) state over an infinite time interval.
- Point availability: The probability that the system is in an operational state at a given point in time.
- Mean cumulative function (MCF): The expected number of system failures over a specified interval.
- Mean time for system to recover after failure.

Consider a diesel power plant. If the diesel engine drives the base load generator to produce electricity, then the reliability measure of interest is asymptotic availability. On the other hand, if it drives a backup generator, then the reliability measure of interest is point or interval availability.

1.3.3.2 Non-Reliability-Related Performance Measures

These are system specific. In Example 1.1, they would be noise and appearance (for small units) and cost to operate and efficiency (for large units). In Example 1.2, they would be fuel efficiency, safety, noise level, quality of ride, etc. In Example 1.3, they would be emission levels, fuel efficiency, lubrication oil consumed, etc. In Example 1.5, they would be safety, speed, punctuality, etc.

1.4 Maintenance

Maintenance actions (in the context of products and plants) can be grouped into two broad categories:

- Preventive maintenance (PM)
- Corrective maintenance (CM).

PM is the set of actions to control the rate of degradation of a system and reduce the likelihood of failure occurrence. It involves tasks such as monitoring relevant variables, collecting and analysing degradation data and initiating appropriate actions. CM is the set of actions to restore a failed system to an operational state. It can involve either repairing or replacing the failed components. PM and CM actions can be done either at component level or at some higher level, depending on the system.

Maintenance of a system involves carrying out several activities as indicated in Fig. 1.1 [adapted from Dunn (1999)].

Performing maintenance costs money, and it can be a significant part of the operating budget of an individual or business. Thus, devising optimal maintenance strategies is critical to minimise running costs. Devising such strategies involves building reliability models to assess the impact of alternative maintenance actions on the degradation process and on the occurrence of failures.

Maintenance in the context of infrastructures involves another category—termed service/operations which depend on the industry sector. This is discussed further in Chap. 2.

1.4.1 Maintenance Outsourcing

Outsourcing of maintenance involves some or all of the maintenance actions (preventive and/or corrective) being carried out by an external service agent under a MSC. The contract specifies the terms of the maintenance and the cost issues. It can be simple or complex and can involve penalty and incentive terms.

There are many different contract scenarios depending on how these activities are outsourced, and these are discussed in Chap. 5.

1.5 Extended Warranties and Maintenance Service Contracts

There is lot of confusion between EWs and MSCs.[9] To properly understand these two terms, we start with a brief discussion of base warranties.

1.5.1 Base Warranties

A base warranty (BW) is often simply called a warranty and is normally associated with products.[10] It is a contractual agreement between a manufacturer and a buyer (customer) that is entered into upon the sale of a product. The contract defines the compensation available to the buyer if the performance of the product is found to be unsatisfactory. It is part of the sale, and its cost is factored into the price of the product.

1.5.2 Extended Warranties

An EW is associated mainly with standard (consumer, industrial and commercial) products. The difference between a BW and an EW is that the latter is entered into voluntarily and is purchased separately—the buyer may even have a choice of

---

9 This is discussed in more detail in Chap. 3.

10 Warranties for plants and infrastructures are more complex and are discussed in Chap. 5.

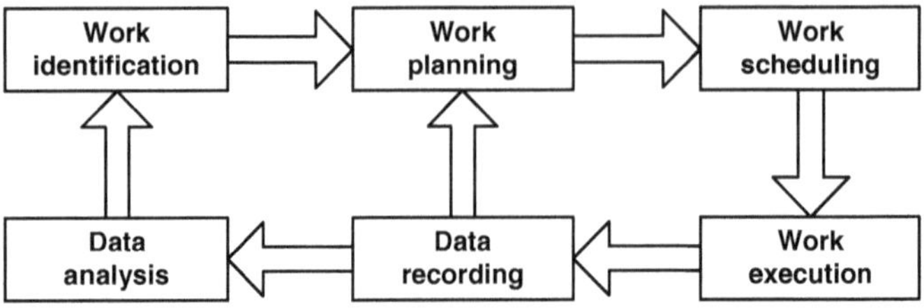

Fig. 1.1 Activities in the maintenance of a system

terms, whereas a BW is part of product purchase and integral to the sale. The terms of an EW can be the same as those for the BW provided by the manufacturer, or they may differ in the sense that the EW may include additional features such as

- Cost sharing,
- Exclusions,
- Cost limits on individual claims or on the total claims made under the warranty,
- Deductibles.

EWs are currently offered on a wide range of products, including automobiles, electronics, appliances and many other items. Often the customer has to buy an EW at the same time as the product is purchased. In some cases, the customer has the option to buy an EW any time before the BW expires or even after the BW expiry date. The customer pays an extra amount, depending on the duration and the terms of the EW.

Originally, EWs were offered only by manufacturers. Currently, for some products, they are offered by other service providers (such as retailers, insurance companies and other parties) rather than by manufacturers. For other products, manufacturers compete with the EW service providers. Two important features of EWs are the following:

- The terms of an EW are decided by the manufacturer or service provider with the customer having limited input.
- An EW covers only the CM costs (associated with failures of components that are not excluded) but not the PM costs or any consequential losses incurred by the owner of the product.

1.5.3 Maintenance Service Contracts

A MSC is similar to EW in that the maintenance of a system is carried out by an external service provider. The MSC defines the period over which the maintenance (PM and/or CM) actions are to be carried out and the payment to be made by the owner of the system to the service provider. These contracts are often associated with the maintenance of products (commercial and industrial), plants and infrastructures.

MSCs can be either (1) standard contracts offered by the service provider, (2) contracts where some customisation of a standard contract is made to meet the specific needs of an owner and (3) contracts initiated and dictated by the owner. Standard (and customised) contracts are mainly for standard products. Owner initiated contracts are usually for the maintenance of complex systems and infrastructures with one or more service providers.

MSCs can include none, or one or more of the following:

- PM only, CM only or both PM and CM,
- Exclusions,
- Deductibles,
- Cost limits,
- Guarantees on system or service performance,
- Incentives (penalties) if the system or service performance exceeds (falls below) a specified level,
- Compensation for some or all of the consequential losses incurred by the owner due to the unavailability of the product (system).

1.6 Lease Contracts

As mentioned earlier, a customer (individual, business or government agency) can decide to lease a system (product, plant or infrastructure) instead of owning it. A lease is a contractual agreement under which the owner of a system (referred to as the "lessor") allows a customer (referred to as the "lessee") to operate the system for a stated period of time and under specified conditions of the LC. There are several types of leases and several reasons for customers deciding to lease rather than own.[11] One reason is that the customer is often not responsible for the maintenance of the leased item.

The terms of the LC are usually decided by the lessor for products (consumer, commercial and industrial) or jointly with the lessee for infrastructures. The terms can include none, or one or more of the following:

- Guarantees on system and/or service performance.
- Incentives (penalties) if the performance exceeds (falls below) some specified level.
- Compensation for some or all of the consequential losses incurred by the owner due to the unavailability of the product (system).

These terms have implications for the maintenance actions of the lessor.

[11] These are discussed in Chap. 9.

1.7 Decision Problems

EWs, MSCs and LCs all involve two or more parties. In the case of an EW or a MSC, there is a system owner and a service provider who provides some or all the maintenance under a contract. In the case of a LC, there is a lessor who owns the system and does the maintenance and a lessee who leases the product (system).[12]

The cost of servicing failures under an EW, MSC or LC depends on several factors (such as system reliability, age of the system at the start of the contract, operating environment and usage mode and intensity). This cost is of interest to both parties, and it plays an important role in their decision-making processes. Each party has to evaluate the different options available and then determine their optimal decision, taking into account the interactions between the different parties and system performance.

We now list some of the decision problems that the parties need to address.

1.7.1 Extended Warranty/Maintenance Service Contract

The two parties are the owner of the system and the service provider.

1.7.1.1 Owner's Perspective

In the case of products, some of the decisions are as follows:

1. Whether to buy an EW or not?
2. How to evaluate whether the EW price is reasonable or not?
3. How to decide on the best EW if there is more than one option?

In the case of plants and infrastructures, some of the decisions are as follows:

1. Should some or all of the maintenance be outsourced?
2. What should be the terms of the MS contract?
3. How to select the best service provider when there is more than one?

1.7.1.2 Service Provider's Perspective

In the case of products, some of the decisions are as follows:

1. Should the service provider offer one or more EW policies?
2. What should be the terms of the EW policies?
3. What are the costs of servicing the different EW policies?

---

[12] There can be many other parties involved for certain types of systems, such as rail networks. This is discussed further later in the chapter.

4. What should be the pricing of the EW policies?
5. How to deal with competition in the EW market?
6. How to plan the servicing logistics?

In the case of plants and infrastructures, some of the decisions are as follows:

1. Should the service provider offer one or more MSCs?
2. What should be the terms of the different MSCs?
3. What are the costs of servicing the different MSCs?
4. What should be the pricing of the different MSCs?
5. How to tender for a MSC?
6. How to deal with competition in the maintenance service market?
7. What should be the optimal number of customers to have?
8. How to plan the servicing logistics?

1.7.2 Lease Contract

The two parties are the lessee and the lessor.

1.7.2.1 Lessee's Perspective

In the case of products, some of the decisions are as follows:

1. How to evaluate alternative LCs?
2. How to decide on the best LC?

In the case of plants and infrastructures, some of the decisions are as follows:

- What should be the terms of the LC?
- How to select the best lessor when there is more than one?

1.7.2.2 Lessor's Perspective

In the case of products, some of the decisions are as follows:

1. Should the number of LCs be one or more?
2. What should be the terms of each LC?
3. What are the costs of servicing different LCs?
4. What should be the pricing of different LCs?
5. How to deal with competition in the lease market?
6. How to plan the servicing logistics?

In the case of plants and infrastructures, some of the decisions are as follows:

1. Should one or more LCs be offered?
2. What should be the terms of the different LCs?

3. What are the costs of servicing the different LCs?
4. What should be the pricing of the different LCs?
5. How to tender for a LC?
6. How to deal with competition in the lease market?
7. What is optimal number of lessees to have?
8. How to plan the servicing logistics?

1.8 Framework and Approach

The systems modelling approach, together with game theory (GT) and agency theory, provides the framework needed to find solutions to these decision problems.

1.8.1 Systems Approach

The systems approach is a multistep process useful for solving problems. The steps involved are as follows:

Step 1: *Problem Definition*
The problem definition depends on the particular context. For EWs, MSCs and LCs, we will be looking at a variety of problems.

Step 2: *System Characterisation*
This involves a characterisation of the salient features of the real world that are relevant to the problem. It is a process of simplification and is done by defining relevant variables and the interactions (e.g. cause–effect relationships) between them. A good understanding of reliability theory is important for carrying out this step.

Step 3: *Model Building*
A model is a representation of the real world. The system characterisation is a descriptive model which highlights the interactions between different variables. A mathematical model links the descriptive model to an appropriate mathematical formulation. Since uncertainty is a significant feature in EWs, MSCs and LCs, probabilistic and stochastic formulations are needed.

Step 4: *Model Analysis and Optimisation*
Techniques from probability theory, stochastic processes and optimisation theory are needed to carry out the analysis and obtain an optimal solution to the problem.

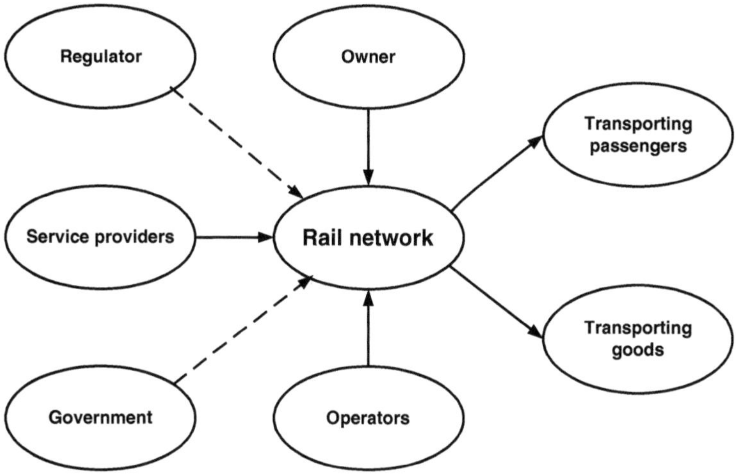

Fig. 1.2 Parties involved in the maintenance of a rail network

1.8.2 Game Theory

A game is a decision-making situation involving two or more parties (called "players" in the game theory (GT) literature) where the objective function of each party is a function of the decision variables of all the parties involved.

In the case of infrastructures, there can be several other parties involved. Figure 1.2 shows the different parties in the context of a rail transport system. Here, the focus is on the track and its owner. The track is used by several different operators (different businesses) who own and operate the rolling stock which carries passengers and moves different types of goods. The service providers are the businesses that provide the maintenance for the track and the rolling stock. Other parties are the regulators who ensure safety standards and governments. If the owner (or an operator) leases the track (rolling stock), then several other parties, such as financial institutions and insurance companies, are also involved.

There are several types of games. Information uncertainty, information asymmetry and other issues make game-theoretic formulations more complex, and there is a vast GT literature which is relevant to the study of EWs, MSCs and LCs. GT is discussed in Chap. 4.

The goals and interests of the various parties are different and are characterised and modelled through their individual objective functions. The parties make their decisions in order to optimise their objective functions while taking into account the interactive nature of these decisions. In the rail network context, GT can provide the framework needed to build models to obtain the optimal decisions for all the parties involved.

1.8.3 Agency Theory

Agency theory (AT) deals with the relationship that exists between two parties (a principal and an agent) where the principal delegates work to the agent who performs that work and a contract defines the relationship. AT is concerned with resolving two problems that can occur in agency relationships.

The first problem arises when the two parties have conflicting goals, and it is difficult or expensive for the principal to verify the actual actions of the agent and whether the agent has behaved properly or not. The second problem involves the risk sharing that takes place when the principal and agent have different attitudes to risk (due to various uncertainties).

The focus of AT is on determining the optimal contract, behaviour versus outcome, between the principal and the agent. Issues such as moral hazard, adverse selection, risks, information asymmetry, costs, monitoring and incentives need to be taken into account, and these are all relevant in the context of maintenance outsourcing.

AT can also be extended to deal with relationships between more than two parties. There is an extensive literature dealing with multiple principal/multiple agent problems and the design of optimal contracts. AT is discussed in Chap. 4.

1.9 Scope and Focus of the Book

EWs, MSCs and LCs involve unreliable systems (products, plants and infrastructure) which degrade with age and/or usage and require maintenance (preventive and corrective). There are certain features that are common to all of these types of contract and others that differ. The common and salient features are as follows: (1) there are two or more parties involved each with different goals and (2) the decisions of each party have an impact on the others. The objectives of the book are as follows:

- To develop a framework to study EWs, MSCs and LCs in a unified manner using concepts from the theories of warranty, reliability and maintenance.
- To identify the key features (goals, decision variables, etc.) from each party's perspective and the interactions between them.
- To give an overview of the literature relating to EWs, MSCs and LCs.
- To review the mathematical models (using concepts from GT and agency theory) that have been proposed to characterise the optimal decision-making processes from the perspectives of the different parties involved.
- To suggest issues and topics for future research in EWs, MSCs and LCs.

The book is aimed at the following three groups:

1. Postgraduate students in engineering and/or management programs: As a text or reference book for a course on servicing unreliable systems dealing with EWs, MSCs and/or LCs.
2. Senior-level managers in industry: As a reference source for understanding the issues and the management of EWs, MSCs and LCs.
3. Researchers in engineering, mathematics, economics and/or management to carry out new research in EWs, MSCs and LCs.

1.10 Structure and Outline of the Book

The book consists of this introductory chapter, followed by four parts (Parts I–IV), with each part consisting of two or more chapters. The four parts are as follows:

- Part I: Background Material (Chaps. 2–4)
- Part II: Extended Warranties and MSC (Chaps. 5–8)
- Part III: LCs (Chaps. 9 and 10)
- Part IV: Management Issues (Chap. 11) and Epilogue (Chap. 12).

A brief description of the contents of Chaps. 2–12 is given below:

Chapter 2 System Degradation and Maintenance
Every system (product, plant or infrastructure) is unreliable in the sense that it degrades and eventually fails. Maintenance is needed to compensate for this unreliability. Any decision-making with respect to maintenance requires a proper understanding of the degradation processes over time and the actions of maintenance from a system life cycle perspective. This chapter looks at the issues involved using concepts from the theory of reliability and of maintenance.

Chapter 3 Modelling and Analysis of Degradation and Maintenance
Mathematical models play an important role in solving decision problems. Modelling is both an art and a science. For some systems, the degradation and failure depend solely only on age, whereas for others, it depends on both age and usage. Since degradation and failures occur in an uncertain manner, one needs one- and two-dimensional stochastic formulations to model the two cases. This chapter deals with models and the modelling process and discusses the issues involved in the context of modelling degradation and maintenance.

Chapter 4 Introduction to Stochastic Optimisation and Game Theory
If there is only one decision-maker (DM), then this DM has an optimisation problem to solve. The presence of two or more DMs, with possibly conflicting objectives, requires a different approach, and then, techniques from GT need to be

used. Issues such as uncertainty, information available to different DMs and attitude risk play an important role in this context. This chapter deals with all of these issues and provides an overview of the quantitative approaches using mathematical models for decision-making.

Chapter 5 EWs/MSCs—An Overview
EWs and MSCs are similar in many respects, but they are also differences. A proper understanding of EWs requires concepts from base warranties (BWs). Similarly, a proper understanding of MSCs requires concepts from outsourcing in general. This chapter deals with these two topics and looks at the different aspects of EWs and MSCs and their similarities and differences.

Chapter 6 EW/MSC Processes
The EW/MSC process can be viewed as a chain involving several stages and the involvement of many different parties—EW/MSC providers (sellers of EWs/MSCs), customers, administrators (responsible for the EWs/MSCs sold), underwriters, insurers, service agents and others (such as regulators and governments). The characterisation of each party and the interactions between the EW/MSC providers and customers define the EW/MSC process. This chapter deals with these issues.

Chapter 7 EW and MSC Cost Analysis
EW and MSC providers generate revenue by selling EW and MS contracts. However, they incur costs in servicing claims over the contract period. The costs associated with servicing a claim are either borne by the provider or shared between the provider and the customer, depending on the terms of the contract. This chapter deals with the cost analysis of EWs and MSCs from both provider's and customer's perspectives.

Chapter 8 Game-Theoretic Models for EW/MSC Decision-Making
Since EW and MSC involve two or more parties, the decision-making by each party is best described using a game-theoretic framework which deals with the interests and objectives of the different parties. Several different models have been proposed in the literature, and this chapter reviews these models.

Chapter 9 Leasing and Maintenance of Leased Assets
The traditional approach for acquiring a system (product, plant or infrastructure) has been to own it either outright by cash payment or conditionally with a deferred payment plan. There is a growing trend towards leasing with or without an option to purchase. The maintenance of leased assets raises some new and interesting issues. The responsibility for the maintenance can be either with the owner (lessor) or with the user (lessee), and they, in turn, can either do the maintenance in-house or outsource it to third-party external service agents. Maintenance decisions need to take into account the terms in the LC. This chapter discusses these issues.

Chapter 10 Models for Lease and Maintenance Decisions

This chapter deals with mathematical models for the two parties (lessor and lessee) to make their optimal decisions (relating to issues such as lease terms and price) and the optimal decisions by the lessor or lessee (depending on the contract).

Chapter 11 Management of EWs/MSCs and LCs

EWs, MSCs and LCs processes are complex processes, and failure to manage them properly can have serious consequences to both customers and service providers. Managing the process is very critical to avoid the potentially costly consequences. This needs to be done from a contract life cycle perspective, which in turn needs to be incorporated into the bigger overall business framework. The framework needed is different for customers and providers as only some of the issues involved are common. Issues such as data and information, qualitative factors that need to be taken into consideration in decision-making, servicing logistics and risk analysis are important in the context of effective management. This chapter deals with these and looks at management from both customer's and service provider's perspectives.

Chapter 12 Epilogue

An evaluation of the literature on EWs, MSCs and LCs indicates that these topics have been studied in a disjointed manner. This book proposes an integrated approach to look at all three areas in a proper, unified manner. However, there are several shortcomings in the literature. This chapter deals with these and highlights some of the issues and topics that need to be looked at in the future.

Reference

Dunn S (1999) Maintenance outsourcing–Critical issues. http://www.plantmaintenance.com/maintenance_articles_outsources.html

Part I
Background Material

Chapter 2
System Degradation and Maintenance

2.1 Introduction

Every system (product, plant or infrastructure) is unreliable in the sense that it degrades and eventually fails. Maintenance is needed to compensate for this unreliability. Any decision-making with respect to maintenance requires a proper understanding of the degradation processes over time and the actions of maintenance from a system life cycle perspective. The life cycle from the manufacturer (or builder) perspective is different from that of the customer (or owner). This chapter looks at these issues using concepts from reliability theory and the characterisation of maintenance actions. It forms the basis for the modelling of the degradation process and maintenance actions which is the focus of the next chapter.

The outline of the chapter is as follows. Section 2.2 deals with system life cycle from both customer (owner) and manufacturer (builder) perspectives. Section 2.3 looks the characterisation of system (product, plant and infrastructure) performance. Section 2.4 deals with product reliability and looks at various issues such as linking component reliability to product reliability and different notions of reliability from a product life cycle perspective. Section 2.5 looks at maintenance and the characterisation of different types of maintenance actions appropriate for products and plants and Sect. 2.6 looks at maintenance of infrastructures.

2.2 System Life Cycle

The life cycle of a system is basically the period of time during which it is in existence, either conceptually or physically, and may be defined in various ways. The life cycle for products differs somewhat from that for plants or infrastructures.

D. N. P. Murthy and N. Jack, *Extended Warranties, Maintenance Service and Lease Contracts*, Springer Series in Reliability Engineering,
DOI: 10.1007/978-1-4471-6440-1_2,

2.2.1 Products

The life cycle for a product (consumer, commercial or industrial) is commonly referred to as the product life cycle (PLC).

2.2.1.1 Manufacturer Perspective

The PLC for a standard consumer durable or an industrial product, from the point of view of the manufacturer, is the time from initial concept of the product to withdrawal of the product from the marketplace. The life cycle involves six stages, as indicated in Fig. 2.1.[1]

The process begins with the idea of building a product to meet some customer requirements, such as performance targets. This is usually based on a study of the market and the potential demand for the product being planned. The next step is to carry out a feasibility study. This involves determining if it is possible to achieve the targets within specified cost limits. This analysis is done in the front-end stage (Stage 1) of Fig. 2.1.[2]

If the analysis indicates that the project is feasible, an initial product design is undertaken. A prototype is then built and tested. It is not unusual at this stage to find that achieved performance levels of the prototype product are below the target values. In this case, further product development is undertaken to overcome the problem. These define the Stages 2 (Design) and 3 (Development) of the PLC as shown in Fig. 2.1. Once these are achieved, the next step is to carry out trials to determine performance of the product in the field and to start a pre-production run. This is required because the manufacturing process must be fine-tuned and quality control procedures established to ensure that items produced have the same performance characteristics as those of the final prototype.

After this, the production and marketing efforts begin. These constitute Stages 4 (Production) and 5 (Marketing) of the PLC shown in Fig. 2.1. The items are produced and sold. Production continues until the product is removed from the market because of obsolescence and/or the launch of a new product. Post-sale support of the product continues at least until expiration of the warranty on the last item sold but can continue beyond this point in terms of spare parts, service contracts, etc. This defines Stage 6 (post-sale) of the PLC.

2.2.1.2 Customer Perspective

From the consumer's viewpoint, the PLC is the time from the purchase of an item to its discarding when it reaches the end of its useful life or is replaced earlier due

[1] The number of stages in the PLC can vary. For more on this, see Murthy et al. (2008).

[2] The Front End stage is also often referred to as the Feasibility stage.

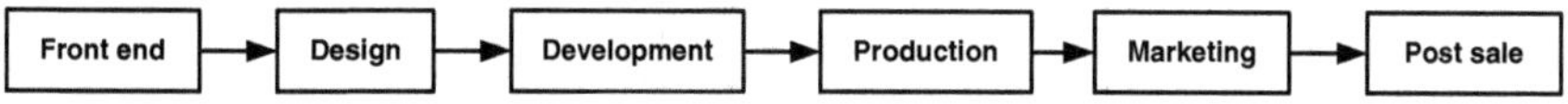

Fig. 2.1 Product life cycle (manufacturer perspective)

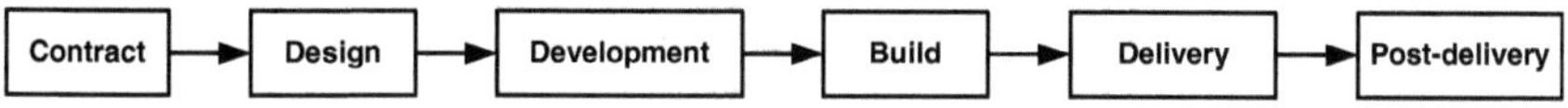

Fig. 2.2 Product life cycle for plants and infrastructures (manufacturer perspective)

to either technological obsolescence or the item being no longer of any use. The life cycle involves the following three phases:

- Purchase.
- Operation and maintenance.
- Discarding (leading to replacement by a new one).

2.2.2 *Plants and Infrastructures*

2.2.2.1 Builder Perspective

The life cycle for a custom built system (product, plant or infrastructure) is slightly different and is as shown in Fig. 2.2. Here, the initial requirements of the plant or infrastructure are specified by the owner and then jointly agreed through discussions leading to a contract that specifies the final agreed requirements. The builder then builds the plant or infrastructure to the specifications stated in the negotiated contract. The process then follows basically the same steps as those for products.

2.2.2.2 Owner Perspective

From the consumer's viewpoint, the life cycle is the time from the initiation of the process and to discarding or upgrading the plant or infrastructure. As such the life cycle involves all the phases (except post delivery) shown in Fig. 2.2 and the following additional phases after the delivery phase:

- Operation and maintenance.
- Discarding or major upgrade (leading to a new life cycle).

2.2.3 *Salvage Value*

The salvage value of a system is the value of the system at the end of its economical or useful life. It is used in accounting to determine depreciation amounts and to determine deductions for taxation purposes. The value can be a best guess of the end value (or determined by a regulatory body such as the Taxation Department). It depends on the state of the system and is influenced by factors such as usage, maintenance and technological obsolescence. It is also referred to as the *residual value*.

2.3 Characterisation of System Performance

Every system (product, plant and infrastructure) is designed for some specified performance as illustrated by the following example:

- Electric bulb (Product): To produce light.
- Engine (Product): To operate to some specified efficiency.
- Power station (Plant): To produce specified output with cost/unit below some specified value.
- Rail system (Infrastructure): To provide passenger service to some specified schedule (frequency and punctuality) at a cost below some specified value.

The performance of a system is a function of the *condition* or *state* of the system. The state of a system, in turn, depends on the state of its elements. We first look at the characterisation of component state and then the characterisation of the state of products, plants and infrastructures.

2.3.1 Characterisation of Component State

The condition of a component (of a product) degrades with time (and usage) and can be characterised through a variable $X(t)$ which represents the *state* of the component. Note that $t = 0$ corresponds to the instant a new component is put into use for the first time. We have three different characterisations with increasing degrees of detail.

2.3.1.1 Characterisation 1 (Binary)

Here, $X(t)$ is binary valued with

- $X(t) = 1$ corresponding to the component being in the *working state* (performance satisfactory or acceptable), and
- $X(t) = 0$ corresponding to component being in the *failed state* (performance is unsatisfactory or unacceptable).

The component starts in the working state and changes to the failed state after a period $\tilde{T}$ as shown in Fig. 2.3. $\tilde{T}$ is the time to failure (or lifetime of the

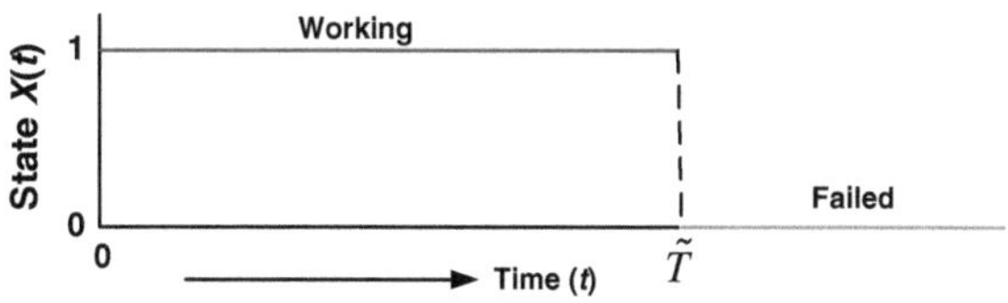

Fig. 2.3 Time to failure (binary characterisation)

component). This is a random variable[3] as the time instant of change from working to failed is uncertain.

A typical example where this characterisation is appropriate is an electric bulb where the state changes from working to failed in a very short time so that it can be viewed as being instantaneous.

2.3.1.2 Characterisation 2 (Finite Number of Levels)

Here, $X(t)$ can assume values from the set $\{1, 2, \ldots, K\}$ with

- $X(t) = 1$ corresponding to component performance being fully acceptable (component is in the good *working state*),
- $X(t) = i,\ 1 < i < K,$ corresponding to component performance being partially acceptable (component is in a working state with a higher value of i implying a higher level of degradation) and,
- $X(t) = K$ corresponding to component performance being unacceptable (component is in the *failed state*).[4]

The time to failure of the component is given by $\tilde{T} = \inf\{t : X(t) = K\}$ as shown in Fig. 2.4. Let $\tilde{T}_i$ denote the duration for the time the component state is $i,\ 1 \le i \le K - 1$. This is a random variable, and as result the time to failure is the sum of $(K - 1)$ random variables.

A typical example where this characterisation is appropriate is the wear in a tire where no wear corresponds to state 1 and complete wear corresponds to state K.

2.3.1.3 Characterisation 3 (Infinite Number of Levels)

This is an extension of the above case with $K = \infty$. $X(t)$ is now a non-decreasing continuous time stochastic process as shown in Fig. 2.5. Here, a higher value of $X(t)$ implies greater degradation, and the component failure time is given by $\tilde{T} = \inf\{t : X(t) = x*\}$.[5]

---

[3] See Appendix A for a definition of a random variable and an introduction to probability theory.

[4] The numbering of states is arbitrary. One can easily reverse the order so that the lower the state the greater the degradation.

[5] In some cases $X(t)$ could be non-increasing with lower values corresponding to greater degradation. In this situation the curve in Fig. 2.5 would be downward sloping.

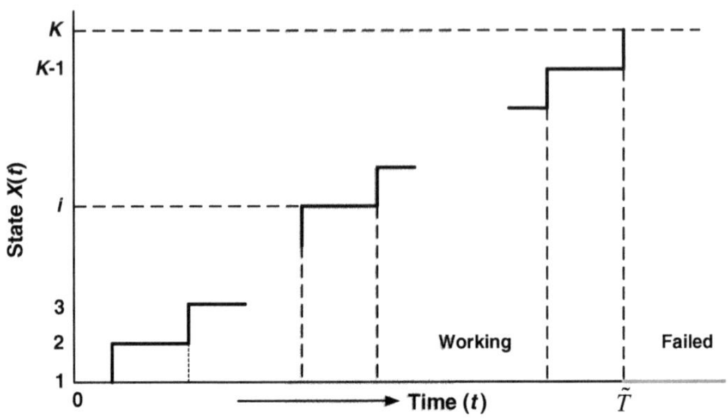

Fig. 2.4 Time to failure (multistate characterisation with finite states)

Fig. 2.5 Time to failure (multistate characterisation with infinite states)

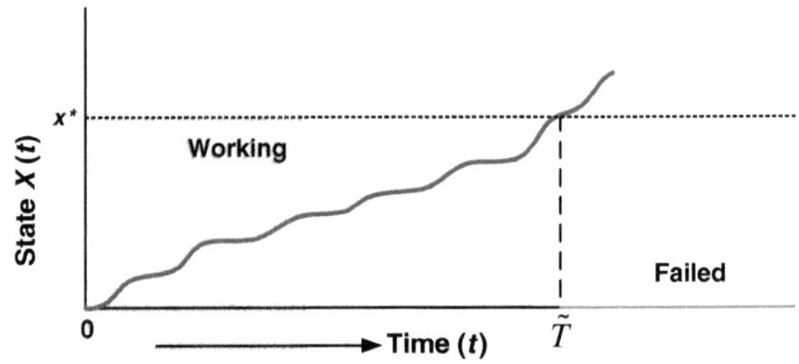

A typical example, where this characterisation is appropriate, is failure due to crack growth in a pipe. The state depends on the crack length, and failure occurs when the crack length reaches some critical value.

Comment: The rate of deterioration of the state depends on factors that impact on the stress (thermal, mechanical, etc.) on the component. The stresses are, in turn, influenced by the load or throughput of the system.

2.3.2 Characterisation of Product (Plant) State

At the product (plant) level, the characterisation of the state is more complex and two approaches can be used. In the first approach, the product (plant) is viewed as a black box and the characterisation is done in a manner similar to the previous subsection with $X(t)$ denoting the state [defined by the output of the product (plant)]. The second approach views the product (plant) in terms of its components. Each component is characterised using Characterisation 1 discussed earlier, and fault tree analysis (FTA) is used to link the product-(plant) level characterisation to the component-level characterisation.

2.3.2.1 Fault Tree Analysis

A fault tree is a logic diagram that displays the relationship between a potential event affecting product (plant) performance and the reasons or underlying causes for this event. The reason may be failures (primary or secondary) of one or more components of the system, environmental conditions, human errors and other factors.

A fault tree illustrates the state of the system denoted the TOP event (binary characterisation—working/failed) in terms of the component states (binary characterisation—working/failed) denoted basic events. The connections are done using gates, where the output from a gate is determined by the inputs to it. A special set of symbols (for gates and basic events) is used for this purpose.[6]

2.3.2.2 Multiunit Plants and Service Facilities

Many industrial plants and service facilities have multiunits—for example a power plant having three units (each with output capacity 50, 100 and 200 MW, respectively), a bus operator having a fleet of K buses. In this case, the system state (from a business-level perspective) is best done in terms of the different levels of output. In the case of the power plant, the different levels correspond to 350 MW (all three units working), 0 MW (all three failed) and six different levels (50, 100, 150, 200, 250 and 300 MW) depending on the number of units (1 or 2) in failed state.

Comment: The rate of deterioration of the state depends on several factors such as the production (throughput) rate, environmental factors and maintenance actions.

2.3.2.3 Fleet

A fleet refers to multiple units of an asset (such as machines, automobiles, ships aircraft, computers, etc.). A fleet can be viewed as a multiunit system, where each unit operates independently and a failure of a unit does not result in the failure of the system but can affect the overall performance (e.g. production capacity) of the system.

---

[6] The extension of FTA to the case where the performance is based on Characterisation 2 is more complex. For further details see, Blischke and Murthy (2000) or Rausand and Høyland (2004).

2.3.3 Characterisation of Infrastructure State

The characterisation of the state of an infrastructure is still more complex for the following reasons:

- There are several types of infrastructure—road, rail, utilities (gas, water sewerage, etc.), concrete structures (dams, buildings, bridges, etc.) and communication networks, etc., and each is different.
- Most infrastructures involve two types of elements—(i) discrete or lumped (similar to a product or plant), and (ii) distributed (with a spatial dimension). The characterisation of the state for the discrete elements is similar to that for products and plants discussed earlier and hence will not be discussed here.
- If one focuses on the distributed elements, the term *quality* is often used to indicate the state or condition. This in turn is defined through terms such as *damage*, *defects*, etc. Also, the characterisation of failure is not so clear. Often failure is defined to occur when the quality falls below some specified norm.
- Often there are several parties involved each with a different objective.
- The performance characterisations for each party are different and each involves a multitude of variables.
- The degradation of the state of the infrastructure is influenced by several factors such as weather, state of the system, usage intensity and output of (or throughput through) the system.
- Each party's performance of interest is different, but they are all functions of the state of the infrastructure.
- Safety also plays a role as poor condition of the asset can lead to dramatic consequences, e.g. in the case of tracks, roads, etc.

We confine our discussion to road infrastructure.

2.3.3.1 Road Infrastructure[7]

Road infrastructure consists of pavements (or "roads") and other items such as traffic signals, signs, etc. There are two types of pavements—rigid and flexible. Rigid pavements consist of a thick concrete top surface. Flexible pavements have a flexible layer on top of the surface.

When a road is built, the surface is dugout down to the designed depth of the intended road. Preparation is carried out on the ground now exposed below (such

[7] The material for the remainder of this section is based on Worm and van Harten (1996). For other issues relating to road maintenance, can be found in Dekker et al. (1998) and Rose and Bennett (1992).

as compaction). The road itself will then be built up above, usually consisting of several layers. The two bottom layers are as follows:

- Subgrade: The ground that is exposed once the ground has been dugout ready to build the road. The top level of this is termed the formation
- Capping: This is a layer added above the subgrade to protect it in new constructions (and often constitutes the formation).

This is followed by four more layers (in ascending order from bottom to top) are

- Sub-base,
- Base,
- Binder Course,
- Surface Course.

The nearer the surface, the profile needs to be more flatter as an uneven surface will be uncomfortable for vehicle occupants and will wear more quickly (as each time a vehicle hits a bump the hammering effects impacts on the surface). These factors are the main reasons for the layered construction of the road. Weight on any unbound material will compact it down with time, as material is forced down and fills gaps. For this reason, during construction of each layer compaction is carried out.

The most commonly used material for use in sub-base and base is an unbound material made from crushed rock, crushed slag, crushed-concrete and recycled aggregates. The binder course helps distribute the load of traffic above onto the base course, which is usually a weaker material. Materials used include open-graded macadam,[8] dense-coated macadam and rolled asphalt. Surface courses are laid in a wide range of bituminous materials, ranging in thickness from 20 to 40 mm. The material selected is dependent on the anticipated traffic intensity. Asphalt pavement is known for its durability and resilience.

The deterioration of a road depends on the materials used in the construction of the road and several other factors. In the case of asphalt pavement, the deterioration is because the materials that make up asphalt begin to break down over time and are affected by elements such as rain, sunlight and chemicals that come into contact with the pavement surface. The liquid asphalt binder that is the "glue" of the pavement begins to lose its natural resistance to water, allowing it to penetrate into and underneath the pavement. Once this happens, the surface can quickly fall prey to a number of different types of deterioration. The premature deterioration of asphalt pavement is usually due to failures in construction and/or human error and includes the following factors:

[8] Compacted broken stone usually bound with tar or asphalt (also referred to as bitumen).

Table 2.1 Groups of damage

Groups of damage	Features of damage
Texture	Ravelling[a], skidding resistance[b]
Evenness	Transverse and longitudinal evenness[c], irregularities[d], roughness[e]
Soundness	Transverse and longitudinal cracks[f], crazing[g], potholes[h], marginal strip, edge damage[i], kerb
Miscellaneous	Water run-off, verge[j]

[a] *Ravelling* loss of aggregate (used in road construction) due to (i) cohesive failure of the bituminous mortar, or (ii) adhesive failure in the adhesive zone

[b] *Skid resistance* characterises the cumulative effects of snow, ice, water, loose material and the road surface on the traction produced by the wheels of a vehicle

[c] *Longitudinal (transverse) evenness* measurement of longitudinal (transverse) profiles for determination of rutting. A rut is sunken track or groove made by the passage of vehicles within pavement layers that accumulates over time

[d] *Irregularity* something irregular, such as a bump in a smooth surface

[e] *Roughness* deviations of surface from true planar surface with characteristic dimensions that affect vehicle dynamics, ride quality, dynamic loads and drainage. It is defined using the International Roughness Index (IRI)—deviations (in metres) per kilometre. Roughness is a function of age, strength, traffic loading, potholes, cracking, ravelling, rutting, environment, etc

[f] *Longitudinal (transverse) cracking* cracks that run along (perpendicular to) the road

[g] *Craze* a fine crack in a surface of the road

[h] *Pothole* an open cavity in road surface with at least 150 mm diameter and at least 25 mm depth

[i] *Edge Damage* loss of bituminous surface material (and possibly base materials) from the edge of the pavement, expressed in square metres per km

[j] *Verge* a strip of grass or other vegetation beside a road

- Insufficient or improperly compacted base below the asphalt.
- Over or under compaction of the asphalt.
- Improper temperature of the asphalt when applied.
- Poor drainage.

Cracks, potholes, edge defects, depressions and corrugations are the significant road defects observed in the field. Traffic, age, road geometry, weather, drainage, construction quality as well construction material and maintenance policy are the major factors that affect the deterioration of a road. The state of a road surface can be described by a state vector where each component corresponds to one of the groups of damage, and these (and the features of damage for each group) are indicated in Table 2.1.

The quality of a road is often described through a function involving one or more of these features. Figure 2.6 illustrates the quality deterioration over time (as defined through some feature such as ravelling) with no maintenance actions.

The *quality standards* (also referred to as *norms*) for a road are derived from the lowest acceptable value for these features. They can be (i) local—for segments (for example 100 m in length) of a road (or lane) or (ii) global—the whole length of the

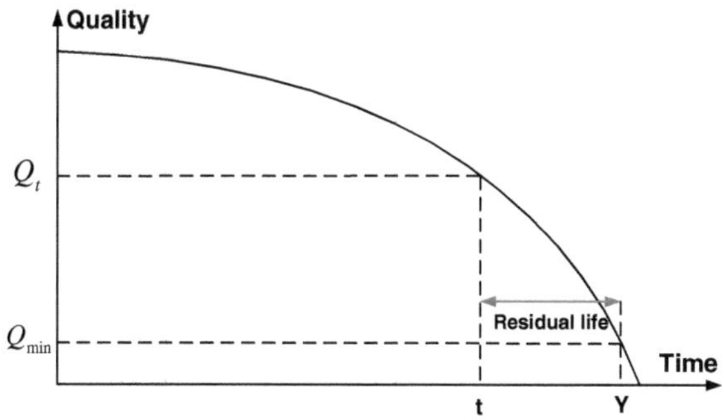

Fig. 2.6 Deterioration of road quality over time

road. In Fig. 2.6 at time t (since the construction of the road), the quality is Q_t and Y denotes the time when the quality reaches the minimum acceptable level $Q_{\min}$ at which instant CM action is needed. The interval $(Y - t)$ provides a window over which PM action can be initiated to avoid the need for CM action.

The performance of road transport is a complex function of the quality of the road. It is a vector that characterises the flow rate (number of cars passing per unit time) which would depend on the number of lanes open for traffic and the quality of ride. These depend on the speed of travel, which in turn, depends on the condition of the road (potholes, roughness to ensure grip) and weather conditions (rain, snow, etc.). From the public perspective, the quality of ride and safety are the important performance measures. The latter is also of importance to the regulators. From a road owner's perspective, cost of maintenance (to ensure the minimum standards for safety) and profits (in the case of toll roads operated by private business enterprises) are two important performance measures.

2.4 Reliability

The reliability of a product (component or some intermediate element) conveys the concept of dependability, successful operation or performance and the absence of failures. It is an external property of great interest to both manufacturer and consumer. Unreliability (or lack of reliability) conveys the opposite. More technical definitions of reliability are the following:

> The ability of an item to perform a required function, under given environmental and operational conditions and for a stated period of time (ISO 8402 1986).
>
> The reliability of a system is the probability that the product (system) will perform its intended function for a specified time period when operating under normal (or stated) environmental conditions (Blischke and Murthy 2000).

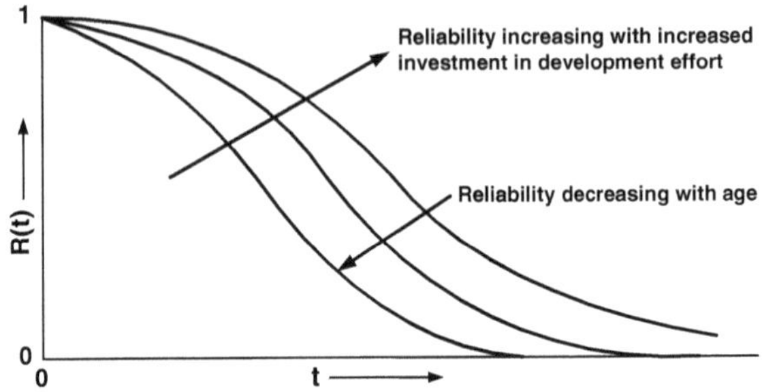

Fig. 2.7 Plots of reliability functions

The reliability is given by a function $R(t;\,\theta)$ with the following properties[9]:

- $R(t)$ is a non-increasing function of t, $0 \le t < \infty$.
- $R(0) = 1$ and $R(\infty) = 0$.

Typical plots of $R(t)$ are shown in Fig. 2.7.

2.4.1 Linking Product and Component Reliabilities

The linking of component reliabilities to product reliability is done through the structure function. This function can be obtained using either FTA (discussed in the previous subsection) or a reliability block diagram (RBD).

2.4.1.1 Reliability Block Diagram

In a RBD, each component is represented by a block with two end points. When the component is in its working state, there is a connection between the two end points. This connection is broken when the component is in a failed state. A product (system) can be represented as a network of such blocks, each with two end points. The product (system) is in the working state if there is a connected path between the two end points. If no such path exists, then the system is in a failed state.

2.4.1.2 Structure Function

A product contains n components, and $X_i(t)$, $1 \le i \le n$, denotes the state of component i at time t, with

[9] θ is the set of parameters for the reliability function. Often we will suppress this and use $R(t)$ instead of $R(t;\theta)$ for notational ease. $F(t) = 1 - R(t)$ is called the failure distribution function and characterises the time to first failure (a random variable).

$$X_i(t) = \begin{cases} 1 & \text{if component } i \text{ is working at time } t \\ 0 & \text{if component } i \text{ is failed at time } t \end{cases} \tag{2.1}$$

Let $X(t) = (X_1(t), X_2(t), \ldots, X_n(t))$ denote the state of the n components at time t, and $X_S(t)$ (a binary variable) denote the state of the system at time t. Then, from FTA or the RBD, one can derive an expression of the form

$$X_S(t) = \phi\left(\underset{\sim}{X}(t)\right), \tag{2.2}$$

which links the component states to the system state. $\phi(\cdot)$ is called the *structure function.*[10]

Let $\underset{\sim}{R}(t) = (R_1(t), R_2(t), \ldots, R_n(t))$ denote the set of reliability functions of the n components of the product and $R_S(t)$ the reliability function for the system. If the component failures are independent, then

$$R_S(t) = \phi\left(\underset{\sim}{R}(t)\right) \tag{2.3}$$

so the system reliability can be expressed in terms of the component reliabilities. When failures are not independent, deriving the expression for the structure function is more complicated.

2.4.2 PLC Perspective: Different Notions of Reliability

From a product life cycle perspective, there are several different notions of reliability. Figure 2.8 (Murthy et al. 2008) shows how these are sequentially linked and the factors that affect them. We briefly discuss four reliability concepts.

2.4.2.1 Design Reliability

At the design stage, the desired product reliability is determined through a trade-off between the cost of building in reliability and the consequences of failures. This trade-off is discussed in detail in Murthy et al. (2008). From this, one derives the reliability specification at the component level. One then evaluates the design reliability.

---

[10] The details can be found in many books on reliability; see, for example, Blischke and Murthy (2000) and Rausand and Høyland (2004).

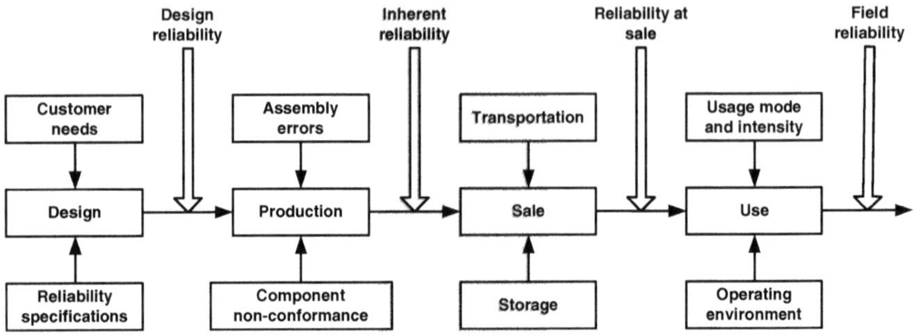

Fig. 2.8 Different notions of reliability (standard product)

2.4.2.2 Inherent Reliability

For standard products produced in volume, the reliability of the produced item can differ from the design reliability because of assembly errors and component non-conformance. The reliability of produced items is the "inherent reliability" of the product.

2.4.2.3 Reliability at Sale

After production, the product must be transported to the market and is often stored for some time before it is sold. The reliability of an item at sale depends on the mechanical load (resulting from vibrations during transport) and impact load (resulting from mishandling) to which it has been subjected, the duration of storage and the storage environment (temperature, humidity, etc.). As a result, the reliability at sale can differ from the inherent reliability. Once an item is sold, it may either be stored for an additional time (if the item has been purchased for later use or is used as a spare), or it may be put into operation immediately. The additional storage time may again affect its reliability.

2.4.2.4 Field Reliability

The reliability performance of an item in operation depends on the length and environment of prior storage and on operational factors such as the usage intensity (which determines the load—electrical, mechanical, thermal and chemical—on the item), usage mode (whether used continuously or intermittently) and operating environment (temperature, humidity, vibration, pollution, etc.) and, in some instances, on the human operator. The reliability performance of an item in operation is often referred to as "field reliability".

2.5 Maintenance of Products and Plants

Maintenance consists of the different functions (or activities) necessary to keep a system in, or restoring it to, an acceptable state (or operating condition). Maintenance involves one or more of the following actions:

- Servicing
- Testing/Inspection
- Removal/Replacement
- Repair/Overhaul
- Modification.

Comment: In the literature, the term "MRO" is used extensively. It is acronym for the following actions:

M Maintenance (minor PM actions)
R Repair (CM actions)
O Overhaul (major PM action).

2.5.1 Corrective Maintenance

The failure of a system is due to the failure of one or more of its components. CM actions are actions to restore a failed system to operational state by rectification actions (repair or replace) on the failed components.

2.5.1.1 Classification of CM Actions

Let x denote the time of first failure. The different types of CM actions and their impact on system reliability are as follows:

Back to New: This involves the replacement of a failed item by a new one. As such the system reliability at time t is given by $R_1(t) = R_0(t - x)$ for $t > x$ where $R_0(\cdot)$ is the reliability of a new system. Note that this is appropriate for maintenance actions at the component level.

Minimal Repair: Here, the reliability of the item is unaffected by the maintenance action. As such, the reliability after repair is the same as that just before failure. This is an appropriate characterisation at the system level if the failure is due to one or few components and either repairing or replacing them has very little impact on the overall system reliability. In this case, the system reliability at time t is given by $R_2(t) = R_0(t)/R_0(x)$ for $t > x$.[11]

[11] This follows from simple argument based on conditional probability (see Appendix A).

Imperfect Repair: Here, the reliability of the item is affected by the repair action. One can define two types of imperfect repairs—case (i) and case (ii). In the former case, the reliability after repair is better than what was just before failure. This characterises the situation where the failure is a major failure requiring the replacement of several components by new ones so that the overall reliability improves. In this case, the system reliability at time t, $t > x$, is given by $R_3(t)$ with $R_2(t) < R_3(t) < R_1(t)$. In the latter case, the reliability after repair is lower than that just before failure. This usually is the effect of poor quality of repair that degrades the reliability so that the system reliability at time t, $t > x$, is given by $R_4(t)$with $R_4(t) < R_2(t)$.

Figure 2.9 shows the impact of the different types of CM actions on the reliability of the system after a failure.

2.5.1.2 Repair Time

In general, the time to carry out a CM action is uncertain and needs to be characterised as a random variable. If the variability in the repair time relative to the mean time to repair is small, then one can treat it as a deterministic quantity (the mean time to repair).[12]

2.5.1.3 Repair versus Replace

When a repairable item fails, there is an option to either repair or replace it by a new (or used) item. The optimal decision is usually based on cost considerations and the impact of the actions on future failures of the item involved.

2.5.2 Preventive Maintenance

PM actions are actions to control system degradation and reduce the likelihood of failure

2.5.2.1 Classification of PM Actions

The different types of PM actions are as follows:

Clock-based maintenance: Here, PM actions are carried out at set times.

---

[12] Typically, the time taken to repair or replace a failed item is often very much smaller than the time between failures (in a statistical sense) so that one can ignore repair times and treat the repairs as being instantaneous for the purpose of modelling of failures over time. This is discussed in Chap. 3.

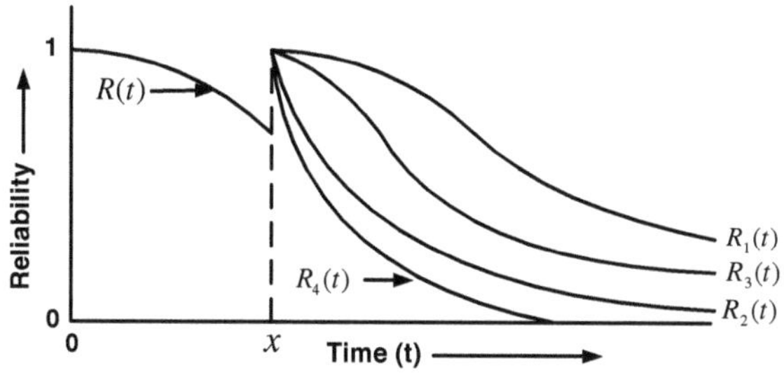

Fig. 2.9 Impact of different types of CM action on system reliability

Age-based maintenance (*ABM*): Here, PM actions for an item (component or higher level element) are based on the age of the item.[13]

Usage-based maintenance (*UBM*): Here, PM actions are based on total output (or usage) of the item since the last PM action.

Condition-based maintenance (*CBM*): Here, PM actions for an item are based on the condition of the item being maintained. This involves monitoring of one or more variables characterising the wear process (e.g. crack growth in a mechanical component).

Opportunity-based maintenance (*OBM*): This is applicable for multicomponent items, where maintenance actions (PM or CM) for a component provide an opportunity to carry out PM actions on one or more of the remaining components contained in the item.

Design-out maintenance (*DOM*): This involves carrying out modifications through re-design of one or more components so that the new components have better reliability characteristics.

Imperfect PM Actions: Here, the reliability characteristics improve after a PM action but not to as-good-as new and are similar to imperfect CM actions.

Overhaul (*Shutdown Maintenance*): In the case of complex products and plants, major overhaul involves dismantling the whole system and replacing components that have deteriorated significantly. The reliability characteristics improve significantly after an overhaul. However, the reliability of the system after overhaul decreases with the number of overhauls as indicated in Fig. 2.10.

Let $R_j(t)$ denote the system reliability after the jth overhaul with t denoting the time subsequent to the overhaul (one is using a local clock that is reset to zero after each overhaul).[14] Note that $j = 0$ corresponds to a new system. Then, we have the following:

- $R_j(t)$ is a decreasing function of t (the effect of degradation)
- $R_{j+1}(t) < R_j(t), j \geq 0,$ implying that the reliability of a system subjected to $(j + 1)$ overhauls is inferior to that subjected to j overhauls.

---

[13] The first two types of PM are also referred to as Time-based maintenance (TBM).

[14] Here the subscripts refer to the number of times an item has been subjected to overhauls and should not be confused with the notation in Sect. 2.4.1 where it refers to reliability of different components.

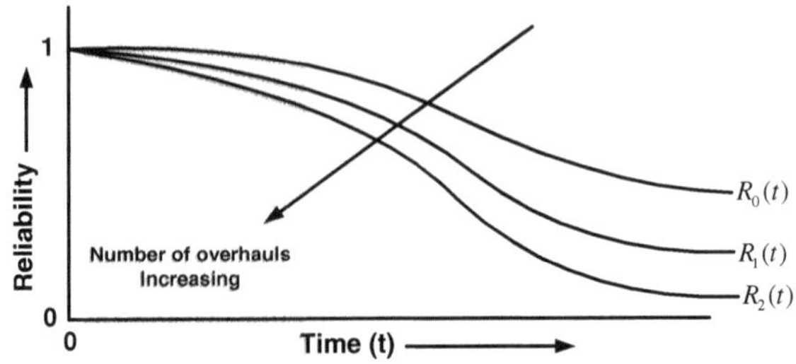

Fig. 2.10 Effect of overhauls on system reliability

2.5.2.2 Time for PM Action

In general, the time to carry out a PM action is predictable and as such it can be characterised as a deterministic quantity.

2.5.3 *Maintenance Costs*

Maintenance costs can be divided into two categories.

2.5.3.1 Direct Costs

The direct costs (which are viewed as part of the maintenance budget) are as follows:

- cost of manpower
- cost of material and spares
- cost of tools and equipment needed for carrying out maintenance actions
- overhead costs
- etc.

2.5.3.2 Indirect Costs

In addition, many other costs are affected either directly or indirectly by maintenance (or, more precisely, by lack of an effective maintenance policy). The costs involved depend on the nature of the business. In the case of a manufacturing operation, some of these costs are as follows:

- Equipment-related
  - accelerated wear because of poor maintenance
  - excessive spare parts inventory
  - unnecessary equipment redundancy
  - excessive energy consumption

- Production-related
  - rework
  - excessive scrap and material losses
  - idle operators due to breakdowns
  - delays in fulfilling orders
- Product-related
  - quality and reliability issues
- Customer-related
  - Customer dissatisfaction
  - Negative word-of-mouth effects.

2.5.4 Some Maintenance Policies

2.5.4.1 Maintenance of Products

Several different maintenance policies for products have been proposed and studied.[15] Examples of some of these policies are the following:

Policy 1 (*Age Policy*): Replace an item (PM action) by a new item when it reaches age v after being put into use or on failure (CM action) should the item fail earlier. This policy is characterised by the decision variable set $\Upsilon \equiv \{v\}$.

Policy 2 (*Block Policy*): Replace an item (PM action) by a new item at set times $t_k = kv$, $k = 1, 2, \ldots$. Failures between PM actions are rectified (CM action) by replacing the failed item by a new one. This policy is also characterised by the decision variable set $\Upsilon \equiv \{v\}$.

Policy 3 (*Periodic Policy*): Replace an item (PM action) by a new one at set times $t_k = kv$, $k = 1, 2, \ldots$. Failures between PM actions are minimally repaired (CM action). This policy is also characterised by the decision variable set $\Upsilon \equiv \{v\}$.

2.5.4.2 Maintenance of Plants

Policies for the maintenance of plants policies involve imperfect PM and overhauls with the above three policies used at the product or component levels. Two such policies are the following.

Policy 4 (*Imperfect PM*): The item is subjected to K imperfect PM actions before it is replaced by a new item. The time instants at which these actions are carried out are given by $\{t_k, 1 \le k \le K\}$ with $t_i < t_j$ for $i < j$. The reduction in the

[15] Nakagawa (2005) deals with the modelling and analysis of several maintenance policies.

failure hazard function during the kth PM action is δ_k. The item is replaced at time t_{K+1}. All failures between PM actions are rectified through minimal repair. This policy is characterised by the decision variable set $\Upsilon \equiv \{K; (t_k, \delta_k), 1 \le k \le K; t_{K+1}\}$.

Policy 5 (*Overhaul*): The item is subjected to the first overhaul after it has been in operation for a period t_0 and then is subsequently subjected to a sequence of overhauls. After the kth overhaul, the system is kept in operation for a period t_k after which it is subjected to an overhaul if $k < K$ or else is replaced by a new unit after being in operation for t_K after the last overhaul. All failures in between overhauls are repaired minimally. This policy is characterised by the decision variable set $\Upsilon \equiv \left\{K; t_j, 0 \le j \le K; \delta_j, 0 \le j \le K-1\right\}$.

2.5.5 Fleet Maintenance

There are several issues that need to be taken into account in the context of fleet maintenance.[16] Some of these are as follows:

- The age and the condition of units in a fleet can vary significantly so that the units are not statistically similar. The main reasons for this include (a) the units are purchased at different time points, (b) the usage of each unit can be quite different and hence the degradation levels of the units with the same age can be quite different, and (c) the ages of constituent components of a unit can be quite different due to the maintenance history. This raises an issue—how to control the "health level" of the fleet by appropriate maintenance and replacement decisions.
- Fleet maintenance needs to coordinate with production (or service) requirements and needs to take into account resource constraints.
- The failure consequence of a unit strongly depends on the configuration of a fleet and the functional requirements assigned to units within the fleet. This implies that the fleet maintenance needs to consider the priority of each unit and devise appropriate maintenance policies.
- The technological evolution of the unit makes maintenance options multidimensional—repair or replacement; if replacement, whether a particular unit should be replaced with a unit with same technology unit or one with more advanced technology. This implies that one needs to take into account technological evolution in the decision-making process when retiring (or replacing) old or degraded units.

Because of the multiunit nature, group and opportunistic maintenance are appropriate for fleet maintenance. Many different policies have been proposed and we discuss a few of them.

[16] For more information, see Cassady et al. (1998).

2.5.5.1 Group Maintenance Policies

One can define three types of group maintenance policies for a fleet.

Type I Policies: Here, the maintenance actions are based on age of the fleet. A group replacement is performed when the fleet reaches an age T.

Type II Policies: Here, the maintenance is based on the number of failed units. If the fleet is monitored continuously, then maintenance actions are initiated when the number of failed units reaches m. At this instant, all failed units are replaced with new ones (CM action) and all functioning units are serviced (PM action) so that they are restored to good-as-new. When the monitoring is not continuous, then the fleet is inspected at discrete time instants and maintenance actions are initiated only if the number of failed units is equal to or greater than m.

Policy 6 (Assaf and Shanthikumar 1987): The system is inspected at discrete time instants. Upon an inspection, the failed units are repaired if the number of failed units is greater than or equal to m; otherwise, they are left idle (failed state). The time to the next inspection is decided based on the number of failed units. The decision variables are m and the state-dependent inspection time instants.

Type III Policies: Here, maintenance action is based on both age and number of failed units.[17] The maintenance actions are initiated (for the continuous monitoring case) when the fleet reaches an age T or at the time instant when the number of failed units reaches m, whichever comes first. All failed units are replaced with new ones (CM action), and all functioning units are serviced (PM action) so that they become good-as-new.

Policy 7 (Park and Yoo 1993): The fleet consists of a group of identical units. Each unit is replaced on failure during the interval $(0, T)$. Beyond this interval, failed units are left idle until the number of failed units reaches a specified number m, when a block replacement is performed. The decision variables of the policy are T and m.

2.5.5.2 Opportunistic Maintenance Policies

Ritchken and Wilson (1990) deal with a fleet of machines in a production line and propose two types of opportunistic maintenance (Type I and II, respectively). In Type I opportunistic maintenance, CM action on a failed unit needs to be performed without any delay and PM actions on non-failed units can be advanced if appropriate and possible. In Type II opportunistic maintenance, failed units can be kept idle (failed state) for some amount of time so that one can postpone CM action to coincide with the first PM opportunity subsequent to the failure. A Type II policy involving only CM actions and two types of failures (minor and major or catastrophic) is the following:

[17] Type I and Type II policies are special cases of Type III policies.

Policy 8 (Sheu and Jhang 1997): The policy involves two intervals—$[0, T-w]$ and $[T-w, T]$. Minor failures are rectified by minimal repairs at any time, and major failures are rectified immediately through replacements in the first interval and are not rectified in the second interval so that the failed units remain idle. Group maintenance is conducted at time T or when the number of failed units reaches $m(\leq n)$, where n is the total number of units whichever comes first. The decision variables of the policy are w, T and m.

Comment: In some cases, one or more units are cannibalised to provide spares for the other units.

2.6 Maintenance of Infrastructures

Maintenance of infrastructures include services such as clearing (snow, any object hindering the operation, etc. in the case of road and rail tracks) and cleaning (routine cleaning of buildings, rolling stock, vegetation growth on the sides of roads and rail tracks, etc.) and fixing (damaged road signs) for safe operations. These are referred to as service/operations and are different from PM and CM actions relating infrastructure per se. PM and CM are infrastructure specific, and we discuss these for road infrastructure. PM actions include inspection to monitor and assess the condition of the infrastructure. Based on the inspection results, e.g. the severity of the fault to traffic and the availability of resources, the decision is made to rectify the fault immediately or it is planned for a later stage considering all the risks to traffic and business, etc. Also, since failure is not so well defined, there is a blurring of PM and CM actions. In general, PM actions are those tasks that can be carried out in a short time period without too much interruption to the normal operation of the infrastructure. In contrast, CM actions take a longer time to complete (possibly running into months) and affect normal operations in a significant manner and are costly.

The main purpose of maintenance actions (PM and CM) is to control infrastructure degradation due to age, usage, load carried and other environmental factors, etc. and restore it to normal operating condition in the case of failure or other faults. In some industry sectors, OPEX (Operating Expenditure) denotes the expenditure associated with service/operations and PM actions and CAPEX (Capital Expenditure) the expenditure associated with CM actions and upgrades, etc. Major maintenance and investment involve a great deal of expenditure but have a direct influence on the financial and operational performance of the infrastructure.[18]

[18] In a 2009 report released by the American Association of State Highway and Transportation Officials (USA) about 50 % of the roads in the USA are in bad condition with urban areas worse.

Table 2.2 Work types for road maintenance

Work class	Work type	Work activity/operation
Routine maintenance	Routine pavement	Patching, edge-repair, crack sealing, spot re-gravelling, shoulders repair, etc.
	Drainage	Culvert repairs, clearing side drains
	Routine miscellaneous	Vegetation control, markings, signs
Periodic maintenance	Preventive treatment	Fog seal, rejuvenation
	Resurfacing	Surface dressing, slurry seal, cape seal, re-gravelling
	Rehabilitation	Overlay, mill and replace, inlay
	Reconstruction	Partial reconstruction, full pavement reconstruction
Special	Emergency	Clearing debris, repairing washout/ subsidence, traffic accident removal, etc.
	Winter	Snow removal, salting, gritting, etc.
Improvement	Widening	Partial widening, lane addition
	Realignment	Horizontal and vertical geometric improvements, junction improvement
	Off-carriageway	Shoulders addition, shoulders upgrading, side drain improvement, etc.
Construction	Upgrading	Upgrading by changing the surface class
	New section	Expanding of an existing section (with more lanes), new section (link)

2.6.1 Road Infrastructure

Pavements are designed for an expected service (design) life that can vary from 10 to 60 years, and for asphalt pavements, the typical life is 40 years. On each lane sector (for a multilane road), the initiating event for maintenance can be of two kinds:

1. End of the technical lifetime of the asphalt,
2. Economic depreciation of the road surface before its technical lifetime is over.

Maintenance is considered in the whole life cost of the road with CM actions at 10, 20 and 30 year milestones, and there is considerable freedom for maintenance planning with 15–20 possible actions per lane sector, from which the best choice has to be made. Maintenance (PM and CM) include many activities, and these are listed Table 2.2 [adapted from Archondo-Callao (2008)].

References

Archondo-Callao R (2008) Bituminous road deterioration, highway development and management (HDM-4). The World Bank, Washington
Assaf D, Shanthikumar JG (1987) Optimal group maintenance policies with continuous and periodic inspections. Manage Sci 33:1440–1452
Blischke WR, Murthy DNP (2000) Reliability: modeling, prediction and optimization. Wiley, New York
Cassady CR, Nachlas JA, Murdock WP (1998) Comprehensive fleet maintenance management. In: Proceedings of the IEEE international conference on systems, man and cybernetics, vol 5. IEEE, New York, pp 4665–4669
Dekker R, Plasmeijer R, Swart J (1998) Evaluation of a new maintenance concept for the preservation of highways. IMA J Math Appl Bus Ind 9:109–156
ISO 8402 (1986) Quality vocabulary. ISO, Geneva
Murthy DNP, Rausand M, Osteras T (2008) Product reliability: performance and specifications. Springer, New York
Nakagawa T (2005) Maintenance theory of reliability. Springer, London
Park KS, Yoo YK (1993) (τ, k) block replacement policy with idle count. IEEE Trans Reliab 42:561–565
Rausand M, Høyland A (2004) System reliability theory: models, statistical methods, and applications, 2nd edn. Wiley, Hoboken
Ritchken P, Wilson JG (1990) (m, T) group maintenance policies. Manage Sci 36:632–639
Rose G, Bennett D (1992) Locating and sizing road maintenance depots. Eur J Oper Res 63:151–163
Sheu SH, Jhang J (1997) A generalized group maintenance policy. Eur J Oper Res 96:232–247
Worm JM, van Harten A (1996) Model based decision support for planning of road maintenance. Reliability Eng Syst Safety 51:305–316

Chapter 3
Modelling and Analysis of Degradation and Maintenance

3.1 Introduction

Models play an important role in solving decision problems. They are used to (i) analyse the effect of changes to decision variables on system performance (for example, the effect of different PM actions on system failures) and (ii) decide on the optimal values of decision variables to achieve some specified objectives (for example, optimum PM to minimise total maintenance costs).

There are many different types of models and our focus will be on mathematical models. Modelling is both an art as well a science. In this chapter, we look at models and the modelling process. The concepts will be used in later chapters to build a variety of models in the context of extended warranties, maintenance service and lease contracts.

For some systems (mainly products and plants), the degradation and failure depends only on age with usage not having any impact. In this case, one uses one-dimensional (1-D) formulations to model failures. However, for other systems, the degradation and failure depend on both age and usage. In this case, one can either use 1-D or 2-D formulations.

The outline of the chapter is as follows. We start with a general discussion on models and the modelling process in Sect. 3.2. Since most of the models use 1-D formulations, we will deal with these in Sects. 3.3, 3.4, 3.5 and 3.6. Section 3.3 deals with the modelling of first failure (for products and plants) and looks at the impact of usage and environment on reliability. In Sect. 3.4, we look at the modelling of maintenance actions and this leads us to Sect. 3.5 which deals with the modelling of failures over time (second and subsequent failures). The modelling and analysis of several maintenance policies is carried out in Sect. 3.6. Section 3.7 deals with 2-D model formulations. We conclude with a brief discussion of the modelling of the degradation and maintenance of infrastructures in Sect. 3.8.

D. N. P. Murthy and N. Jack, *Extended Warranties, Maintenance Service and Lease Contracts*, Springer Series in Reliability Engineering,
DOI: 10.1007/978-1-4471-6440-1_3,

3.2 Models and Modelling Process

3.2.1 Models

A model is a representation of the real world that is relevant to the problem of interest. A mathematical model is an abstract representation involving a mathematical formulation. When uncertainty is a significant feature of the real world (as is the case, for example, with the time to failure of an item), then concepts from probability and statistics, as well as data from the real world, play an important role in linking the model to reality, as indicated in Fig. 3.1.

3.2.2 Modelling Process

One can use two different approaches.

(i) Black-Box Approach
In the black-box approach, the underlying physical mechanisms responsible for failure occurrence are ignored. The model formulation is selected based solely on the data available. Product-related data are the failure times and service times[1] while other data are operational data (such as usage intensity, operating environment). This modelling approach is also termed *data-based* or *empirical* modelling.

(ii) White-Box Approach
Here, the mechanisms leading to degradation and failure are modelled using relevant theories (for example, different theories such as corrosion, wear, overstress—for component failures). This approach to modelling is also called *physics-based* or the *theory-based* approach.

3.2.3 Black-Box Approach to Modelling

The black-box approach to modelling is an iterative process involving several steps, as indicated in Fig. 3.2.[2]

---

[1] Service time refers to the duration in the working state for a non-failed item.

[2] There are many books that discuss the modelling process in detail; see for example, Murthy et al. (1990) and the references cited therein.

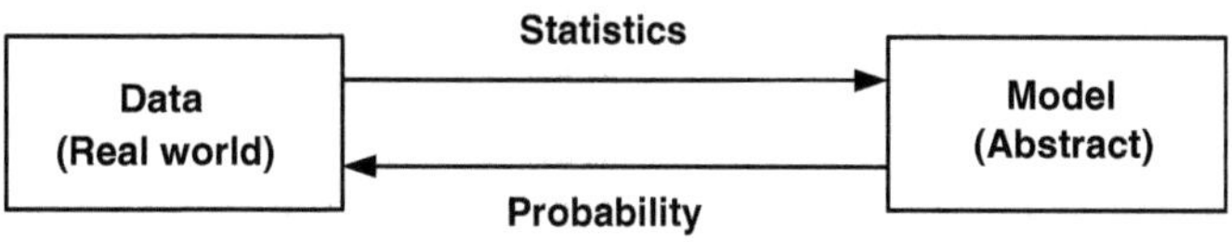

Fig. 3.1 Link between real world and model

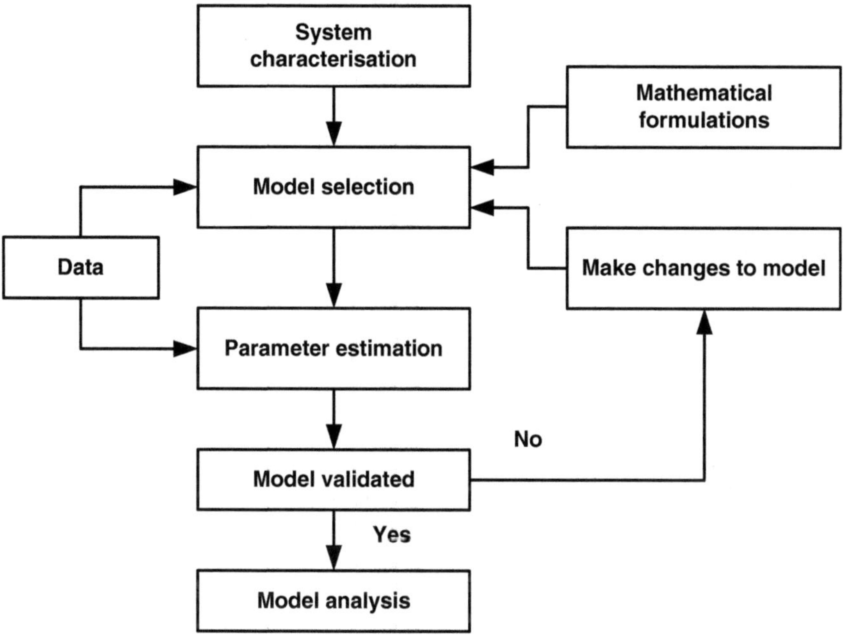

Fig. 3.2 Modelling process

We now discuss the key steps in the modelling process. These principles will be applied to reliability and maintenance modelling.

Step 1 System Characterisation
Characterisation of a system details the salient features of the system that are relevant to the problem under consideration. This generally involves a process of simplification. The variables used in the system characterisation and the relationships between them are problem dependent.

Step 2 Model Selection
The type of mathematical formulation to be used depends on the system characterisation and the approach used. This is discussed further in the next subsection.

Step 3 Parameter Estimation
The model will involve one or more unknown parameters, and numerical values for these are needed. These values are obtained by means of a statistical methodology called parameter estimation. The approach used depends on the type and amount of data available.

Step 4 Model Validation
Validation involves testing whether or not the model selected (along with the assigned parameter values) imitates the real world sufficiently adequately to yield a meaningful solution to the problem of interest. The approach used can vary from a visual comparison between model predictions and observed data to statistical methods such as hypothesis testing for goodness of fit.

Step 5 Model Analysis
Several different approaches can be used to conduct analysis of the model. These include analytical methods (which yield closed form results as functions of the model parameters), computational methods and simulation.

3.2.4 Classification of Maintenance Models

In Sect. 1.2.4, we defined a multilevel decomposition of an engineered object (product, plant or infrastructure) viewed as a system. The modelling for maintenance can be done at any level. The engineered object is one element with many other elements at the business level. In maintenance modelling, we have three levels of models.

3.2.4.1 Component-Level Models

Here, one uses either the white-box (if there is a well-developed theory) or the black-box approach to build the model. In the black-box approach, the modelling involves probability distribution functions and point process formulations. In the white-box approach, one models the degradation of a component through some variable (such as wear, crack growth) using a stochastic differential equation (if time is treated as a continuous variable) or a difference equation (if time is treated as discrete variable) formulation. These models are useful for (CBM) of components.

3.2.4.2 System-Level Models

Here, one uses the black-box approach and the degradation and failures are modelled using different types of stochastic formulations. The formulations involve stochastic processes with discrete state (with two or more states) and time treated as either continuous or discrete.

3.2.4.3 Business-Level Models

Here, the engineered object is viewed as one of many other various elements which include business- or societal-related issues (such as finance, marketing, impact on customers, business performance). The modelling of the object depends on the context. For example, for life cycle costing, one models the cost of maintenance on a yearly basis.

Comment: Component-level models based on white-box approach are often referred to as *microscopic models* and business-level models are referred to as *macroscopic models.*

3.3 Modelling First Failure (1-D Formulations)

The modelling of first failure can be done at the system level, component level or any intermediate level. In Sect. 2.3.1, we discussed three approaches to characterising performance degradation. The first two approaches (binary and finite multistate characterisation) are used in black-box modelling while the third approach (infinite multistate characterisation) is used in white-box modelling. The black-box approach uses distribution functions from probability theory for modelling first failure and point process formulations for modelling subsequent failures.[3] The white-box approach uses stochastic differential equations for modelling first and subsequent failures.

The time to first failure, $\tilde{T}$, for an item (system, component or some intermediate level) is a non-negative continuous random variable. This is characterised by a distribution function $F(t; \theta)$ (also called a *cumulative distribution function* or CDF), which gives the probability that the item fails at or before time t. The CDF is given by

$$F(t; \theta) = P\{\tilde{T} \leq t\}. \tag{3.1}$$

Comment: θ denotes the set of parameters of the distribution function. For notational ease, the dependence on θ is often suppressed and $F(t)$ is used instead of $F(t; \theta)$. We follow this convention in the remainder of the chapter.

When the failure distribution function $F(t)$ is differentiable, its derivative $f(t) = \mathrm{d}F(t)/\mathrm{d}t$ is called the failure density function.

The reliability function $R(t)$ (sometimes denoted by $\bar{F}(t)$),[4] is defined to be the probability that the item survives beyond time t, so that

$$R(t) = P\{\tilde{T} > t\} = 1 - F(t). \tag{3.2}$$

---

[3] Appendix A [B] reviews material from probability theory [stochastic processes] that is relevant for reliability modelling.

[4] We will use both notations throughout the book.

The conditional probability that the item will fail in the interval $[t, t + \delta t)$, given that it has not failed prior to t, is given by

$$F(t + \delta t|t) = \frac{F(t + \delta t) - F(t)}{R(t)}. \tag{3.3}$$

The *hazard function* (or failure rate function) $h(t)$ associated with $F(t)$ is defined as

$$h(t) = \lim_{\delta t \to 0} \frac{F(t + \delta t|t)}{\delta t} = \frac{f(t)}{R(t)}. \tag{3.4}$$

The hazard function $h(t)$ can be interpreted as the probability that the item will fail in $[t, t + \delta t)$, given that it has not failed prior to t. In other words, it characterises the effect of age on item failure more explicitly than $F(t)$ or $f(t)$. The hazard function can have many different shapes (such as constant, increasing, decreasing, bathtub, roller coaster and many more) depending on the form of the distribution function and its parameters.

The *cumulative hazard function*, $H(t)$, is defined as

$$H(t) = \int_0^t h(t')\mathrm{d}t'. \tag{3.5}$$

It can easily be shown that

$$R(t) = 1 - e^{-H(t)} \tag{3.6}$$

Comment: Characterising the time to failure can be achieved either through the distribution, density or hazard function since they are all equivalent and any one function can be derived from any of the others.

3.3.1 Distribution (Density) Functions for Modelling

Many different distribution (density) functions have been used in modelling the time to first failure. The distribution function $F(t)$ must have the property that $F(t) = 0$ for $t \leq 0$. Some of the well-known basic distribution functions with this property used extensively in modelling first failure are the following[5]:

---

[5] Expressions for the various distributions mentioned in this subsection can be found in Appendix A.

1. Exponential distribution
2. Gamma distribution
3. Weibull distribution.

Many other distributions that are derived from basic distributions (some with $F(t) > 0$ for $t \leq 0$) through transformations have also been used extensively since they exhibit more complex patterns for the hazard function. These include the following:

1. Inverse Gaussian (Wald) distribution
2. Log-normal distribution
3. Three-parameter Weibull distribution
4. Extended Weibull distribution
5. Modified Weibull distribution
6. Exponentiated Weibull distribution.

Distributions involving two or more basic distributions allow for a still more diverse range of shapes for both density and hazard rate functions (for example, bimodal shapes for the density function and roller coaster shapes for the failure rate). As a result, they are extremely useful for modelling complex data which cannot be adequately modelled by a single basic or derived distribution. Of particular importance are the following three forms.

1. Mixtures Models
2. Competing risk Models
3. Multiplicative Models.

Comment: Mixture models (competing risk models) are appropriate for modelling the effect of component non-conformance (assembly errors). For more details, see Murthy et al. (2003).

3.3.2 Modelling the Effect of Usage and Environment

As discussed in Sect. 2.4.2, there are several notions of reliability. Let $R_0(t)$ denote the inherent reliability. The field reliability differs from this due to factors such as usage mode and intensity. We discuss the modelling of these effects in this section.

3.3.2.1 Usage Mode

Products are often used intermittently, resulting in a usage pattern such as that shown in Fig. 3.3. Intermittent usage involves a cyclic change from the "Operating" state to the "Idle" state in an uncertain manner. Here, $\tilde{T}_{1j}$ denotes the time in operating state and $\tilde{T}_{0j}$ the time in the idle state during the jth cycle.

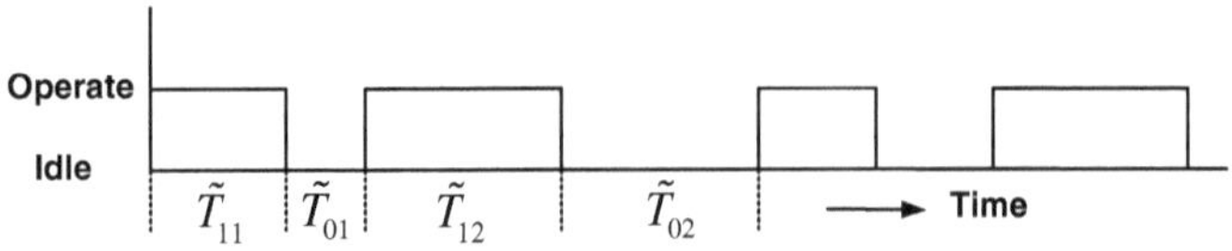

Fig. 3.3 Intermittent usage time history

Products are designed for some nominal usage mode, usage intensity and operating environment to ensure the desired reliability $R_0(t)$ when operated continuously. Let $R_i(t)$ be the reliability when it is used intermittently. Linking the two functions requires building a model for the usage pattern. Blischke et al. (2011) deal with a model where the usage pattern is modelled by a two-state continuous-time Markov chain.

3.3.2.2 Usage Intensity and Operating Environment

Products are designed for some nominal usage intensity (for example, in the case of a washing machine, this corresponds to the number of washes per week and/or size of loads washed). Usage intensity can vary considerably across the customer population. When the usage intensity is higher [lower] than the nominal usage intensity, the degradation (due to higher wear and/or increased stresses on the components) is faster [slower]. As a result, the field reliability can be lower or higher than the design reliability.

The same is true with the operating environment (for example, road conditions in the case of an automobile, operating temperature in the case of an electronic product).

Both of the above factors affect the stresses on the components (electrical, mechanical and/or thermal) of the product and, in turn, the reliability. Let s_0 denote stress under nominal (design) condition and s the actual stress on an item in field. Define $\hat{s} = s/s_0$. Let $R_e(t)$ denote the field reliability (which takes into account the influence of the operating environment) and this differs from $R_0(t)$, the design reliability. The two well-known models linking field reliability to design reliability are the accelerated failure time (AFT) Model (Nelson 1990) and the proportional hazards (PH) Model (Kumar and Klefsjo 1994).

AFT Model

Let $\tilde{T}_s$ denote the time to failure under stress s and $\tilde{T}_0$ the failure time under nominal stress. The AFT model assumes the following

$$\tilde{T}_s = \tilde{T}_0 \phi(\hat{s}) \tag{3.7}$$

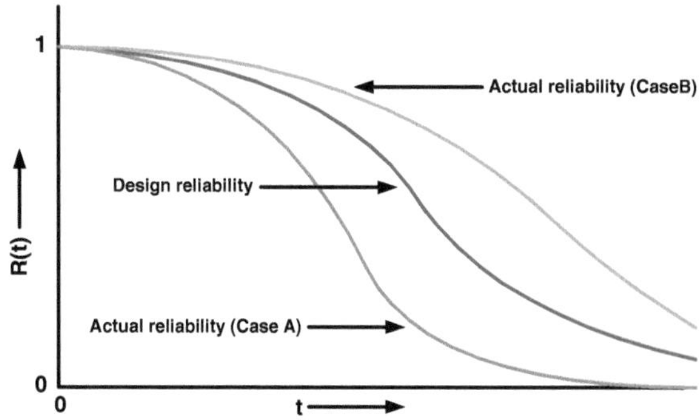

Fig. 3.4 Design and actual (field) reliabilities

where $\phi(\hat{s})$ is a non-negative and monotonically increasing function with

$$\phi(\hat{s}) \quad \begin{cases} > 1 & \text{when } \hat{s} < 1 \\ = 1 & \text{when } \hat{s} = 1 \\ < 1 & \text{when } \hat{s} > 1 \end{cases} \tag{3.8}$$

As a result, $R_e(t)$ has the same form as $R_0(t)$ and the scale parameters of the two are linked by a relationship similar to that in (3.7). The scale parameter for $R_e(t)$ decreases (increases) as $\hat{s}$ increases (decreases). Figure 3.4 shows the effect of $\phi(\hat{s})$ on the field reliability, with case A corresponding to $s > s_0$ ($\hat{s} > 1$) and case B corresponding to $s < s_0$ ($\hat{s} < 1$).

PH Model
Let $h_e(t)$ [$h_0(t)$] denote the hazard function associated with $R_e(t)$ [$R_0(t)$]. The PH model assumes that

$$h_e(t) = h_0(t)\phi(\hat{s}) \tag{3.9}$$

where $\phi(\tilde{s})$ is as in the AFT Model. As a result,

$$R_e(t) = [R_0(t)]^{\phi(\hat{s})}. \tag{3.10}$$

3.4 Modelling Maintenance Actions (1-D Formulations)

As discussed in Sect. 2.5, maintenance actions can be grouped into two categories—corrective and preventive.

3.4.1 Corrective Maintenance Actions

We need to differentiate items (product, component or anything intermediate) which are repairable and non-repairable. In the case of non-repairable item, the

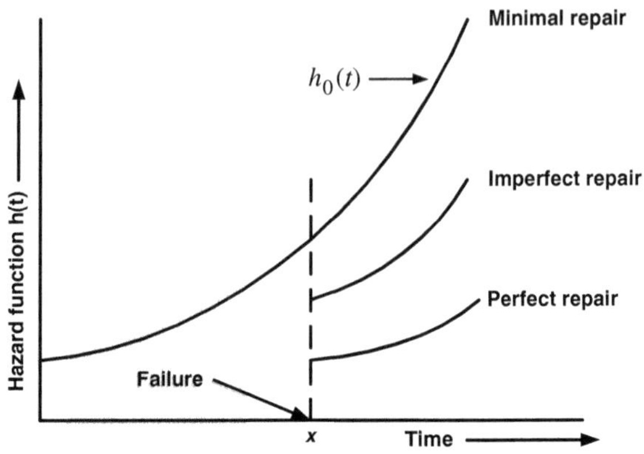

Fig. 3.5 Effect of different CM actions on hazard function

only option is to replace by a new item, whereas in the case of repairable item, one can choose between different types of repair (such as minimal, imperfect) or replace by new.

Let $F_0(t)$ denote the failure distribution function for a new item and $h_0(t)$ the associated hazard function. Let x denote the age at the first failure. Let $F(t)$ and $h(t)$ denote the failure distribution function and hazard function after the repair/replacement has been performed.

3.4.1.1 Replace by New Item

In this case, the time to next failure, using calendar clock,[6] is given by a distribution function

$$F(t) = F_0(t - x) \tag{3.11}$$

or by

$$F(t) = F_0(t) \tag{3.12}$$

using local clock which is reset to zero after replacement.

3.4.1.2 Minimal Repair[7]

Under minimal repair, the reliability of the item is unaffected by the repair action. As a result, the hazard function after repair is the same as if the failure had not occurred and this results in

[6] We will be using calendar clock unless specifically some other clock (such as local, age) is indicated.

[7] The concept of minimal repair was first proposed by Barlow and Hunter (1961).

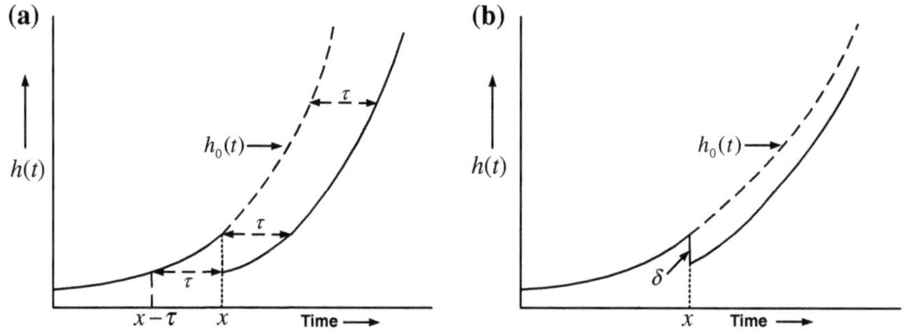

Fig. 3.6 Two models for imperfect repair. **a** Reduction in age. **b** Reduction in hazard function

$$h(t) = h_0(t),\ t > x. \tag{3.13}$$

3.4.1.3 Imperfect Repair

In some situations, the reliability characteristic of a repaired item is better than that under minimal repair but not as good as that for a new item. This type of repair is referred to as "imperfect repair" and the hazard function after repair satisfies the inequality $h_0(t - x) < h(t) < h_0(t)$, $t > x$ as shown in Fig. 3.5.

Two ways of modelling imperfect repair are as follows[8]:

Reduction in Age (Virtual age)
The effect of CM action is modelled through a reduction $\tau(0 < \tau < x)$ in the age.[9] The virtual age of the item at time t is given by $(t - \tau)$ for $t > x$. As a result, the hazard function after repair is given by

$$h(t) = h_0(t - \tau),\ t > x, \tag{3.14}$$

as shown in Fig. 3.6a.

Reduction in Hazard Function
Here, a CM action results in a reduction in the intensity function. The effect of CM on the intensity function is given by $h(x^+) = h_0(x^-) - \delta$ where δ is the reduction resulting from the CM action at time x. As a result, the hazard function after repair is given by

$$h(t) = h_0(t) - \delta,\ t > x, \tag{3.15}$$

[8] For more on imperfect repair, see Pham and Wang (1996).

[9] See Kijima (1989), Doyen and Gaudoin (2004) for more details.

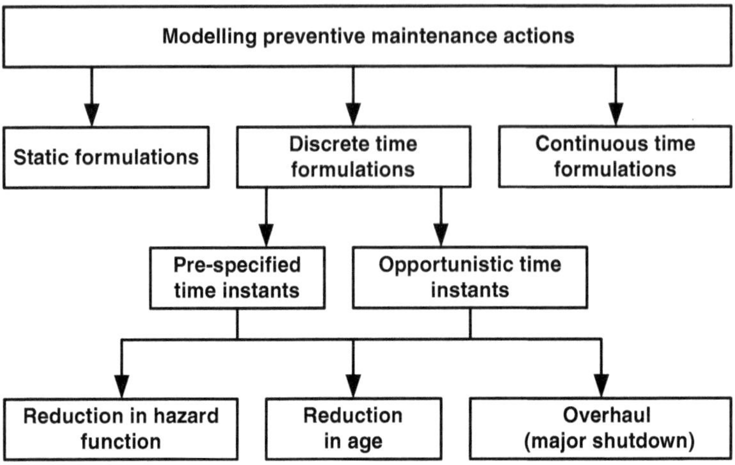

Fig. 3.7 Modelling of PM actions

as indicated in Fig. 3.6b. δ depends on the level of CM effort and needs to be constrained to satisfy the following inequality:

$$0 \le \delta \le h_0(x^-) - h_0(0). \tag{3.16}$$

3.4.2 Preventive Maintenance Actions

There are three different ways of modelling PM actions, depending on the kind of formulations used, as indicated in Fig. 3.7.

3.4.2.1 Static Formulations

In the static formulation, PM effort is modelled as a parameter which captures the different actions (or level of PM action) in an aggregated manner. The OEM recommends PM level u_0 and the actual level that the owner decides is u with $0 \le u \le u_0$.

Impact on hazard function
The hazard function is affected by the level of PM as indicated in Fig. 3.8. The hazard function with OEM recommended level of PM is given by $h(t, u_0)$, and with PM level $u(0 < u \le u_0)$, the hazard function is given by $h(t, u)$.

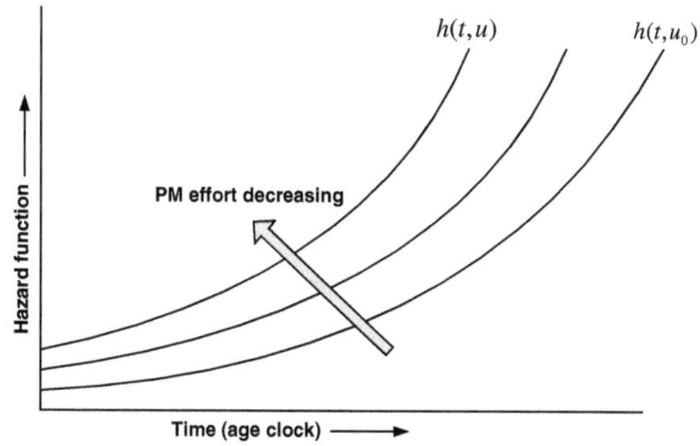

Fig. 3.8 Impact of PM on hazard function (static formulation)

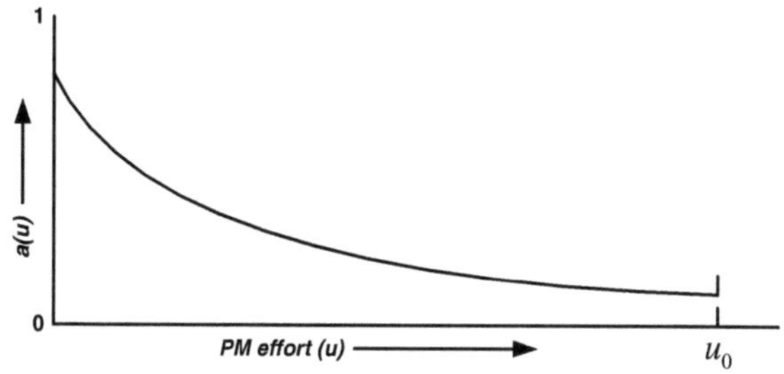

Fig. 3.9 Effect of PM level on salvage value (static formulation)

It is important to note the following:

- $\partial h(t, u)/\partial t > 0$ implying the ageing effect on the hazard function.
- $\partial h(t, u)/\partial u < 0$ implying an increased hazard function with reduced PM.

Impact on Salvage value

The salvage value of an item depends on its age as well as its condition, which in turn, depends on the level of PM. Let the salvage value of an item of age t be $\upsilon(t)$ and effect of PM level u on the salvage value can be modelled in many different ways. A simple (deterministic) model is the following:

$$\frac{\mathrm{d}\upsilon(t)}{\mathrm{d}t} = -a(u)\upsilon(t), \quad \upsilon(0) = C_p \tag{3.17}$$

where C_p is the purchase price. Note that the rate of decrease in the salvage as a fraction of the salvage value given by $[\mathrm{d}\upsilon(t)/\mathrm{d}t]/\upsilon(t) = -a(u)$ where $a(u)$ is a decreasing function of the PM level as indicated in Fig. 3.9 implying that an item maintained with a higher PM level has a higher salvage value over time.

3.4.2.2 Point Process Formulations

Often, the time to carry out a PM action is relatively small compared to the time between PM actions. In this case, PM actions can be modelled as points along the

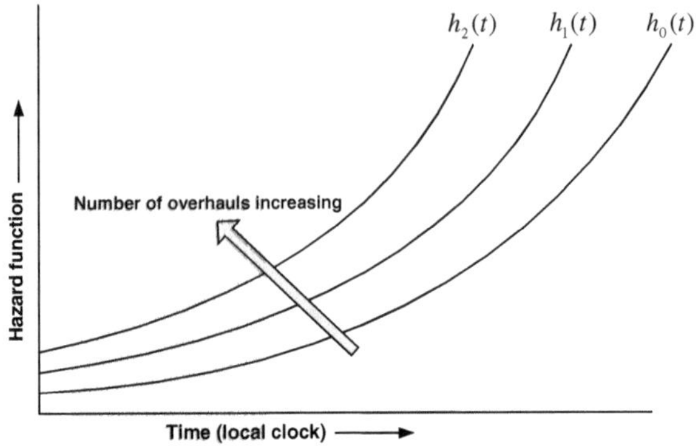

Fig. 3.10 Modelling the effect of overhauls (major shutdowns)

time axis. Such formulations have been used extensively to model many different types of PM actions.[10] The time instant for a PM action can be either deterministic (based on calendar or age clock) or uncertain (also referred to as an opportunistic PM action). The effect of a PM action is to improve the reliability of the item after the PM. In the non-opportunistic case, PM actions can be one of the following three types.

1. Replace by new.
2. Reduction in age [Virtual age] and reduction in hazard function.
3. Overhaul [Major shutdown PM]—the level of reduction being dependent on the components replaced.

For the first two types, modelling of the effect of PM action on the hazard function is similar to that for the CM case discussed earlier and hence will not be discussed any further.

Overhaul (Major Shutdown Maintenance)
An overhaul involves a complete dismantling of the system and replacing all the components that have deteriorated significantly. The hazard function after the jth ($j \geq 1$) overhaul is given by $h_j(t)$ with $h_0(t)$ being the function for a new system. Typical shapes for these functions are as shown in Fig. 3.10 where we use local clock which is reset to zero after each overhaul.

To note are the following:

1. $h_j(t)$, $j \geq 0$, is an increasing function of t implying that the system degrades with age.
2. $h_{j+1}(t) > h_j(t)$ for $j \geq 0$ implying that each overhaul improves the hazard function but there is progressive deterioration.

Comment: If the system is restored back to as good as new after each overhaul, then $h_j(t) = h_0(t)$, $j \geq 1$. This implies that an overhaul can be either perfect or imperfect depending on the maintenance actions.

---

[10] Nakagawa (2005) discusses several models based on this formulation.

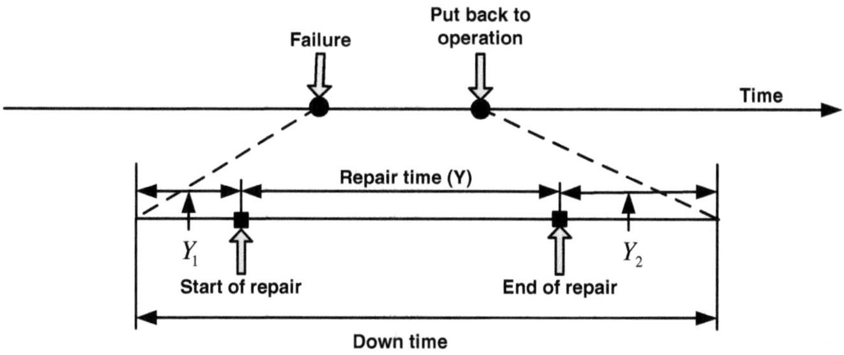

Fig. 3.11 Downtime and repair time

3.4.2.3 Continuous-Time Formulations

Many complex plants require PM actions to be performed at fairly short intervals. A discrete-time formulation results in a complex model (the curse of dimensionality). In this case, PM actions are better modelled through a continuous function $u(t)$ which changes over time. These types of models are used in CBM to determine PM actions based on continuous data collection and analysis.[11]

3.4.3 Repair Times and Downtimes

Downtime is characterised by two events—(1) failure of the item and (2) the item being put back into operation after repair/replacement. The time between failure and the item being returned to operation is usually larger than the actual repair time which is characterised by two events—(1) start of repair and (2) end of repair. As shown in Fig. 3.11, downtime = repair time $(Y) + Y_1 + Y_2$.

3.4.3.1 Modelling Repair Times

Repair time is comprised of several time periods—investigation time (time needed to locate the fault), the time needed to carry out the actual repair and testing time after repair. It can also include the waiting times that can result because of lack of spares or because of other failed items awaiting rectification actions. This time is dependent on the inventory of spares and the staffing of the repair facility.

Some of these times can be predicted precisely whereas others (e.g. time to carry out the actual repair) can be highly variable, depending on the item and the

[11] For more on CBM, see Williams et al. (1994).

type of failure. The easiest approach is to aggregate all the above-mentioned times into a single repair time Y modelled as a random variable with a distribution function $F_R(y) = P\{Y \leq y\}$. We assume that $F_R(y)$ is differentiable and let $f_R(y) = \mathrm{d}F_R(y)/\mathrm{d}y$ denote the density function and $\bar{F}_R(y) = 1 - F_R(y)$ the probability that the total repair time will exceed y. Analogous to the concept of the failure rate function, one can define a repair rate function $\rho(y)$ given by

$$\rho(y) = \frac{f_R(y)}{\bar{F}_R(y)} \tag{3.18}$$

$\rho(y)\mathrm{d}y$ is interpreted as the probability that the repair will be completed in $[y, y + \delta y)$, given that it has not been completed in $[0, y)$. In general, $\rho(y)$ will be a decreasing function of y [see, Mahon and Bailey (1975)], indicating that the probability of a repair being completed in a short time increases with the duration of the repair time. In other words, $\rho(y)$ has a "decreasing repair rate", a concept analogous to that of a decreasing failure rate.[12]

If the variability in the repair time is small in relation to the mean time for repair, then one can approximate the repair time as being deterministic.

3.4.3.2 Modelling Downtimes

As indicated in Fig. 3.11, the down time is comprised of three components. If maintenance is to be carried out on site, then Y_1 is the travel time and Y_2 is negligible. If a failed item has to be transported to a central repair facility, then both these variables are nonzero. Again, these times can be either predicted precisely or uncertain. One can aggregate all the three times and model the downtime by a distribution function similar to that for the repair time.

3.4.4 Modelling Maintenance Costs

Some of the PM and CM costs can be predicted precisely whereas others (e.g. the cost of the time to carry out the actual repair) can be highly variable, depending on the item and the type of failure. The easiest approach is to aggregate all the above-mentioned costs into a single cost $\tilde{C}$ modelled as a random variable with a distribution function $F_C(c) = P\{\tilde{C} \leq c\}$. We assume that $F_C(c)$ is differentiable with density function $f_C(c) = \mathrm{d}F_C(c)/\mathrm{d}c$ and $\bar{F}_C(c) = 1 - F_C(c)$ is the probability that the aggregated cost will exceed c. If the cost variability is small (relative to the average cost), then it can be ignored, and in this case, the cost is modelled as a

[12] Kline (1984) suggests that the log-normal distribution is appropriate for modelling the repair times for many different products.

deterministic quantity. Obviously, deterministic modelling is much easier than probabilistic modelling.

For the analysis of some maintenance policies, one only needs to use average values. In such cases, one models the uncertain costs through these average values and in this case the modelling is simpler.

3.4.4.1 PM Costs

In general, the uncertainty in the cost of carrying out a PM action is very small so that it can be ignored and the costs can be treated as deterministic.

Replace by new: This is the cost associated with replacement by a new unit and can be treated as being deterministic. We denote the cost by C_p.

Imperfect PM: The cost of an imperfect PM depends on the level of PM action performed (modelled through the reduction in the virtual age or in the hazard function). The reduction is characterised through variables τ or δ as discussed in Sect. 3.4.1. The cost can also depend on the age of the item (a) at the time of the PM action and/or the number of times (j) the item has been subjected to previous imperfect PM actions.

If the cost depends only on the age of the item and the level of imperfect PM, then it modelled by a function $C_p(\tau, a)$ or $C_p(\delta, a)$ with the various first partial derivatives >0 implying that the cost increases as the item ages and/or as the level of PM action increases. If age has no significant effect, then the cost is modelled using a simpler function $C_p(\tau)$ or $C_p(\delta)$.

Overhaul: The cost of an overhaul increases with the number of times the item has been overhauled. If there is high uncertainty (due to the parts that need to be replaced), then the cost needs to be modelled probabilistically. In this case, the cost of the kth overhaul is denoted by $\tilde{C}_O(k)$ with the expected value $C_O(k) = E[\tilde{C}_O(k)]$, $k \geq 1$. If the variability is not significant, then one only needs to specify the function $C_O(k)$, $k \geq 1$.

3.4.4.2 CM Costs

Replace by new: This is the cost of replacing a failed item by a new item and can be treated as being deterministic. We denote this cost by C_f with $C_f \geq C_p$.

Repair: If the variability in the cost is large, we need to model it probabilistically. In this case, we denote it by $\tilde{C}_r$ with the distribution function $F_r(\cdot)$. Let c_r denote the expected value ($c_r \equiv E[\tilde{C}_r]$). If the variability is insignificant, then the modelling is done using c_r.

Comment: If the repair cost depends on the age (a) of the component, then the expected cost $c_r(a)$ is an increasing function of a.

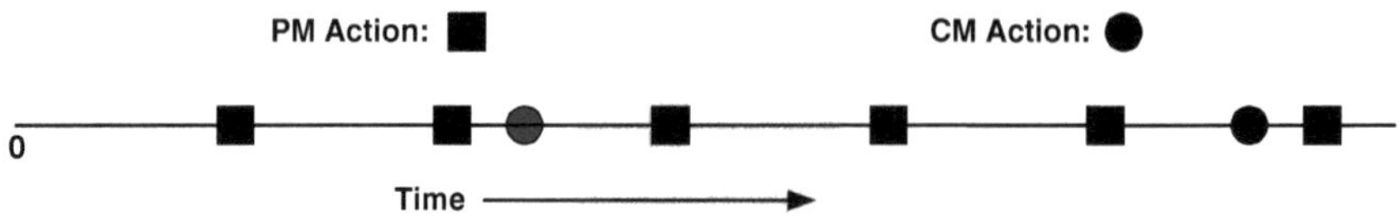

Fig. 3.12 CM and PM events along the one-dimensional time axis

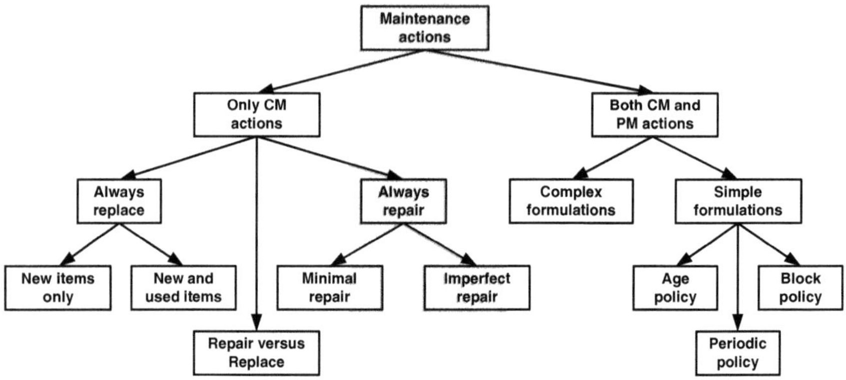

Fig. 3.13 Different scenarios

3.5 Modelling Subsequent Failures (1-D Formulations)

Subsequent failures depend on the CM and PM actions performed and as such they can be viewed as random events (as failures occur in an uncertain manner) along the time axis as shown in Fig. 3.12.

The modelling of the events involves stochastic point processes. The type of[13] formulation needed depends on the CM and PM actions, and one needs to consider several different scenarios as indicated in Fig. 3.13.

In the modelling of subsequent failures and the cost analysis, we assume the following:

1. Failures are detected immediately.
2. The time to carry out a PM or CM action is negligible and assumed to be zero.[14]
3. Only expected costs are considered.

One can relax these assumptions but the model formulation and analysis becomes more complex.

---

[13] For details of formulation and analysis of the two processes (NHPP and renewal) can be found in Appendix B.

[14] This is justified as, in general, the time for a repair/replacement $\ll$ time between events (CM or PM actions). However, if downtime is needed for determining penalty costs, then it needs to be modelled. However, it can be ignored for modelling subsequent failures as its impact is, in general, negligible.

3.5.1 One-Dimensional Point Processes

In a one-dimensional point process formulation, the time between events, the number of events in an interval and the probability of an event occurring in a short interval are all random variables. As a result, the characterisation of a point process can be done in three different (equivalent) ways as indicated below:

C-1 (Time between events): This is characterised by X_i, $i = 1, 2, \ldots$, the time between event i and $(i - 1)$ for $i > 2$ and X_1 being the time instant for the first event measured from time $t = 0$.

C-2 (Count of events over an interval): This is characterised by $N(t)$ the number of events over the interval $[0, t)$ and $N(t_2, t_1) = N(t_2) - N(t_1)$ denoting the number events over $[t_1, t_2)$.

C-3 (Intensity function): Here, the probability of an event occurring in a short interval $[t, t + \delta t)$ is given by $\mu(t)\delta t$ and the probability of two or more events occurring is $o(\delta t^2)$. $\mu(t)$ is also referred to as the *intensity function.*

Depending on the context, one of these characterisations can be much simpler to use than the others. In the context of modelling subsequent failures, two types of point processes are of particular importance and they are discussed below.

3.5.1.1 Non-homogeneous Poisson Process

This type of process can be characterised either using C-2 or C-3. Using the former, for an NHPP $\{N(t), t \geq 0\}$, the probability of j events occurring over an interval $[s, s + t)$ is given by

$$p_n(s+t, s) = P\{N(t+s) - N(s) = j\} = \frac{e^{-\left\{\int_s^{s+t} \lambda(t')\mathrm{d}t'\right\}}\{\int_s^{s+t} \lambda(t')\mathrm{d}t'\}^j}{j!} \tag{3.19}$$

for $j \geq 0$ and for all s and $t \geq 0$. $\lambda(t)$ is the intensity function (characterisation C-3). The cumulative intensity function is given by

$$\Lambda(t) = \int_0^t \lambda(t')\mathrm{d}t' \tag{3.20}$$

The expected number of events in $[0, t)$, $M(t)$, is given by

$$M(t) = E[N(t)] = \Lambda(t) \tag{3.21}$$

Two forms of intensity function that have been used extensively in modelling failures over time are the following:

- The *Weibull (Power-law) intensity function* is given by

$$\lambda(t) = \frac{\beta}{\alpha}\left(\frac{t}{\alpha}\right)^{\beta-1}, \alpha > 0, \beta > 0, t \geq 0 \tag{3.22}$$

with $\Lambda(t) = (t/\alpha)^{\beta}$.

- The *log linear intensity function* is given by[15]

$$\lambda(t) = \exp(\gamma_0 + \gamma_1 t), -\infty < \gamma_0, \gamma_1 < \infty, t \geq 0 \tag{3.23}$$

with $\Lambda(t) = [\exp(\gamma_0)][\exp(\gamma_1 t) - 1]/\gamma_1$.

Comments

- If $\beta = 1$ $[\gamma_1 = 0]$, the power-law [log linear] NHPP becomes the homogeneous poisson process (HPP).
- If $\beta < 1$ $[\gamma_1 < 0]$, the power-law [log linear] intensity is a strictly decreasing function of age.
- If $\beta > 1$ $[\gamma_1 > 0]$, the power-law [log linear] intensity is an increasing function of age.

3.5.1.2 Renewal Process

An *ordinary renewal process* is characterised very easily through C-1. For an *ordinary renewal process,* the inter-event times are independent and identically distributed with an arbitrary distribution function $F(t)$.

A *delayed renewal process* is similar to the ordinary renewal process with the following important difference—the time to the first event, is a non-negative random variable with distribution function $F(t)$ and the time intervals between subsequent events are independent and identically distributed random variables with a distribution function $\tilde{F}(t)$ which is different from $F(t)$.

The expected number of renewals in $[0, t)$, $M(t)$, is given by the integral equation

$$M(t) = F(t) + \int_0^t M(t-x) f(x) \mathrm{d}x \tag{3.24}$$

[15] This is also known as the exponential law or the Cox-Lewis intensity function.

3.5.1.3 ROCOF

$N(t)$ denotes the count of events (failures, CM or PM actions) in a point process over $[0, t)$. The mean function for $N(t)$, often referred to as the *mean cumulative function* (MCF), is denoted by $\psi(t) = E[N(t)]$. From characterisation C-3 of a point process, we have over the interval $[t, t + \delta t)$ one of three things happening:

- No failure with probability $1 - \mu(t)\delta t$
- One failure with probability $\mu(t)\delta t$
- Two or more failures with probability $o(\delta t^2)$

As a result, the expected number of failures over $[t, t + \delta t)$ is given by

$$\psi(t + \delta t) - \psi(t) = 0\{1 - \mu(t)\delta t\} + 1\{\mu(t)\delta t\} + o(\delta t^2) = \mu(t)\delta t + o(\delta t^2) \tag{3.25}$$

Diving both sides by δt and then taking the limit as $\delta t \to 0$, we have

$$\mu(t) = \frac{d\psi(t)}{dt}. \tag{3.26}$$

$\mu(t)$ is the *rate of occurrence of failures* (ROCOF) and is the derivative of the MCF.[16]

In the case of the NHPP, $\psi(t) = \Lambda(t)$ given by (3.20) and $\mu(t) = \lambda(t)$ the intensity function. In the case of the renewal process, $\psi(t) = M(t)$ given by (3.24) and $\mu(t) = dM(t)/dt$.

3.5.1.4 Renewal Points and Cycles

Let t_i, $i = 0, 1, \ldots$, be an increasing sequence of points in time. This defines a sequence of renewal points (for a point process $\{N(t), t \geq 0\}$) if $P\{\xi(N(t)), t \geq t_i\}$ is the same for all $i \geq 0$, where $\xi(N(t))$ is function of $N(t)$. The time interval between two adjacent renewal points defines a renewal cycle and is characterised by the fact that $P\{\xi(N(t)), t_i \leq t < t_{i+1}\}$ is the same for all $i \geq 0$. In the context of maintenance modelling, $\psi(\xi(t))$, $t_i \leq t < t_{i+1}$, defines the cycle cost CC and $(t_{i+1} - t_i)$ defines cycle length CL. Both CC and CL are random variables. The expected values of these variables (ECC = E[CC] and ECL = E[CL]) play an important role in obtaining expressions for the asymptotic expected cost rates of maintenance policies.

---

[16] For more on MCF and ROCOF, see, Ascher and Feingold (1984), Rigdon and Basu (2000).

3.6 Modelling and Analysis of Maintenance Actions

In this section, we look at the modelling and analysis of some of the maintenance actions shown in Fig. 3.13. In the case where there is only CM and no PM, the counting process is $\{N_f(t), t \geq 0\}$ which counts the number of failures (and so CM actions) over [0, t). When both PM and CM actions are present, we have three different counting processes—$\{N_f(t), t \geq 0\}$ as defined above, $\{N_p(t), t \geq 0\}$ which counts the number of PM actions over [0, t) and $\{N(t) \equiv N_f(t) + N_p(t), t \geq 0\}$ which counts the total number of events (each PM and CM being an event). We assume the following:

- Time for repair or replacement is negligible and hence is treated as being instantaneous.
- All items used in replacements are statistically similar.
- The failures are all independent.

We use the following notation:

- $\psi_f(t|\Upsilon)$ $[\psi_p(t|\Upsilon)]$: MCF for $N_f(t)$ $[[N_p(t)]$
- $\tilde{J}(t|\Upsilon)$: Maintenance cost over [0, t)—a random variable
- $J(t|\Upsilon)$: Expected maintenance cost over $[0, t) = E[\tilde{J}(t|\Upsilon)]$
- $J_\infty(\Upsilon)$: Asymptotic maintenance cost per unit time $J_\infty(\Upsilon) = \lim_{t\to\infty} \frac{J(t;\Upsilon)}{t} = \frac{\text{ECC}}{\text{ECL}}$

Comment: Υ is the set of decision variables.

Case 1: *No PM and Always Replace*
Every failure results in the replacement of the failed item by a new item and is a renewal point. As such $\{N_f(t), t \geq 0\}$ is a renewal process with MCF given by (3.24). The expected maintenance cost over [0, t) is given by

$$J(t) = C_f M(t). \tag{3.27}$$

The expected cycle length ECC is the mean time to failure and the expected cycle cost is ECL $= C_f$.

Case 2: *No PM and Always Minimally Repair*
In this case $\{N_f(t), t \geq 0\}$ is a non-homogeneous poisson process (NHPP) with intensity function $\lambda(t) = h(t)$, the hazard function associated with $F(t)$.[17] The MCF is given by $\Lambda(t) = H(t)$. The expected maintenance cost over [0, t) is given by

$$J(t) = C_r H(t). \tag{3.28}$$

[17] For a proof of this, see Nakagawa and Kowada (1983)

Case 3: *Age Policy*
In this case, note that every replacement is a renewal point for the processes $\{N_f(t), t \geq 0\}$ and $\{N_p(t), t \geq 0\}$. The MCF $\psi_f(t)$ can be obtained using a conditional approach (see, Appendix A) where the conditioning is done on the time to first item failure, T_1. Note that if $T_1 < v$, then the failure count increases by one, and otherwise, there is no change in the count. As a result,

$$\psi_f(t; v|T_1 = x) = \begin{cases} 1 + \psi_f(t - x; v) & \text{if } x < v \\ \psi_f(t - v; v) & \text{if } x \geq v \end{cases} \tag{3.29}$$

On removing the conditioning, we have

$$\psi_f(t; v) = \int_0^v [1 + \psi_f(t - x; v)]f(x)\mathrm{d}x + \psi_f(t - v; v)\bar{F}(v). \tag{3.30}$$

This is a renewal type integral equation (see, Appendix B) and a computational approach is needed to evaluate $\psi_f(t; v)$ from this equation.[18] Using a similar approach, $\psi_p(t; v)$, the expected number of PM replacements over $[0, t)$ is given by

$$\psi_p(t; v) = \int_0^v \psi_p(t - x; v)f(x)\mathrm{d}x + [1 + \psi_p(t - v; v)]\bar{F}(v). \tag{3.31}$$

The expected total maintenance cost over $[0, t)$ is given by

$$J(t; v) = C_f\psi_f(t; v) + C_p\psi_p(t; v). \tag{3.32}$$

It is easily shown that the expected cycle length and expected cycle cost are given by

$$\mathrm{ECL} = \int_0^v xf(x)\mathrm{d}x + v\bar{F}(v) \text{ and } \mathrm{ECC} = C_fF(v) + C_p\bar{F}(v). \tag{3.33}$$

The asymptotic expected maintenance cost per unit time is given by[19]

$$J_\infty(v) = \frac{\mathrm{ECC}}{\mathrm{ECL}} = \frac{C_fF(v) + C_p\bar{F}(v)}{\int_0^v xf(x)\mathrm{d}x + v\bar{F}(v)}. \tag{3.34}$$

---

[18] See Blischke and Murthy (1994) for more details.

[19] This expression is used as the objective function to determine the optimal decision variable v if the goal is to minimise the asymptotic cost per unit time.

Case 4: *Block Policy*
In this case, every PM action is a renewal point and all failures between these renewal points occur according to renewal process. As a result, for $jv < t \le (j+1)v, j \ge 0$, we have

$$\psi_f(t; v) = jM(v) + M(t - jv) \tag{3.35}$$

where $M(t)$ is the renewal function associated with $F(t)$ and which is given by (3.24). The first term in (3.35) represents the expected number of failures (CM actions) over the first j intervals (an interval is the period between two PM actions) and the second term represents the expected number of failures (CM actions) over $(jv, t]$.

The number of PM replacements is a deterministic quantity as the PMs occur at time instants $t = jv, j \ge 1$. As a result, the expected total maintenance cost over $[0, t)$ for $jv < t \le (j+1)v$ is given by

$$J(t; v) = C_f \psi_f(t; v) + jC_p. \tag{3.36}$$

The asymptotic expected maintenance cost per unit time is given by

$$J_\infty(v) = \frac{\text{ECC}}{\text{ECL}} = \frac{C_f M(v) + C_p}{v}. \tag{3.37}$$

Case 5: *Periodic Policy*
In this case, every PM action is a renewal point and failures between renewal points occur according to an NHPP with intensity function given by the hazard function associated with $F(t)$. As a result, for $jv < t \le (j+1)v$, we have

$$\psi(t; v) = j \int_0^v h(x)\mathrm{d}x + \int_0^{t-jv} h(x)\mathrm{d}x = jH(v) + H(t - jv). \tag{3.38}$$

The first term in result (3.37) represents the expected number of failures (CM actions) over the first j intervals. The second term represents the expected number of failures (CM actions) over $(jv, t]$. As with the Block policy, the number of PM replacements is a deterministic quantity with. As a result, the expected total maintenance cost over $[0, t)$, $jv < t \le (j+1)v$ is given by

$$J(t; v) = c_r \psi_f(t; v) + jC_p. \tag{3.39}$$

The asymptotic expected maintenance cost per unit time is given by

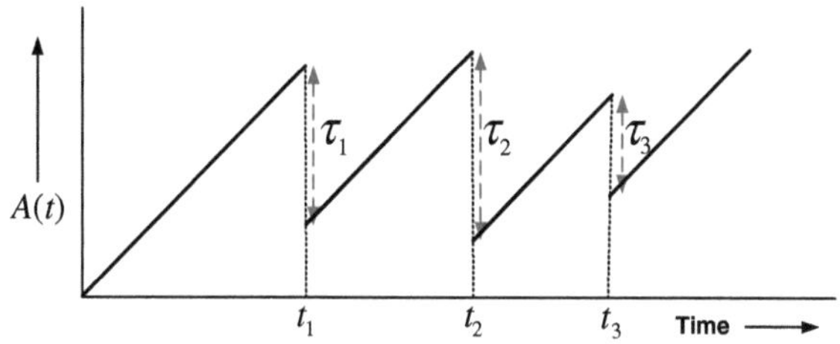

Fig. 3.14 Plot of virtual age $A(t)$

$$J_{\infty}(v) = \frac{\text{ECC}}{\text{ECL}} = \frac{c_r H(v) + C_p}{v}. \tag{3.40}$$

Case 6: *Imperfect PM and Minimal Repair* [*Reduction in age*]
This involves the concept of virtual age which increases linearly with time and every PM action results in a reduction in the virtual age. The ROCOF is a function of the virtual age. Let $A(t)$ denote the virtual age of the item at time t and t_i, $i \geq 1$, denote the time instants at which PM actions are carried out. After the ith PM action, the reduction in the virtual age is τ_i so that virtual age is given by $A(t) = t - \sum_{j=0}^{i} \tau_j$, for $t_i < t \leq t_{i+1}$, $i \geq 0$ with $\tau_0 = 0$ and $t_0 = 0$. Figure 3.14 shows a plot of the virtual age $A(t)$ as a function of time.

As a result, the ROCOF is given by the function

$$\mu(t|\Upsilon) = h(A(t)) = h\left(t - \sum_{j=0}^{i} \tau_j\right), \quad t_i < t \leq t_{i+1}, i \geq 0. \tag{3.41}$$

where $\Upsilon \equiv \{k; (t_1, \tau_1), (t_2, \tau_2), \ldots, (t_k, \tau_k)\}$. The reduction in the virtual age at the jth PM action is constrained by the relationship

$$0 \leq \tau_j < t_j - t_{j-1}, \; j \geq 1, \tag{3.42}$$

This implies that the item can never be restored to as good as new condition.

The cost of a PM action depends on the reduction in age. Let $C_p(\tau)$ denote this cost with $\mathrm{d}C_p(\tau)/\mathrm{d}\tau > 0$. The expected cost of each CM action is c_r. The expected total maintenance cost over $[0, t)$, $t_j < t \leq t_{j+1}$, $j = 0, 1, \ldots, k$ is given by (with $t_0 = 0$)

$$J(t; \Upsilon) = c_r \int_0^t \mu(t; \Upsilon)\mathrm{d}t + \sum_{j=1}^{k} C_p(\tau_j) \tag{3.43}$$

where the first term represents the expected total cost of all the CM actions and the second term represents the total cost of the k PM actions.

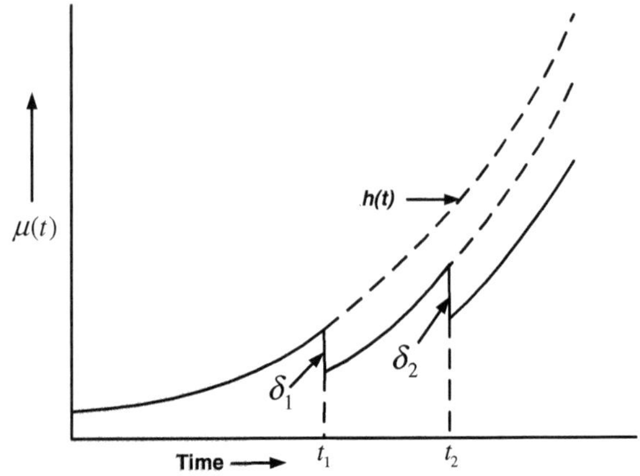

Fig. 3.15 Plot of ROCOF with reductions due to PM actions

Case 7: *Imperfect PM and Minimal Repair* [*Reduction in ROCOF*]
Here, with each PM action, there is a reduction in the ROCOF so that after the ith PM, the ROCOF is given

$$\mu(t; \Upsilon) = h(t) - \sum_{j=1}^{i} \delta_j, \; t_i < t \leq t_{i+1}, \tag{3.44}$$

where $Y \equiv \{k; (t_1, \delta_1), (t_2, \delta_2), \ldots, (t_k, \delta_k)\}$. δ_j is the reduction in the ROCOF at the jth PM and is constrained by the relationship

$$h(0) \leq \sum_{j=1}^{i} \delta_j < h(t_i) \tag{3.45}$$

for $t_i < t < t_{i+1}$. This ensures that the value of the ROCOF is never less than the failure rate for a new item. Figure 3.15 shows a plot of the ROCOF.

Comment: When the time to carry out a PM action cannot be ignored, then the ROCOF is not defined over the periods when PM actions are carried out.

The cost of a PM action depends on the reduction in the ROCOF. Let $C_p(\delta)$ denote this cost with $\mathrm{d}C_p(\delta)/\mathrm{d}\delta > 0$. The expected cost of each CM action is C_f. The expected total maintenance cost over $[0, t)$, $t_j < t \leq t_{j+1}$, $j = 0, 1, \ldots, k$ is given by (with $t_0 = 0$)

$$J(t; \Upsilon) = c_r \int_0^t \mu(t; \Upsilon)\mathrm{d}t + \sum_{j=1}^{k} C_p(\delta_j) \tag{3.46}$$

where the first term represents the expected total cost of all the CM actions and the second term represents the total cost of the k PM actions.

Case 8: *Major Overhaul*
In this case, the ROCOF of the system after the *j*th ($j \geq 1$) overhaul is given by $\mu_j(t)$ where $j = 0$ corresponds to a new system. We model the effect of PM on the ROCOF as follows:

1. For each $j \geq 0$, $\mu_j(t)$ is an increasing function of t (with $t = 0$ after each overhaul) implying that the item always degrades with age after each overhaul.
2. $\mu_j(t) < \mu_{j+1}(t)$ for $j \geq 0$ implying that each overhaul improves the ROCOF, but there is progressive deterioration.

Let t_1 denote the age at which the first overhaul occurs and t_j, $j = 2, 3, \ldots$ denote the duration between the completion of *j*th overhaul and the start of the $(j+1)$st overhaul. Let $C_p(j)$, $j \geq 1$ denote the cost of the *j*th overhaul. The total maintenance cost over $[0, t)$ depends on the number of overhauls carried out, the duration for which the system has been in operation since the last overhaul and the number of CM actions carried out. Let k and $y(t)$, ≥ 0, denote the number of overhauls carried out and the duration in operation since the last overhaul. Let $J(t|k;\, t_i,\, 1 \leq i \leq k)$ denote the total expected maintenance cost, and it is given by

$$J(t|k;t_i,1\leq i\leq k) = c_r\left[\sum_{j=0}^{k}\{\int_0^{T_j}\mu_j(t)\mathrm{d}t\} + \int_0^{y(t)}\mu_j(t)\mathrm{d}t\right] + \sum_{j=1}^{k} C_p(j) \tag{3.47}$$

where the terms in the square brackets represent the total expected cost of CM actions and the last term is the cost of PM actions.

3.7 Two-Dimensional Formulations

For many items, the failure is a function of both the age (T) and usage (U) at failure are random variables. The notion of usage depends on the item as illustrated the following examples:

- Automobile: The distance travelled until failure the first failure.
- Photocopier: The number of copies made until the first failure.
- Machine tool: The number of components machined until the first failure.

In this case, one can define *usage rate* ($Z = U/T$) as the output per unit time until failure. In the case of an automobile, it could represent the distance travelled per week, month or year and so on.

3.7.1 First Failure

The time to first failure is a random point in a two-dimensional plane with age and usage being the two coordinates. The data available for modelling can be either complete or incomplete. In the case of complete data, the age and usage at first failure for all n items is known. In the case of incomplete data, for failed items, we have the age and usage, and for non-failed items, we might or might not know the service time and/or usage for the remaining.

There are three different approaches to modelling such data. The underlying formulation depends on the approach used and is as follows:

Approach 1: This approach assumes a constant usage rate for an item and the rate varies from item to item. Usage rate is modelled as a random variable Z with distribution function $G(z) = P\{Z \leq z\}$ and density function $g(z)$.[20] The time to first failure, conditional on the usage rate, is given by the conditional failure distribution function $F(t|Z = z)$.

Normally, products are designed for some nominal usage rate z_0. As the usage rate increases (decreases), the rate of degradation increases (decreases) and this, in turn, accelerates (decelerates) the time to failure. As a result, the reliability decreases [increases] as the usage rate increases (decreases).

Let $\bar{F}_0(t)$ $[\equiv 1 - F_0(t)]$ denote the base survivor function when the usage rate is the nominal value z_0. Conditional on the usage rate, the time to first failure is modelled by a survivor function

$$\bar{F}(t|z) = \bar{F}_0(t\tilde{z}^{\gamma}) \tag{3.48}$$

where $\tilde{z} = z/z_0$ and $\gamma > 1$.

Approach 2: In this approach, the two scales, usage u and time t, are combined to define a *composite scale* v and the time to first failure is modelled by a distribution function $F_V(v)$.

Approach 3: The time to first failure is modelled by a bivariate distribution function $F(t, u)$. The density, survivor and hazard functions associated with this are given by $f(t,u), \bar{F}(t,u)$ and $h(t, u)$, respectively. The bivariate failure distribution function $F(t, u)$ is given by

$$F(t,u) = P\{T \leq t, U \leq u\} \tag{3.49}$$

$F(t, u)$ must be such that $E[U|T = t]$ is a non-decreasing function of t in order to ensure that on the average usage increases with time.

The density function associated with $F(t, u)$ (provided the function is differentiable) is given by

$$f(t,u) = \frac{\partial^2 F(t,u)}{\partial t \partial u} \tag{3.50}$$

[20] For notational ease, we omit the parameters of the functions.

The survivor function is given by

$$\bar{F}(t,u) = P\{T > t, U > u\} \tag{3.51}$$

The hazard function associated with $F(t, u)$ is given by

$$h(t,u) = \frac{f(t,u)}{\bar{F}(t,u)} \tag{3.52}$$

with $h(t, u)\delta t\delta u$ defining the probability that the first system failure will occur in the rectangle $[t, t+\delta t) \times [u, u+\delta u)$, given that $T > t$ and $U > u$.

A variety of 2-D distributions have been used in modelling failures.[21]

3.7.2 Subsequent Failures

Subsequent failures depend on the CM and PM actions used. If failed items are replaced by new items and the replacement times are negligible, then subsequent failures can be modelled by a two-dimensional renewal process.[22] The concept of minimal repair is not fully developed and is a topic for new and further research.[23]

3.8 Modelling Infrastructure Degradation and Maintenance

Most infrastructures consist of discrete elements (such as pumps in gas or water system, signalling devices in rail and road systems, power plants in energy system) and distributed elements (such as pipes in gas or water system, tracks in rail and road systems, generators, turbines in energy system). The maintenance modelling of discrete elements is similar to that for products or plants discussed in the earlier sections.

For the distributed elements, one can break the elements into a number of small sections and treat each as a discrete element or treat the whole as a system involving both time and spatial coordinates.

The modelling needs to take into account the effect of usage (traffic volume, amount of fluid pumped, etc.), operating conditions (flow rate and pressure in a gas or water pipe network, axle load in the case of rails, etc.). Another complicating factor is the environment (snow affecting road and rail operations, blocked drains

---

[21] See Johnson and Kotz (1972), Hutchinson and Lai (1990) for more on 2-D distributions. Murthy et al. (2003) discuss a variety of 2-D Weibull distributions useful in modelling failures.

[22] For more on 2-D renewal processes, see Hunter (1974a, b, 1996).

[23] For further discussion, see Baik et al. (2004, 2006).

affecting the subsoil of rail and road tracks). These all make the modelling of degradation and failures a challenge and as a result the modelling of maintenance a bigger challenge.

The bulk of the existing models use a discrete-state—discrete time or discrete state—continuous-time stochastic formulation (with a large number of states) to model the degradation, and PM and CM actions are modelled by changes to the state.

References

Ascher H, Feingold H (1984) Repairable system reliability. Marcel Dekker, New York

Baik J, Murthy DNP, Jack N (2004) Two-dimensional failure modelling and minimal repair. Naval Res Logistics 51:345–362

Baik J, Murthy DNP, Jack N (2006) Erratum: two-dimensional failure modelling with minimal repair. Naval Res Logistics 53:115–116

Barlow RE, Hunter L (1961) Optimum preventive maintenance policies. Oper Res 8:90–100

Blischke WR, Murthy DNP (1994) Warranty cost analysis. Marcel Dekker, New York

Blischke WR, Karim MR, Murthy DNP (2011) Warranty data collection and analysis. Springer, London

Doyen L, Gaudoin O (2004) Classes of imperfect repair models based on reduction of failure intensity or virtual age. Reliab Eng Syst Saf 84:45–56

Hunter JJ (1974a) Renewal theory in two dimensions: basic results. Adv Appl Probab 6:376–391

Hunter JJ (1974b) Renewal theory in two dimensions: asymptotic results. Adv Appl Probab 6:376–391

Hunter JJ (1996) Mathematical techniques for warranty analysis. In: Blischke WR, Murthy DNP (eds) Product warranty hand book. Marcel Dekker, New York

Hutchinson TP, Lai CD (1990) Continuous bivariate distributions, emphasizing applications. Rumsby Scientific Publishing, Adelaide

Johnson NL, Kotz S (1972) Distributions in statistics: continuous multivariate distributions. Wiley, New York

Kijima M (1989) Some results for repairable systems with general repair. J Appl Probab 26:89–102

Kline MB (1984) Suitability of the lognormal distribution for corrective maintenance repair times. Reliab Eng 9:65–80

Kumar D, Klefsjo B (1994) Proportional hazards model: a review. Reliab Eng Syst Saf 29:177–188

Mahon BH, Bailey RJM (1975) A proposed improved replacement policy for army vehicles. Oper Res Q 26:477–494

Murthy DNP, Page NW, Rodin Y (1990) Mathematical modelling. Pergamon Press, Oxford, England

Murthy DNP, Xie M, Jiang R (2003) Weibull models. Wiley, New York

Nakagawa T (2005) Maintenance theory of reliability. Springer, London

Nakagawa T, Kowada M (1983) Analysis of a system with minimal repair and its application to a replacement policy. Eur J Oper Res 12:253–257

Nelson W (1990) Accelerated testing. Wiley, New York

Pham H, Wang H (1996) Imperfect maintenance. Eur J Oper Res 94:425–438

Rigdon SE, Basu AP (2000) Statistical methods for the reliability of repairable systems. Wiley, New York

Williams JH, Davies A, Drake PR (1994) Condition based maintenance and machine diagnostics. Chapman and Hall, London

Chapter 4
Introduction to Stochastic Optimisation and Game Theory

4.1 Introduction

Decision problems may be classified in different ways. They may be either *static* or *dynamic*. In the static situation, decisions have to be made only once whereas a dynamic problem involves multiple decisions over time. Decision problems may also be either *deterministic* or *stochastic*. In a deterministic problem, uncertainty is assumed to be insignificant and is ignored whereas, in a stochastic problem, the effect of uncertainty is included in the modelling, and this is done through the presence of random variables.

If there is only one decision-maker (DM), then this DM has an *optimisation* problem to solve. The presence of two or more DMs, with possibly conflicting objectives, requires a different approach, and then, techniques from *game theory* (GT) may be used. The information available to a DM when a decision is made is also very important. More information should produce better decisions, but the effect of different amounts of information among DMs also needs to be considered.

This chapter provided an overview of the quantitative approaches used for decision-making. Deterministic optimisation is discussed in Appendix C, and an understanding of this topic is essential in order to follow this chapter. Section 4.2 deals with stochastic optimisation problems. The game theoretic approach to decision-making is introduced in Sect. 4.3. Section 4.4 deals with games involving two DMs, and this discussion is extended to more than two DMs in Sect. 4.5. GT techniques will be used in Chap. 8 to analyse the strategic behaviour of EW/MSC providers and customers. Finally, agency theory (AT) which is relevant for the design of MSCs is described in Sect. 4.6.

D. N. P. Murthy and N. Jack, *Extended Warranties, Maintenance Service and Lease Contracts*, Springer Series in Reliability Engineering,
DOI: 10.1007/978-1-4471-6440-1_4, © Springer-Verlag London 2014

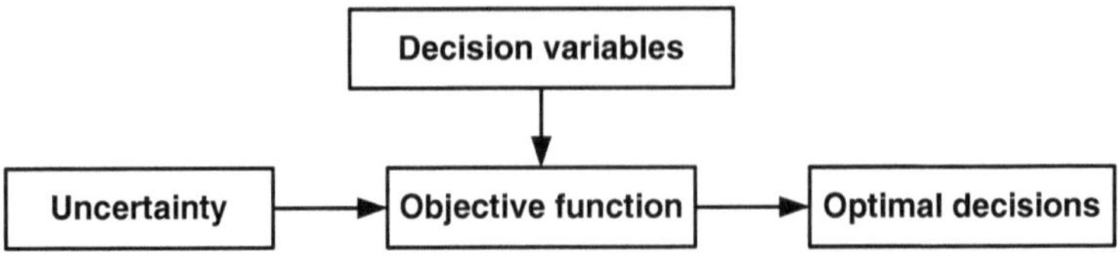

Fig. 4.1 Stochastic optimisation problem

4.2 Stochastic Optimisation

Most real-world optimisation problems involve uncertainty. This uncertainty can affect both the pay-off to the DM and other outcomes and is often said to be due to "*Nature*". The general structure of an optimisation problem with uncertainty (a stochastic optimisation problem) is shown in Fig. 4.1.

Let V denote the monetary outcome which is a random variable. The DM's objective function needs to account for this randomness, and one way of doing this is to use the expected value of the monetary outcome—$E[V]$. However, this function does not take into account the variability in the monetary outcome or the DM's attitude to risk. One way of capturing these concepts is through a utility function $U(V)$.

$U(V)$ is a measure of a DM's preferences for different pay-offs and is always increasing in V, so $U'(V) = \mathrm{d}U(V)/\mathrm{d}V > 0$. The risk attitude of the DM is captured by the second derivative $U''(V) = \mathrm{d}^2U(V)/\mathrm{d}V^2$. A strictly concave (convex) shape—$U''(V) < 0$ $(U''(V) > 0)$ represents the utility function of a risk-averse (risk-loving) DM, whereas a linear utility function $(U''(V) = 0)$ represents risk-neutral behaviour.

For a given utility function $U(V)$, the *Arrow-Pratt* risk aversion function is given by $r(V) = -U''(V)/U'(V)$. Risk-averse (risk-loving) behaviour corresponds to $r(V) > (<)\ 0$, while risk-neutral behaviour corresponds to $r(V) = 0$.

A particular utility function that will be used for decision modelling in later chapters is

$$U(V) = \frac{1}{\gamma}\left(1 - e^{-\gamma V}\right). \tag{4.1}$$

For this utility function, the risk aversion function $r(V) = \gamma$ is a constant and if $\gamma > 0$, then the DM's level of risk aversion increases as γ increases.

A rational DM should choose the values of the decision variables $\underline{x}$ in order to maximise the objective function given by $E[U(V)]$ the expected utility of the random pay-off earned. This is known as the principle of *expected utility maximisation*. An equivalent objective function is C_V, the *certainty equivalent* of the random pay-off, which is defined implicitly by

$$U(C_V) = E[U(V)]. \tag{4.2}$$

Thus, the DM is indifferent between receiving V or the fixed amount C_V. For low levels of risk aversion, the approximation for the certainty equivalent is given by

$$C_V \approx E[V] - \frac{r(E[V])}{2}\text{Var}[V]. \tag{4.3}$$

4.2.1 Static Optimisation

In a static stochastic optimisation problem, the pay-off or reward to the DM depends on the particular values chosen for the set of decision variables Υ and the realised values of one or more random variables. If there is only one random variable Ω involved, then the pay-off is given by $v = g(\Upsilon, \Omega)$.

If Ω is a discrete random variable with probability mass function $p(\omega)$, then the DM's objective function to be maximised is the expected utility

$$J(\Upsilon) = \sum U(g(\Upsilon, \omega))p(\omega), \tag{4.4}$$

whereas, if Ω is continuous with probability distribution function $P(\omega)$, then the objective becomes

$$J(\Upsilon) = \int U(g(\Upsilon, \omega))\mathrm{d}P(\omega). \tag{4.5}$$

4.2.2 Dynamic Optimisation

In a stochastic optimisation problem (see Appendix C for details of the deterministic dynamic optimisation scenario), the state transitions and the returns are both functions of random variables. In the discrete time case, the decision points (stages) occur at times $t = 0, 1, 2, \ldots, N-1$, and uncertainty is modelled in each stage by introducing the random variables $\Omega_0, \Omega_1, \ldots, \Omega_{N-1}$. We assume that Υ_t is the set of decision variables that the DM has to select at stage t.

The state S_t at time t is uncertain due to the effect of the random variables Ω_i, $0 \leq i < t$. The equation for the transformation of the state variable is now given by the stochastic difference equation

$$S_{t+1} = \phi_t(S_t, \Upsilon_t, \Omega_t). \tag{4.6}$$

The total pay-off/reward earned by the DM over the time horizon of length N (assuming no terminal reward) is the random variable

$$V = \sum_{t=0}^{N-1} \phi_t(S_t, \Upsilon_t, \Omega_t). \tag{4.7}$$

For a given value of S_0, the objective of the DM is to find the set of decisions (the optimal policy) which maximise $E[U(V)]$. The expectation in this function is evaluated with respect to the joint distribution of the random variables $\Omega_0, \Omega_1, \ldots, \Omega_{N-1}$. If these random variables are independent, then the computation of the expectation is much easier.

S_0 can be either deterministic or random, and there are two types of solution to the problem. In an *open-loop* solution, the choice of the optimal values of Υ_t, $0 \le t < N$, is based solely on the value of the initial state $S_0 = s_0$ at $t = 0$. In a *closed-loop* or *feedback* solution, the optimal choice of Υ_t is made at time t taking into account the present state $S_t = s_t$. Open-loop solutions are always easier to compute but are inferior to their closed-loop counterparts.

4.3 Game Theory

We now discuss decision problems involving two or more decision-makers (DMs), where each DM has his/her own objective function. The pay-offs to each DM now depend on the particular values of the decision variables chosen by the other DMs and are also affected by uncertainty. Thus, the optimal decisions taken by the DMs are interdependent. The general structure of a problem with two DMs is shown in Fig. 4.2.

The framework required to characterise optimal decision-making in problems with at least two DMs is provided by GT. A *game* consists of three elements: The *players* (the DMs who participate in the game), their *strategies* (the plans for each player describing what they will do in any situation) and the *pay-offs* they receive for all combinations of strategies.

In any game, an *action* is the decision that a player makes at a particular point in the game, whereas a strategy specifies what actions the player will take at each point in the game. A *solution concept* is a technique that is used to predict the outcome (*equilibrium*) of the game. It identifies the strategies that the players are actually likely to play in the game.

GT problems may be classified into a number of different ways. The timing of actions by the players and also the number of periods during which games are played lead to different solution approaches. In some games, the players may choose their actions simultaneously, so that no player knows exactly what the others have done when they make a decision. Alternatively, in games with sequential timing, the players choose their actions in predetermined order. These two situations are termed *Nash* games and *Stackelberg* games, respectively.

Some games take place during a single time period, whereas others occur over multiple time periods and the actions taken by the players in each period affect the

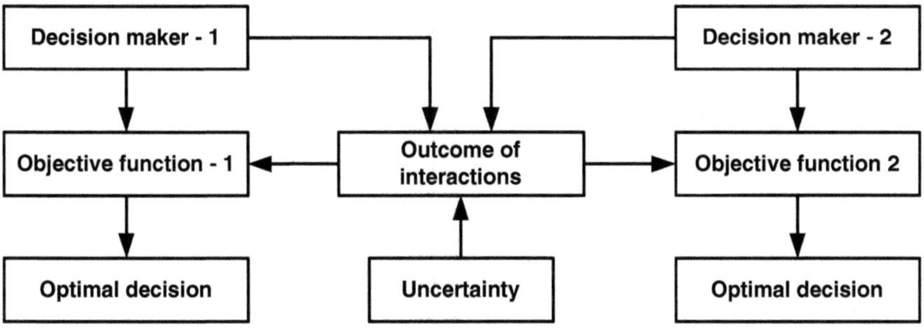

Fig. 4.2 Decision problem structure for two DMs

actions and rewards of the players in subsequent periods. These two situations are termed *static* games and *dynamic* games, respectively.

To describe a game, it is also important to specify the *information* available to each player. In a game with *complete information*, all elements of the structure of the game are known to all players whereas in games with *incomplete information*, some players may have private information. In a game with *perfect information*, all the players know exactly what has happened in the game prior to choosing an action. *Imperfect information* implies that at least one of the players is unaware of the full history of the game. Another important assumption of GT is that the players will always act rationally (choose their best actions/strategies).[1]

Finally, games may be either *cooperative* or *non-cooperative*. In a cooperative game, the players communicate with each other to coordinate their strategies and, most importantly, make binding agreements. This type of game can be formulated as a multiobjective optimisation problem. In a non-cooperative game, the players may communicate, but binding agreements are not made.

We begin by discussing *two-player non-cooperative games* and then move on to deal with games with more than two players.

4.4 Two-Player Games

4.4.1 Static Games

The players involved in the static games are denoted P_1 and P_2, and the sets of possible actions for these two players are Υ_1 and Υ_2, respectively. These action sets may be either finite or infinite. The objective functions (expected utility functions for pay-offs) for the two players are $J_1(x;y)$ and $J_2(y;x)$ for $x \in \Upsilon_1, y \in \Upsilon_2$. Note that both of these functions may also contain parameters which are fixed and so cannot be controlled by the players.

[1] Applications of GT in finance, accounting, operations management and other business areas can be found in Chatterjee and Samuelson (2001), Osborne (2002), and Watson (2008).

Fig. 4.3 Nash game decision structure

4.4.2 Nash Games

We assume that each player selects a single action without knowing the particular action chosen by their rival. This effectively means that the two players P_1 and P_2 choose their actions simultaneously and so have equal decision-making power. This power configuration is shown in Fig. 4.3. $P_1 \leftrightarrow P_2$ indicates that P_1 and P_2 make their decisions simultaneously.

In such a game, the players' strategies are just the single actions they choose, so the terms actions and strategies will be used interchangeably. The most well-known and widely used solution concept for this static game is called *Nash equilibrium* (NE). An NE is a set of strategies (strategy profile) for the two players such that no player has an incentive to change their strategy unilaterally, given the strategy chosen by the other player. More formally, the strategy profile (x^*, y^*) is an NE if

$$\begin{aligned} J_1(x^*; y^*) \geq J_1(x; y^*) \quad &\text{for all } x \in \Upsilon_1, \text{ and} \\ J_2(y^*; x^*) \geq J_2(y; x^*) \quad &\text{for all } y \in \Upsilon_2. \end{aligned} \tag{4.8}$$

An NE can be found using *best response* functions. $P_1's$ best response $\mathrm{BR}_1(y)$ to a given action $y \in \Upsilon_2$ chosen by P_2 is the value of x which maximises $J_1(x; y)$ so

$$\mathrm{BR}_1(y) = \underset{x \in \Upsilon_1}{\operatorname{argmax}}\, J_1(x; y). \tag{4.9}$$

Similarly, $P_2's$ best response $\mathrm{BR}_2(x)$ to a given action $x \in \Upsilon_1$ chosen by P_1 is the value of y which maximises $J_2(y; x)$ so

$$\mathrm{BR}_2(x) = \underset{y \in \Upsilon_2}{\operatorname{argmax}}\, J_2(y; x). \tag{4.10}$$

For an NE, both players' actions must be best responses to each other so the NE strategy profile (x^*, y^*) is the solution of

$$x^* = \mathrm{BR}_1(y^*) \text{ and } y^* = \mathrm{BR}_2(x^*). \tag{4.11}$$

A Nash game may have 0, 1 or more NE. In some games where there are multiple NE, both players will prefer one particular outcome (NE) to the others, and so this will stand out as the "right" prediction for how the game will actually be played. This outcome is then said to *Pareto dominate* the other(s).

Finite Action Sets for P_1 and P_2

If $\Upsilon_1 = \{x_1, x_2, \ldots, x_m\}$ and $\Upsilon_2 = \{y_1, y_2, \ldots, y_n\}$,then all the details of the Nash game can be displayed in a table (matrix) format with m rows and n columns.

Each row corresponds to a possible action x_i for P_1 each column corresponds to a possible action y_j for P_2, and the cells contain the objective function values $J_1(x_i; y_j)$ and $J_2(y_j; x_i)$ for the two players.

In this case, the best response of each player can be found by inspection. $\mathrm{BR}_1(y_j)$ is identified by underlining the largest objective function value(s) for P_1 in the jth column of the table, and $\mathrm{BR}_2(x_i)$ is identified by underlining the largest objective function value(s) for P_2 in the ith row of the table. An NE for the game is indicated by the cell(s) (x_i^*, y_j^*) in the table where both objective function values have been underlined.

Continuous Action Sets for P_1 and P_2

If Υ_1 and Υ_2 are sets of non-negative real numbers and $J_1(x; y)$ and $J_2(y; x)$ are both differentiable and concave, then the best response functions for each player are found from the two respective first-order conditions

$$\partial J_1(x; y)/\partial x = 0, \text{ and } \partial J_2(y; x)/\partial y = 0. \tag{4.12}$$

The first condition is solved for x in terms of y to give $x = \mathrm{BR}_1(y)$ and solving the second condition for y in terms of x gives $y = \mathrm{BR}_2(x)$. The NE strategy profile(s) (x^*, y^*) occur, where these two functions intersect.

4.4.3 Stackelberg Games

We now assume that P_1 chooses an action $x \in \Upsilon_1$ and then P_2 observes x and chooses an action $y \in \Upsilon_2$. P_1 is termed the "*leader*" with P_2 the "*follower*". P_1 has more decision-making power than P_2, and this is indicated in Fig. 4.4. Power is defined to be a player's ability to move first in the game. $P_1 \to P_2$ indicates that P_1 makes a decision before P_2.

The *backward induction* method of solution for this two-stage *Stackelberg* game is as follows.

Stage 2: Given the action x previously chosen by P_1, $P_2's$ problem is to find the value of y that maximises $J_2(y; x)$. The solution to this problem is the best response function

$$\mathrm{BR}_2(x) = \operatorname*{argmax}_{y \in \Upsilon_2} J_2(y; x). \tag{4.13}$$

Thus, P_2 responds optimally to $P_1's$ action.

Stage 1: P_1 anticipates what P_2 will do in Stage 2, so $P_1's$ problem in this part of the game is to solve the problem

$$\max_{x \in \Upsilon_1} J_1(x, \mathrm{BR}_2(x)). \tag{4.14}$$

Fig. 4.4 Stackelberg game decision structure

If x^* is the optimal solution to (4.14), then the outcome of the game is that P_1 chooses x^* and P_2 chooses $\mathrm{BR}_2(x^*)$.

This solution method can be applied when the actions sets of the two players are either finite or infinite.

4.4.4 Dynamic Games

If a game is played more than once (over multiple time periods), then it is termed a *dynamic* or multiperiod game. In a two-player "*state-dependent*" dynamic game, there is an explicit link between periods with the current actions taken by each player impacting on the present and future pay-offs of both players. Actions for both players need to be determined during each period in which the game is played and there is a state equation which determines how the game evolves and how future pay-offs for both players are affected. In each period, the actions of the players may be made either simultaneously or sequentially.

Dynamic games can be analysed using either discrete or continuous time models. In the discrete case, a dynamic programming approach can be used to obtain the optimal solution, while optimal control theory is needed for a continuous time solution. For further details about dynamic games, see Basar and Olsder (1995).

4.5 Multiplayer Games

We now focus on three-player, non-cooperative static games with the players involved being denoted P_1, P_2 and P_3 In this case, there are many possible decision scenarios (power structures), some of which are shown in Fig. 4.5. Games with more than three players produce even more scenarios.

In Fig. 4.5, if P_i has more decision-making power than P_j, then this is represented by $P_i \to P_j$. This means that P_j makes a decision only after observing the decision made by P_i. $P_i \leftrightarrow P_j$ indicates that P_i and P_j have equal decision-making power and so make their decisions simultaneously.

Scenario (i) is a three-stage Stackelberg game which can be solved using an extension of the method described in Sect. 4.4.3. In scenario (iii), P_1 and P_2

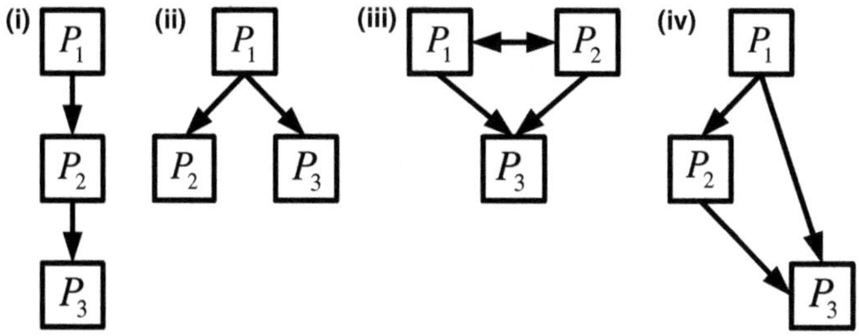

Fig. 4.5 Alternative decision structures in three-player static games

simultaneously choose actions $x \in \Upsilon_1$ and $y \in \Upsilon_2$. P_3 observes these choices, and then in Stage 2 of the game, chooses their action $z \in \Upsilon_3$. The objective functions of the three players are $J_1(x; y, z)$, $J_2(y; x, z)$ and $J_3(z; x, y)$, respectively. The backward induction solution to this two-stage *Nash–Stackelberg* game is as follows.

Stage 2: Given the actions x and y previously chosen by P_1 and P_2 and $P_3's$ best response function is

$$\mathrm{BR}_3(x, y) = \operatorname*{argmax}_{z \in \Upsilon_3} J_3(z; x, y). \tag{4.15}$$

Stage 1: P_1 and P_2 play a Nash game, both anticipating what P_3 will do in Stage 2. Their best response functions are

$$\begin{aligned} \mathrm{BR}_1(y) &= \operatorname*{argmax}_{x \in \Upsilon_1} J_1(x; y, \mathrm{BR}_3(x, y)) \text{ and} \\ \mathrm{BR}_2(x) &= \operatorname*{argmax}_{y \in \Upsilon_2} J_2(y; x, \mathrm{BR}_3(x, y)). \end{aligned} \tag{4.16}$$

The NE (x^*, y^*) for this stage of the game is the solution of $x^* = \mathrm{BR}_1(y^*)$ and $y^* = \mathrm{BR}_2(x^*)$. The outcome of the complete game is that P_1 chooses x^*, P_2 chooses y^* and P_3 chooses $z^* = \mathrm{BR}_3(x^*, y^*)$.

Note: P_1 and P_2 could be two independent EW providers competing to sell an EW to P_3 the customer.

4.6 Agency Theory

AT attempts to explain the relationship that exists between two parties (a principal and an agent), where the principal delegates work to the agent who performs that work under a contract. This is exactly the case in an MSC scenario.

AT is concerned with resolving two problems that can occur in principal–agent relationships. The first is the agency problem that arises when the two parties have conflicting objectives, and it is difficult or expensive for the principal to verify what

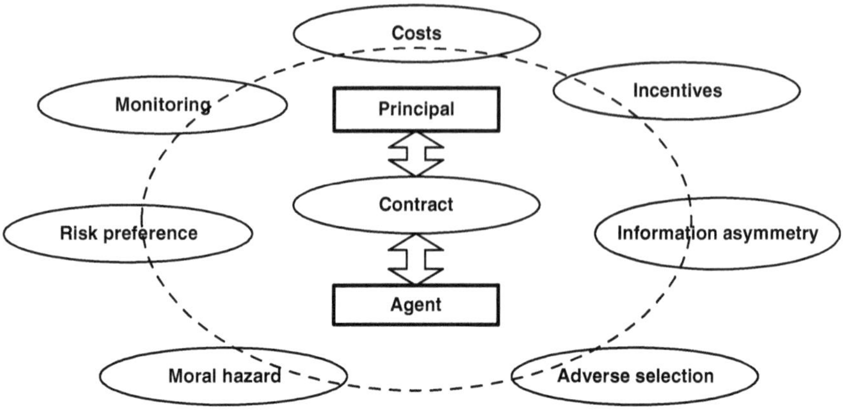

Fig. 4.6 Issues in agency theory

the agent is actually doing and whether the agent has behaved appropriately or not. The second is the problem involving the risk sharing that takes place when the principal and the agent have different attitudes to risk (due to various uncertainties). Each party may prefer different actions because of their different risk preferences.

The different issues that are involved in AT are indicated in Fig. 4.6.

These issues are

- *Moral hazard*: This refers to the agent's possible lack of effort in carrying out the delegated tasks and the fact that it is difficult for the principal to assess the effort level that the agent has actually used.
- *Adverse selection*: This refers to the agent misrepresenting their skills to carry out the tasks and the principal being unable to completely verify this before deciding to hire them. One way of avoiding this is for the principal to contact people for whom the agent has previously provided service.
- *Monitoring*: The principal can counteract the moral hazard problem by closely monitoring the agent's actions.
- *Information asymmetry*: The overall outcome of the relationship is affected by several uncertainties, and the two parties will generally have different information to make an assessment of these uncertainties.
- *Risk*: This results from the different uncertainties that affect the outcome of the relationship. The risk attitudes of the two parties may differ, and a problem occurs when they disagree over the allocation of the risk.
- *Costs*: Both parties incur various kinds of costs. These will depend on the outcome of the relationship (which is influenced by various types of uncertainty), acquiring information, monitoring and on the administration of the contract.
- *Contract*: The key factor in the relationship between the principal and the agent is the contract which specifies what, when and how the work is to be carried out and also includes incentives and penalties for the agent. This contract needs to be designed taking account of all the issues involved.

4.6.1 Principal–Agent Models

GT can be used to analyse the interaction between a principal and an agent. The objective of the game is to determine the structure of the optimal contract, behaviour versus outcome, between the two parties. In the simplest GT model, there is a two-stage Stackelberg game with the principal acting as the leader and the agent the follower. In Stage 1 of the game, the principal offers a contract to the agent with specific terms. In Stage 2, the agent decides whether to accept or reject this contract. Rejection of the contract ends the game. Acceptance means that the agent then chooses a "work or effort level" for the contract period from a set of alternatives. During the contract period, the effort used by the agent is combined with the effect of other uncertainties to determine the pay-off for the principal (e.g. total profit earned) and the resulting payment to the agent at the end of the period. This completes the game. See Watson (2008) for details of the GT model analysis for this problem.

4.6.2 Extended Principal–Agent Problems

There is a large amount of literature dealing with the design of contracts for multiple principal/multiple agent problems see for example, Macho-Stadler and Perez-Castrillo (2001) and Laffont and Martimort (2002). All the AT issues that needed to be considered in the single principal/single agent problem are still relevant in the extended case.

References

Basar T, Olsder GJ (1995) Dynamic Non-cooperative game theory, 2nd edn. Academic, New York

Chatterjee K, Samuelson WF (eds) (2001) Game theory and business applications. Kluwer Academic Publishers, Norwell

Laffont JJ, Martimort DM (2002) The theory of incentives: the principal-agent model. Princeton University Press, New Jersey

Macho-Stadler I, Perez-Castrillo JD (2001) An introduction to the economics of information. Oxford University Press, Oxford

Osborne MJ (2002) An introduction to game theory. Oxford University Press, Oxford

Watson J (2008) Strategy: an introduction to game theory, 2nd edn. W.W. Norton & Company, New York

Part II
Extended Warranties and Maintenance Service Contracts

Chapter 5
EWs/MSCs: An Overview

5.1 Introduction

In the warranty literature there confusion regarding the usage of the term "extended warranty" (EW). In the case of standard consumer products, customers prefer this term whereas providers of EWs have used a plethora of terms including maintenance service contracts (MSCs). EWs and MSCs are similar in many respects but there are also differences. A proper understanding of EWs requires concepts from base warranties (BWs). Similarly a proper understanding of MSCs requires concepts from outsourcing in general. This chapter starts deals with these two topics, looks at the different aspects of EWs and MSCs and their similarities and differences.

The outline of the chapter is as follows. We start with a brief discussion of BWs and their different aspects in Sect. 5.2. This is followed by a general discussion of EWs in Sect. 5.3 where we highlight some of the key issues. Section 5.4 gives a brief introduction to outsourcing and Sect. 5.5 deals with maintenance outsourcing where we focus on the key elements of MSCs and the similarities and differences between EWs and MSCs. In Sect. 5.6 we present some real EWs and MSCs for consumer and industrial products. Section 5.7 looks at MSCs in the context of infrastructures.

5.2 Base Warranty

As mentioned in Chap. 1, a BW is integral to the sale of a product and the customer does not pay anything extra for it. Most standard products (consumer, commercial and industrial) are sold with either a one- or two-dimensional BW. The two most common types are the free replacement warranty (FRW) and the pro rata warranty (PRW) policies. The terms of the BW policy are formulated by the manufacturer. In contrast, the warranty terms for custom built and complex expensive products are jointly decided by the manufacturer and the customer and

D. N. P. Murthy and N. Jack, *Extended Warranties, Maintenance Service and Lease Contracts*, Springer Series in Reliability Engineering,
DOI: 10.1007/978-1-4471-6440-1_5, © Springer-Verlag London 2014

can include reliability performance guarantees which require the manufacturer to improve reliability should the targets be not met. These are referred to as reliability improvement warranty (RIW) policies.

5.2.1 Standard Products

One-Dimensional BWs: A one-dimensional BW policy is characterized by an interval defined in terms of a single variable—time or age.[1] The two most common warranties are the following.

Policy 1: Non-renewing FRW Policy

The seller agrees to repair or provide replacements for failed items free of charge up to a time W from the time of the initial purchase. The warranty expires at time W after purchase.

Policy 2: Non-renewing PRW Policy

The seller agrees to refund an amount $\alpha(T)C_s$ if the item fails at age T prior to time W from the time of purchase, where C_s is the original sale price and $\alpha(T)$ is a non-increasing function of T, with $0 < \alpha(T) < 1$.

Two-Dimensional BWs: A two-dimensional BW is characterized by a region in a two-dimensional plane, usually with one axis representing time or age and the other representing item usage. The most common are the following two policies with a rectangular warranty region.

Policy 3: Two-dimensional Non-renewing FRW Policy.

The seller agrees to repair or provide a replacement for failed items free of charge up to a time W or up to a usage U, whichever occurs first, from the time of the initial purchase. W is called the warranty period and U the usage limit. The warranty region is a rectangle given by $[0, W) \times [0, U)$.

Comment: If the usage is heavy, the warranty can expire well before W, and if the usage is very light, then the warranty can expire well before the limit U is reached. Should a failure occur at age T with usage X, it is covered by warranty only if T is less than W and X is less than U. If the failed item is replaced by a new item, the replacement item is warranted for a time period $W - T$ and for usage $U - X$. Nearly all car manufacturers offer this type of policy, with usage corresponding to distance driven.

Policy 4: Two-dimensional Non-renewing PRW Policy

The seller agrees to refund the buyer a fraction of the original sale price if $T < W$ and $X < U$ at failure. The fraction refunded is a function of $W - T$ and/or $U - X$.

---

[1] The variable can also be usage—for example, number of copies made in the case of photocopiers and number of hours flown in the case of jet engines.

5.2.2 Custom Built and Complex Products

The basic idea of a RIW is to extend the notion of a basic consumer warranty (usually the FRW) to include guarantees on the reliability of the item and not just on its immediate or short-term performance. This is particularly appropriate in the purchase of complex, repairable equipment that is intended for relatively long use. The purpose of a RIW is to negotiate warranty terms that will motivate a manufacturer to continue improvements in reliability after the product is delivered.

Under a RIW, the manufacturer's fee is based on his/her ability to meet the warranty reliability requirements. These often include a guaranteed mean time between failures (MTBF) as a part of the warranty contract. The following is an illustrative example:

Policy 5: RIW Policy [Gandara and Rich (1997)].

Under this policy, the manufacturer agrees to repair or provide replacements free of charge for any failed parts or units until time W after purchase. In addition, the manufacturer guarantees the MTBF of the purchased item to be at least M. If the computed MTBF is less than M, the manufacturer will provide, at no cost to the buyer (1) engineering analysis to determine the cause of failure to meet the guaranteed MTBF requirement (2) engineering change proposals (3) modification of all existing units in accordance with approved engineering changes, and (4) consignment spares for the buyer to use until such time as it is shown that the MTBF is at least M.

5.2.3 Study of BWs

BWs have been studied from three different perspectives—(1) customer (individual, business, or government agency) (2) manufacturer (or distributor, retailer, and so forth) and (3) societal (including legislators, consumer affairs groups, the courts, and public policy decision-makers, etc.).

5.2.3.1 Customer's Perspective

As indicated in Chap. 1, from the customer's point of view, the main role of a BW in product purchase transactions is *protectional*—it provides a means of redress if the item, when properly used, fails to perform as intended or as specified by the manufacturer. A second role is *informational*—a product with a relatively longer warranty period signals a more reliable and longer lasting item than one with a shorter warranty period.

5.2.3.2 Manufacturer's Perspective

From the manufacturer's point of view a BW also serves a protectional role. A warranty contract specifies the use, and conditions of use, for which the product is intended and provides for limited coverage or no coverage at all in the event of misuse of the product. Another role is promotional—as buyers often infer a product to be more reliable when a long BW is offered. As such, the warranty serves as an effective advertising tool and it has become an instrument, similar to product performance and price, used in competition with other manufacturers in the marketplace.

5.2.3.3 Societal Perspective

Civilized society has always taken a dim view of the damage suffered by its members that is caused by someone or some activity, and it has demanded a remedy or retribution for offences against it. Consequently, manufacturers are required to provide compensation for any damages resulting from failures of an item. This has serious implications for manufacturers of engineered objects. Product-liability laws and warranty legislation are signs of society's desire to ensure fitness of products for their intended use and compensation for failures. In the USA during the last century, the Congress passed a sequence of Acts (the Uniform Commercial Code, the Magnuson-Moss Warranty Act, the TREAD Act, and so on).

5.2.3.4 Different Aspects

There are many aspects to a warranty and these have been studied by researchers from diverse disciplines. Some of the warranty issues that have been studied include the following:

1. Historical: origin and use of the notion
2. Legal: court action, dispute resolution, product liability
3. Legislative: Magnusson-Moss Act; Federal Trade Commission, Warranty requirements in government acquisition (particularly military) in the USA and the latest EU legislation
4. Economic: market equilibrium, social welfare
5. Behavioural: buyer reaction, influence on purchase decision, perceived role of warranty, claims behaviour
6. Consumerist: product information, consumer protection
7. Engineering: design, manufacturing, quality control, testing
8. Statistics: data acquisition and analysis, data-based reliability analysis
9. Operations Research: cost modelling, optimization
10. Accounting: tracking of costs, time of accrual

11. Marketing: assessment of consumer attitudes, assessment of the marketplace, use of warranty as a marketing tool, warranty and sales
12. Management: integration of many of the previous items, determination of warranty policy, warranty servicing decisions
13. Societal: public policy issues.

Consequently, the BW literature is very large[2] and Blischke and Murthy (1996) integrate the many different issues that have been addressed. Four topics from BWs, that are relevant in the context of EWs later on, are the following:

5.2.3.5 Warranty Cost Analysis

Whenever an item is returned under warranty, the manufacturer incurs various costs (handling, material, labour, facilities, etc.) and these costs are random (unpredictable) quantities. The following three types of cost are of importance to both customers and manufacturers:

1. Warranty cost per unit sale
2. Life cycle cost per unit sale
3. Life cycle cost over repeat purchases.

Blischke and Murthy (1994) discuss models to determine these costs for many different types of BWs.

5.2.3.6 Warranty and Marketing

The interaction between consumers and manufacturers defines the market for a product. For most products (such as consumer durables, industrial and commercial products), a manufacturer will have several competitors who are producing similar products and attempting to sell them to a given set of consumers, so that the market (for the product) is competitive. For some specific products (mainly industrial and commercial products), the manufacturer has no competitor so that the market is monopolistic rather than competitive. The market outcome depends on the interactions between several variables. On the manufacturer side, the variables include price, promotion, warranty etc. On the consumer side, product choice (no purchase/purchase; which of the competing brands to purchase) depends on several variables such as product features, perceived risk, brand, reputation, etc.

Warranties are seen as reducing perceived performance risk by providing protection against product defects leading to failures within the warranty period.

---

[2] See Djamaludin et al. (1996) for a bibliography listing over 1,500 papers up to 1996. Reviews of the later literature on warranty can be found in Thomas and Rao (1999) and Murthy and Djamaludin (2002).

Financial risk to the consumer is also reduced, as the repair costs to rectify failures occurring under warranty are covered by the manufacturer.

Blischke and Murthy (1996) discuss these issues in more detail.[3]

5.2.3.7 Warranty Management

Warranty management needs to be done at two different levels—strategic and operational. Strategic Management deals with decision-making with regard to all aspects of the product from an overall business viewpoint and over the product life cycle, which is the period from initial conception to manufacture and marketing to product obsolescence. As such, this is a long time frame and the decision-making needs to take into account the uncertain nature of the impact of external factors (for example, the economy, competitors actions, etc.) and some internal factors (for example, outcome of research and development). Warranty decisions must be integrated with decisions relating to technical issues such as design, development and manufacturing, and to commercial issues such as marketing, price, sales, revenue, etc. so as to ensure that the business objectives—profits, return on investment, market share, and so forth—are achieved, while at the same time providing adequate assurance to customers and ensuring customer satisfaction. Operational management deals with the implementation and execution of actions needed to achieve the business goals. It involves monitoring and making the changes needed over shorter time intervals. For more details of warranty management, see Brennan (1994) and Murthy and Blischke (2000, 2005).

5.2.3.8 Warranty Logistics

Warranty logistics deals with all the issues relating to warranty servicing and has an impact on the warranty costs. The manufacturer's ability to service a warranty is affected by the geographical distribution of customers and by the level of their demand for prompt response. The manufacturer needs a dispersed network of service facilities that store spare parts and provide a base for field service. This service delivery network requires a diverse collection of human and capital resources and careful attention must be paid to both the design and the control of the service delivery system. This involves several strategic and operational issues. The strategic issues are (1) the number of service centres and their location (2) the capacity and manning for each service centre (to ensure desired response time for customer satisfaction), and (3) whether to own these centres or outsource them so that the service is carried out by an independent agent. The tactical and operational issues are (1) transportation of the material needed for warranty servicing (2) spare parts inventory management (3) scheduling of jobs and (4) optimal repair/replace decisions. Murthy et al. (2004) discuss this topic in detail.

---

[3] More recent papers dealing with pricing are Huang et al. (2007) and Zhou et al. (2009).

5.3 Extended Warranty

An EW is a similar concept to a BW The difference between a BW and an EW is that the latter is entered into voluntarily and is purchased separately—the customer may even have a choice of terms for an EW, whereas a BW is part of product purchase and is integral to the sale.

Confusion in Terminology
According to Mancuso[4]

> Consumers seem to prefer the term EW. But industry professionals prefer the term service contract, even when they work for companies with the word warranty in their name.

He remarks that describing something as an extension of the manufacturer's warranty is inviting trouble.

> The word warranty only applies to the underlying manufacturer's product warranty, which came with the product. That's what Legal would say to us. If I went in and said, 'We're extending the warranty,' they'd say, 'No, you're not!' Warranty comes from the manufacturer. It ends, and we're asking, 'Would you like a service contract?' They're two distinctly different elements.

There is no consistency in the terminology used in industry. In the automobile industry alone there are 35 different terms used.[5]

5.3.1 Key Elements of an EW

An EW may contain some or all of the elements listed below.

---

[4] Warranty Week January 21, 2010.

[5] The terms used are: service agreement; extended warranty; service contract; maintenance agreement; after-market warranty; extended service plan; vehicle protection plans; extended vehicle coverage; extended auto warranty; vehicle service agreement; extended vehicle service contract; car service contract; vehicle maintenance contract; extended car warranty; extended service contract; vehicle extended warranty; aftermarket warranty; car extended warranty; auto extended warranty; automobile service contract; vehicle service contract; mechanical breakdown insurance; extended service coverage; extended vehicle warranty; auto service contract; extended automobile warranty; automotive extended warranty; motor vehicle service agreement; automotive service contract; power-train extended warranty; vehicle service protection; mechanical breakdown protection plan; service contracts for vehicles; auto extended service contract; automotive service plan.

5.3.1.1 EW providers

EW providers can be

- Manufacturers
- Retailers
- Third parties—insurance companies, credit card providers, etc.

5.3.1.2 Purchase Date and Duration

Often the customer has to purchase an EW at the time the product is purchased. Sometimes the customer has to the option to purchase the EW before the BW expires. In either case the EW starts from the time the BW expires.

In the case of a 1-D EW policy, the duration refers to additional time period W_1 of coverage provided by the EW. In the case of a 2-D EW policy the duration includes the additional time period W_1 and usage limit U_1 provided by the EW.

5.3.1.3 Terms

The terms define what the EW covers in relation to labour and material.

- Labour—full, partial or not covered
- Material—components or parts covered.

With full coverage (for both labour and material) the customer incurs no additional cost during the period of the EW. With partial coverage the cost to the customer depends on the terms of the EW policy.

5.3.1.4 Transferability

This defines whether the EW is transferrable or not should the customer decide to sell the product before the EW expires.

5.3.1.5 Exclusions and Limits

The exclusions and limits refer to claims over the EW period and include the following:

- Transport or freight costs excluded and paid by the customer
- Parts of the product not covered
- Limits are placed on the total number of claims
- Cost limits—limit on each claim, limit on total claims.

5.3.1.6 Price

- Purchase prices of different EW options
- Deductibles—the customer pays a certain fixed amount for each claim.

5.3.1.7 Special Requirements

- Regular preventive maintenance (PM) actions that need to be carried out during the EW period for the EW to be valid
- Nominated agents (e.g. retailers) authorised to carry out the PM actions
- Procedure for making a claim—restricted to a particular repairer.

5.3.2 Three Perspectives

As was the case with a BW, the customer's (an individual, business, or government agency) point of view of an EW is different from that of the EW provider (a manufacturer, retailer or third party). Another perspective is the societal point of view, including that of legislators, consumer affairs groups, the courts, and public policy decision-makers.

5.3.2.1 Customer Perspective

From the customer's point of view, the main role of an EW is assurance for a period after the BW expires. Specifically, the warranty assures the buyer that a faulty item will either be repaired or replaced at no cost or at reduced cost. This is important as the cost to repair a failed item can be high. As such, an EW is like an insurance to cover the high repair costs. In the case of consumer products it provides "peace of mind" which has been exploited by EW providers in their marketing efforts. Two other factors that sometimes influence a customer's decision to buy an EW are the following:

- Without an EW the customer needs to find a repair facility to get a failed item fixed. This is avoided with the purchase of an EW for the duration of the warranty
- The option to choose a particular response and service time when there are several EWs on offer with different response and service times.

5.3.2.2 EW Provider Perspective

EWs are a major source of revenue for many manufacturers and retailers. Over twenty years ago, Sears reported in excess of $1 billion in revenues from EWs

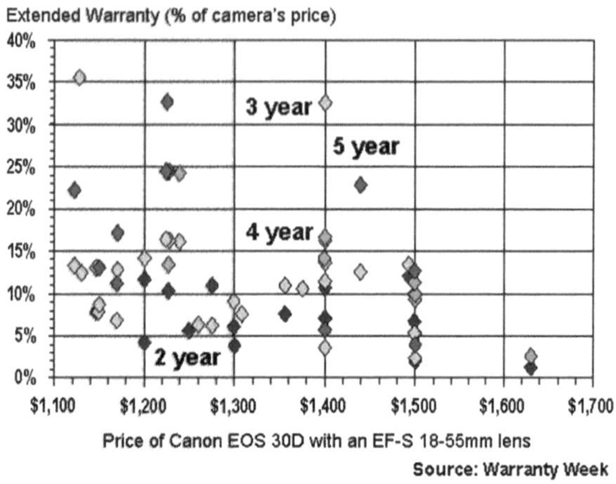

Fig. 5.1 EW price (as percentage of sale price) for four different EW periods

alone[6] and they accounted for over 50 % of profits for some major appliance store chains.[7] The major focus of EW providers is to maximise their profits.

The percentage of consumers buying EWs varies across product categories—from 20 % on products such as automobiles to 75 % on products such as home electronics and appliances.[8] For a given product brand the price charged by EW providers can vary considerably—for example, in the case of the EW for the Canon EOS 30D camera sold in the USA both the sale price and price of the EW (as a percentage of sale price) varied considerably. The figures for four different EW periods are given in Fig. 5.1.[9]

Other benefits are:

- EWs provide a unique mechanism (for both manufacturer and retailer) to build customer loyalty and encourage repeat product purchasing
- EWs help the manufacturer keep in touch with customers long after the expiry of the BW
- EWs create brand-authorised spare parts and allied services
- The servicing of EWs provides valuable information about product reliability that is useful for R&D and Design activities.

---

[6] *San Francisco Chronicle*, January, 1992.

[7] *Business Week*, January 14, 1991.

[8] Padmanabhan and Rao (1993), *PC World*, March 2003, *Wall Street Journal*, November 12, 2002, *Automotive News*, November 26, 2001.

[9] *Warranty Week*, October 24, 2006.

5.3.2.3 Societal Perspective

In the case of EWs, rip-offs can arise in numerous ways, including:

- Overcharging for policies
- Non-payment of valid claims
- Skimping on coverage.

Some retailers and dealers charge relatively high prices (compared to the price of the product the policies cover) because they have a monopoly of opportunity and a monopoly of information.[10]

There have been legislations passed in the USA and UK to protect customers' interests and reduce the exploitation by some of the EW providers.

5.3.3 Some Simple EW Policies

5.3.3.1 One-Dimensional Policies

The warranty coverage for an EW (in the non-renewing case) is to time $W + W_1$, with W_1 being the duration of the EW and W the duration of the BW. The terms of the EW can be the same as those of the BW provided by the manufacturer for a new product (in which case there is no additional cost to the customer), or they may differ in the sense that the EW may include additional features. We list a few EW policies which contain such additional features.

Policy 6: Cost Sharing EW Policies.

Under the cost sharing EW the customer and the service agent (SA) share the repair cost. The basis for sharing leads to several different scenarios;

Policy 6(*a*): Specified parts excluded (SPE).

Let I denote the set of components that are included and $\bar{I}$ the set of components excluded. The SA rectifies all failures of components belonging to the set I at no cost to the customer. The cost of rectifying failures of components belonging to the set $\bar{I}$ is borne by the customer.

Policy 6(*b*): Lump sum cost sharing (LCS).

The cost of repairing a failure is borne by both the customer and the SA. The function characterising the cost sharing can differ depending on the policy. Figure 5.2 shows one such function where the fraction of the cost borne by the customer increases once within the EW period.

Policy 6(*c*): Material or labour cost sharing (MLS).

There are two possible situations. In the first case, the customer pays for the material needed to repair a failure and the SA pays for the labour cost. In the second case, the reverse arrangement applies.

---

[10] In the case of 1-D warranties some EW providers mislead the public by claiming that the warranty period is $W + W_1$ when it is actually W_1.

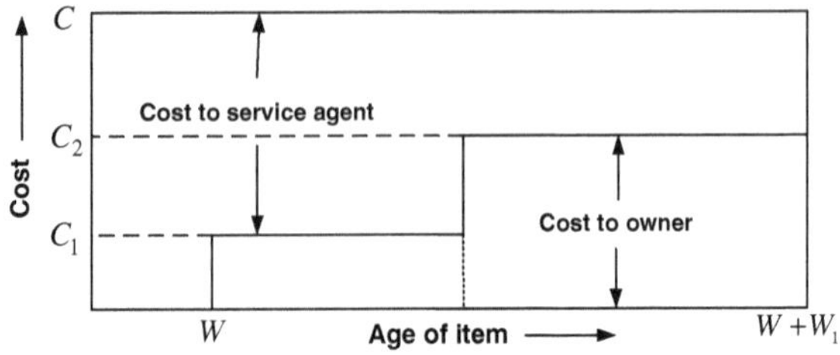

Fig. 5.2 An illustrative example of a cost sharing EW

Policy 7: Cost limit warranty (CLW) Policies

The cost limits can be on each individual claim or on total claims over the EW period.

Policy 7(*a*): Limit on individual cost (LIC).

If the cost of a rectification is below a specified limit c_I then the cost is completely borne by the SA. If the cost exceeds this limit, then the customer pays the excess—the cost of rectification less c_I.

Policy 7(*b*): Individual cost deductible (ICD).

For each claim under an EW the customer pays an amount c_E to the SA. As a result, the SA makes money on an EW claim if the cost of repair is less than c_E and incurs a cost (given by the difference between the actual cost and c_E should the cost of rectification exceed the limit).

Policy 7(*c*): Limit on total cost (LTC).

Under this policy the EW expires when the total rectification cost to fix claims under the EW exceed a limit c_T. Note that in this case the EW can cease before W_1.

5.3.3.2 Two-dimensional EW Policies

As in the 1-D case, several different 2-D EW policies can be formulated involving cost sharing, limits, exclusions, etc. When the EW is purchased at the sale of a product the warranty region is bigger than that for the BW. In the case where the warranty region is a rectangle it is given by $[0, W + W_1) \times [0, U + U_1)$ as indicated in Fig. 5.3.

When an EW is bought just before the BW expires then there can be two scenarios. The first is similar to that discussed above so that parts covered by the EW have a total age limit $W + W_1$ and usage limit $U + U_1$ irrespective of the age and usage when BW expires. In the second, the EW is a rectangle given by $[0, W_1) \times [0, U_1)$ as indicated in the Fig. 5.4.

5.3.4 Study of EWs

In contrast to BWs the literature on EWs is limited and can be broadly grouped into four categories.

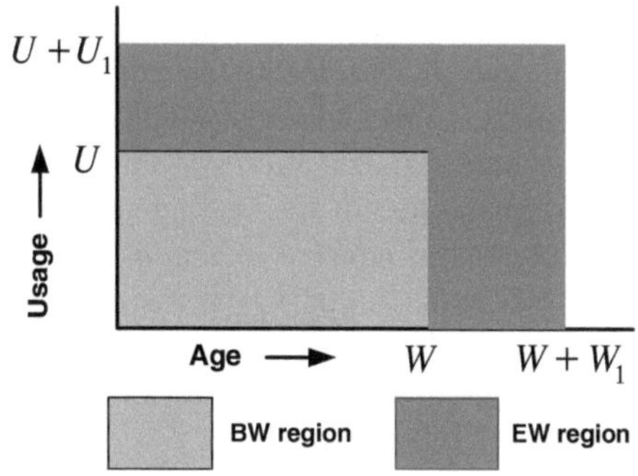

Fig. 5.3 BW and EW regions for an EW purchased at product sale

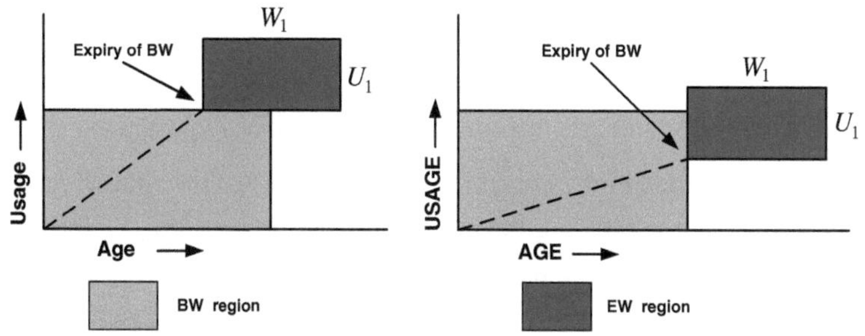

Fig. 5.4 EW regions dependent on the situation at the expiry of the BW

5.3.4.1 Operational Research

Here the focus is on estimating the EW costs from both the EW provider and customer perspectives and optimal customer decisions are based on the cost analysis. The costs can be

- Cost per unit sale and/or time
- Life cycle cost.

Other issues include such as the terms (price, warranty coverage, etc.) offered by EW providers and the maintenance actions carried out by the customer during the EW period and their implications for the optimal decisions. These will be reviewed in Chap. 8.

5.3.4.2 Economics

The EW research in economics is at the microeconomic level and deals mainly with EW market related issues. The EW market is the outcome of interactions between EW providers (manufacturer and others) and customers purchasing EWs.

The focus is on the economic efficiency of the EW Market.[11] Inefficiency occurs due to distortions created either by the market (actions of EW providers and/or customers) and/or government actions (through legislation) or inactions. For the market to be economically efficient information plays a critical role. Asymmetry in the information that different parties in the market possess can lead to market inefficiency due to the problems of *adverse selection*[12] and *moral hazard.*[13] The difference between these two terms is that adverse selection is caused by hidden information whereas moral hazard is the result of hidden actions which are either unobservable or costly to observe. Some examples of hidden information in the context of EW markets are the following:

- The inability of the EW provider to service EW claims either due to lack of expertise, or an unsound financial state so that bankruptcy can take place before the EW ceases. This situation is known to the EW provider but is not communicated to potential customers and can lead to adverse selection by customers.
- The customer's maintenance effort and usage mode which might not be revealed to the EW provider and can lead to adverse selection by providers.

Some hidden actions in the context of EW markets are the following:

- The EW provider not doing the EW servicing properly and the customer being unable to observe this—EW provider moral hazard.
- The customer not investing in the due maintenance effort and care and the EW provider being unable to observe this—customer moral hazard.

There are two other EW issues that are dealt with in the economic literature. Warranties can signal product quality to consumers when quality (reliability) is unobservable. This is called '*signalling*', with a longer warranty assumed to signal a better product. When consumers are heterogeneous, offering different price/warranty combinations to the market and allowing consumers to self-select increases the EW providers' profit and this process is referred to as '*screening*'. There are several papers that focus on screening taking into account moral hazard and adverse selection resulting from information asymmetry.

---

[11] In economics, the term *economic efficiency* refers to the use of resources so as to maximize the production of goods (products and services). A situation can be called *economically efficient* if:

- No one party can be made better off without making some other worse off (commonly referred to as Pareto efficiency).
- No additional output can be obtained without increasing the amount of inputs.
- Production of goods proceeds at the lowest possible per-unit cost.

[12] In economic theory adverse selection refers to a class of problems where pre-contractual opportunism by parties possessing private information leads to inefficiency in the operation of a market. Hollis (1999) deals with the effect of adverse selection on market outcome.

[13] In economic theory moral hazard is a situation where the behaviour of one party may change to the detriment of another after the transaction has taken place.

Heterogeneity in the customer population can be due to one or more of the following:

- Valuation of the product—some customers value a working item more than others and are willing to pay extra for an EW that provides a faster service [see, Lutz and Padmanabhan (1998) and Huysentruyt and Read (2010)].
- Attitude to risk: Risk-averse customers are willing to pay more for an EW compared to less risk-averse customers. [see, Padmanabhan and Rao (1993)].
- Usage: Product usage (e.g. km/year travelled in the case of a car; copies made per week in the case of a photocopier) can vary considerably [see, Padmanabhan (1995) and Hollis (1999)].
- Income: Customer income also varies across the population and, in general, those with higher income are more likely to purchase an EW than those with lower income [see, Lutz and Padmanabhan (1994)].

The bulk of the EW literature in economics is dominated by insurance theory which assumes that customers are more risk-averse than EW providers and EWs are a form of insurance to compensate for product failures.[14] The bulk of the papers have very stylised models in a non-dynamic setting with a warranty being viewed as monetary compensation.

Accounting for EWs is another important issue from the service provider perspective. Graves and Levitin (1990) discuss this.

5.3.4.3 Marketing

The focus of the marketing literature is on the following two topics:

1. Design of EW policies: The design of an EW policy includes terms and price and the aim is to make it more appealing to customers. Day and Fox (1985) conduct a qualitative study of consumer perceptions and decision making with regards EWs. Most customers view EWs as being overpriced and a way for EW providers to make huge profits. Fox and Day (1998) suggest the use of conjoint analysis[15] to design better EW policies which make them more appealing. They suggest two ways of doing this—the first is to provide a rebate (where the customer is given a refund at the end of the EW period should there be no warranty claims) and the second is by deductibles (where the customer pays a fixed amount to get each claim made under the EW serviced). This latter case

[14] Two other theories of warranty are—(1) the signalling theory (warranty serving as a signal of product quality) and (2) the incentive theory (to effectively address the double moral hazard issues).

[15] Conjoint analysis is a measurement technique that has been widely used by market researchers for new product development across many different product and service categories. For more details, see Green and Srinivasan (1978).

allows for the option of lowering the price of an EW and to make it more appealing to customers.[16]

2. Channel coordination: This concerns the different channel arrangements that a manufacturer can use to sell EWs (e.g. direct to customers or through a retailer). Desai and Padmanabhan (2004) consider the impact of these different arrangements on EW sales. This topic is discussed further in Chap. 8.

5.3.4.4 Consumerist and Legislative

Most customers view an EW as insurance. Their perception of repair far exceeds actual repair experience as they over estimate the cost of repair as well as the probability of failure. As a result they pay a price which is well in excess of the fair insurance price and in many industries (for example, consumer electronics) EWs have been highly profitable to manufacturers—see Padmanabhan (1996) and the UK Competition Commission Report (2003). According to *Consumer Reports*, EWs are not needed except in a few cases. Others (such as *Warranty Week*) say they provide good value at a reasonable price. EW legislation aims to address this problem. The new laws governing EWs in UK include the following:

1. Retailers must display the price of EWs alongside the price of the relevant products in both the storefront and in any advertisements,
2. Customers must be told of their right to cancel the EW contract within 45 days and to expect a full refund if no claims have been made during that time,
3. Customers must be informed in writing that the EW being offered to them at the time of sale remains available on the same terms for 30 days, and
4. Customers must be informed in writing that alternatives exist, both from third party EW providers and the product manufacturer, and perhaps even from their existing household insurance provider.

5.4 Outsourcing

Businesses producing goods (products and/or services) need to come up with new solutions and strategies to develop and increase their competitive advantage. Outsourcing is one of these strategies that can lead to greater competitiveness (Embleton and Wright 1998). It can be defined as a managed process of acquiring goods from an external agent under a contract rather than doing it in-house. The agent charges a fee

---

[16] More recent papers dealing with consumer perception are Maronick (2007) and Albaum and Wiley (2010); designing and price—Brooks and White (1996) and Hartman and Laksana (2009); adoption of EW—Bouguerra et al. (2012); options to consumers—Lam and Lam (2001); flexible warranty—Jack and Murthy (2007); purchase—Chen et al. (2009).

and in exchange the business (henceforth called the *customer* and recipient of the goods) is provided with the goods at a guaranteed quality or service level.

Most contracts stipulate specific, measurable metrics called *Service level agreements* (SLAs). These depend on the goods involved. Often SLAs also have penalties associated with not meeting the specified metrics, and sometimes rewards as incentives for exceeding the metric. Needless to say, there is a multitude of ways of constructing outsourcing agreements.

5.4.1 Reasons for Outsourcing

The conceptual basis for outsourcing (Campbell 1995) is as follows:

1. Domestic (in-house) resources should be used mainly for the core competencies of the company.
2. All other (support) activities that are not considered strategic necessities and/or whenever the company does not possesses the adequate competences and skills should be outsourced (provided there is an external agent who can carry out these activities in a more efficient manner).

There are a number of reasons that drive businesses to outsource. The list of reasons include

- Reduce costs: Sometimes achieved through lower wages costs, but also achieved through economies of scale when the external agent provides the goods to multiple businesses.
- Improve service: This often requires better educated or skilled people which either is not available in-house or not economical to have.
- Obtain expert skills: An external agent is often a business that is allegedly an expert in the delivery of the goods under consideration and thus should be able to do it better than the customer.
- Improve processes: For complex processes often external sources have expertise with similar processes that is needed to improve the process.
- Improve focus on core activities: Outsourcing frees management from having to worry about the inner-workings of a non-core activity. The customer focuses on the internal core competencies, and the others are outsourced.

Comment: Unfortunately, many businesses do not look at all these factors and often the primary reason for outsourcing is to reduce their costs.

5.4.2 Problems with Outsourcing

Outsourcing may not be appropriate for some businesses. Some of the reasons for this are the following.

- The business may be too small to effectively outsource.
- The culture within the business may not be appropriate for outsourcing.
- Other reasons (such as confidentiality) may limit or prevent the business's ability to outsource.
- The changes needed to the organisational structure make it difficult.

5.4.3 *Issues in Outsourcing*

Issues that need to be addressed before deciding on outsourcing are the following:

1. Is there a well-defined set of achievable business objectives?
2. Does outsourcing make sense?
3. Is the organisation ready?
4. What are the outsourcing alternatives?
5. What activities should be outsourced?
6. How should the best external agents be selected?
7. What are the negotiating tactics for contract formation?
8. How to decide on the fee?[17]
9. How to decide on incentives and/or penalties in the contract?
10. What systems are needed for effective monitoring?
11. What are the potential risks?

Agency theory (discussed in Sect. 4.6) provides the framework to discuss these issues. The business that seeks goods from an external source is the Principal and the provider of the goods is the Agent.

5.5 Maintenance Outsourcing

Most businesses tend not to view maintenance as a core activity and have moved towards outsourcing it. For these businesses, it is no longer economical to carry out the maintenance in house. There are a variety of reasons for this including the need for a specialist work force and diagnostic tools that often require constant upgrading. In these situations, it is more economical to outsource the maintenance (in part or total) to an external agent through a service contract. Campbell (1995) gives details of a survey where it was reported that 35 % of North American companies had considered outsourcing some of their maintenance.

[17] The fee can take many forms—based on the transaction, labour hour, cost per unit, cost per project, annual cost, cost by service levels, etc.

The advantages of outsourcing maintenance are as follows:

1. Better maintenance due to the expertise of the service agent.
2. Access to high-level specialists on an “as and when needed” basis.
3. Fixed cost service contract removes the risk of high costs.
4. Service providers respond to changing customer needs.
5. Access to latest maintenance technology.
6. Less capital investment for the customer.
7. Managers can devote more resources to other facets of the business by reducing the time and effort involved in maintenance management.

However, there are some disadvantages and these are indicated below.

1. Dependency on the service provider.
2. Cost of outsourcing.
3. Loss of maintenance knowledge (and personnel).
4. Becoming locked into a single service provider.

For very specialised (and custom built) products, the knowledge to carry out the maintenance and the spares needed for replacement need to be obtained from the original equipment manufacturer (OEM). In this case, the customer is forced into having a MSC with the OEM and this can result in a non-competitive market. In the USA, Section II of the Sherman Act (Khosrowpour 1995) deals with this problem by making it illegal for OEMs to act in this manner.

When the maintenance service is provided by an agent other than the OEM often the cost of switching prevents customers from changing their service agent. In other words, customers get “locked in” and are unable to do anything about it without a major financial consequence.

As a result, it is very important for businesses to carry out a proper evaluation of the implications of outsourcing their maintenance. If done properly, outsourcing can be cheaper than in-house maintenance and can lead to greater business profitability.

5.5.1 Different Scenarios for Maintenance Outsourcing

Maintenance of a product or system involves carrying out three sequentially linked activities as indicated in Fig. 5.5. The activities are

- Work Planning (D-1): **What** (components) need to be maintained?
- Work Scheduling (D-2): **When** should the maintenance be carried out?
- Work Execution (D-3): **How** should the maintenance be carried out?

There are three different scenarios (S-1, S-2 and S-3) depending on which of these activities are outsourced and they are shown in Table 5.1.

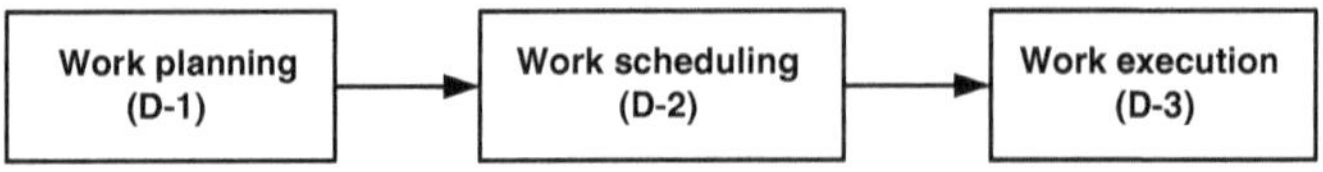

Fig. 5.5 Maintenance activities

Table 5.1 Different maintenance outsourcing scenarios

Scenarios	Decisions	
	Customer	Service agent
S-1	D-1, D-2	D-3
S-2	D-1	D-2, D-3
S-3	–	D-1, D-2, D-3

In scenario S-1, the SA is only providing the resources (workforce and material) to execute the work. This corresponds to the minimalist approach to outsourcing. In scenario S-2, the SA decides on **how** and **when** and **what** is to be done is decided by the customer. Finally, in scenario S-3 the SA makes all three decisions.

5.5.2 *Maintenance Service Contracts*

A MSC is a legal document that is binding on both parties (the business or customer and the service agent) and it needs to deal with technical, economic and other issues.

5.5.2.1 Technical Issues

There is a growing trend towards *functional guarantee contracts*. Here the contract specifies a level for the output generated from equipment, for example, the amount of electricity produced by a power plant, or the total length of flights and number of landings and take-offs per year. The SA has the freedom to decide on the maintenance needed (subject to operational constraints) with incentives and/or penalties if the target levels are exceeded or not.[18] However, these contracts need to take into account restrictions such as usage intensity, operating conditions, etc.

5.5.2.2 Economic Issues

There are a number of alternative contract payment structures as indicated below:

- Fixed or Firm price.
- Variable Price.

[18] For more on this, see Kumar and Kumar (2004a).

- Price ceiling incentive.
- Cost plus incentive fee.
- Cost plus award fee.
- Cost plus fixed fee.
- Cost plus margin.
- Other issues are cost deductibles and cost limits (for individual and total claims).

Each of these price structures represents a different level of risk sharing between the business (customer) and the SA.

5.5.2.3 Other Issues

Some other issues are as follows:

Requirements: Both parties might need to meet some stated requirement. For example, the customer needs to ensure that the usage intensity and operating loads of the asset do not exceed the levels specified in the contract. These can lead to greater degradation (due to higher stresses on the components) and higher servicing costs to the service agent. Similarly, the SA needs to ensure proper data recording.

Contract Duration: This is usually fixed with options for renewal at the end of the contract.

Moral hazard (*Cheating*): In maintenance outsourcing cheating by both owner and SA are issues that need to be addressed. Cheating by the owner occurs when the nominated usage is higher than the actual usage and the SA is not able to observe this. Similarly, cheating by the SA occurs when the actual maintenance is below the nominated maintenance and the owner cannot observe this. Information, monitoring and penalties/incentives can reduce and eliminate the potential for cheating.

Dispute Resolution: This specifies the avenues to follow when there is a dispute. The dispute can be resolved by going to a third party (e.g. an arbitration tribunal or a court).

Unless the contract is written properly and relevant data (relating to the equipment and collected by the service agent) are analysed properly by the customer the long-term costs and risks will escalate.

5.5.3 Key Elements of a MSC

A MSC document contains some or all the elements listed below.

- Parties involved—SA supplier of service and customer (recipient of the service), their names and addresses, etc.
- Definitions—glossary of frequently occurring words in the document.
- Description of the service (maintenance actions, materials, labour, etc.).

- Performance levels.
- Delivery of the service (single or multi locations).
- Term—start date and period of agreement.
- Pricing details (these can vary considerably from contract to contract).
- Pricing adjustment (e.g. annual increases linked to inflation or some other index).
- Payment details—annual, monthly, after each service, etc.
- Responsibilities of the SA—details of services to be performed and SLAs if applicable.
- Responsibilities of the customer—usage of product or system.
- Indemnification and insurance.
- Bankruptcy.
- Confidentiality.
- Force majeure.
- Dispute and arbitration process.
- Termination.
- Renegotiation/renewal.

5.5.4 Two Perspectives

There are two parties (players)—the customer (recipient of the maintenance service) and the MS Provider (the SA providing the maintenance service). There are three different scenarios (1–3) depending whether both are equally dominant or one is more dominant (leader) than the other (follower) as indicated in Table 5.2.

The decision making process for both parties depends on the particular scenario and this is discussed in more detail in Chap. 8.

5.5.5 Classification of MSCs

Maintenance requires materials, parts and labour to carry out the various activities discussed in Sect. 5.5.1. As a result there are several different kinds of MSCs. These can be broadly grouped into three types as indicated below.[19]

---

[19] Martin (1997) uses a different way of classifying MSCs. It also involves three types as indicated below:

1. *Work Package Contract*: The customer performs all planning and scheduling and the SA carries out the execution. This corresponds to Scenario S-1 and Type II in our classification.
2. *Performance contract*: This corresponds to Type III in our classification.
3. *Facilitator contract*: This corresponds to a lease contract in our definition and is discussed in Chap. 9.

Table 5.2 Three different scenarios

Scenario	Customer	MS provider
1	Leader	Follower
2	Follower	Leader
3	Neither leader nor follower	

Type I: SA only responsible for supply of material and parts (includes reconditioned parts).

Type II: SA responsible for material and parts + and carrying out some or all maintenance.

Type III: SA is responsible for complete maintenance + operations.

Comment: Type III contracts are also referred to as *functional guarantee contracts* which were discussed in Sect. 5.5.2.

5.5.6 Comparison of MSCs and EWs

As mentioned earlier there is some confusion in the literature regarding the terms EW and MSC. There are lots of similarities but also some differences as indicated in Table 5.3.

5.5.7 Study of Maintenance Outsourcing and MSCs

The literature on MSCs is large and can be divided into three categories—general, customer perspective, and industry sector. For the second and third categories the literature deals with a variety of topics. We give a small illustrative sample of the literature.

5.5.7.1 General[20]

- Justification for outsourcing: Campbell (1995) and Levery (2002).
- Critical issues: Dunn (1999).
- Enhancing appeal: Fox and Day (1998).
- Learning effects: Tarakci et al. (2009).
- MO and evolving technologies: Tseng et al. (2009).

[20] Maintenance outsourcing survey results, available at: www.plant-maintenance.com/maintenance_articles_outsources.html.

Table 5.3 Comparison of EWs and MSCs

Factors	EW	MSC
Product type	Standard products (consumer, commercial and industrial)	Standard products Custom built products/systems Infrastructure
Contract formulation	OEM	MS provider MS provider + customer
Relationship to BW	Similar Different (more restrictions)	Different
Time of purchase	At product sale Before BW expires	Any time after the BW (or EW) expires
Customisation to meet customer needs	Choosing between few options with no customisation Limited customisation (for industrial and commercial products)	Level of customisation can vary to meet the different customer needs
Complexity of contract	Low–medium	Medium–high
Initiator	EW provider	Customer
Process of selection	Simple	Simple (for standard contracts) Complex involving auctions, tendering, etc. (for complex systems and infrastructure)

5.5.7.2 Customer Perspective

- Decision models: de Almeida (2001, 2005, 2007).
- Selection of MS provider: Bertolini et al. (2004), Brito et al. (2007).
- Competition: Karmarkar and Pitbladdo (1995).
- Cost of MS: Jensen and Stonecash (2009), Datta and Roy (2010).
- Demand: Bryant and Gerner (1982).
- Implications for design and reliability: Guajardo et al. (2012), Laksana and Hartman (2010).
- Management: Sundarraj (2004), Bollapragada et al. (2007).
- Market channels: Chen et al. (2008), Desai and Padmanabhan (2004), Li et al. (2012), Tarakci et al. (2006).
- Market segmentation: Bolton and Myers (2003).
- Mass customisation: Dausch and Hsu (2003).
- Pricing: Bowman and Schmee (2001), Huber and Spliner (2012).

5.5.7.3 Industry Specific

- Aircraft: Bowman and Schmee (2001), Smith and Bachman (2008).
- Defence: Ng et al. (2009), Ng and Nudurupati (2010).

- Industrial equipment and systems: Stremersch et al. (2001); Dausch and Hsu (2003), Kumar and Kumar (2004a, b), Kumar et al. (2004), Markeset and Kumar (2003a, b), Panesar and Markeset (2008).
- Mining: Kumar and Kumar (2004a).
- Mission critical and infrequent restoration: Kim et al. (2010).

5.6 Some Illustrative Examples of EWs and MSCs

We discuss a few EWs and MSCs from different industry sectors. These were obtained from the internet websites of the businesses and further details of some of them are given in Appendix D.

5.6.1 EWs for Consumer Products

Case 5.1 (*Manufacturer's EW for Electrical and Electronic Products [Sony Corporation]*)
Sony Corporation, commonly referred to as Sony, is a Japanese multinational corporation and one of the leading manufacturers of electronics products for the consumer and professional markets.

An EW purchased for a Sony product bought in Australia or New Zealand from a Sony Authorised Dealer contains details of the following five elements

1. EW Services
2. Making a claim
3. Repairs
4. EW Term duration
5. Limitations and exclusions to EW coverage.

Each element contains several items and the details are given in Appendix D.

Case 5.2 (*Retailer's EW for Electrical and IT Products [Harvey Norman]*)
Harvey Norman is a large Australian-based retailer of electrical, computer, furniture, entertainment and bedding goods. It is effectively a franchise and the main brand is owned by Harvey Norman Holdings Limited.

The brochure to market Harvey Norman EWs for electrical and IT products is given in Appendix D. As can be seen customers can choose EWs varying from 2 to 4 years and they must be bought within 14 days of the purchase of an item.

Case 5.3 (*Manufacturer's Warranty for Cars [Chrysler]*)
The Chrysler Corporation is a multi-national company producing a range of cars around the world.[21] Chrysler Service Contracts issued for new cars vary in duration from 3 to 7 years and are available with maximum covered distances of 36,000–100,000 miles. The four different types of EWs offered are:

- Powertrain Care.
- Powertrain Care Plus.
- Added Care Plus.
- Maximum Care.

The details of the components covered are given in Appendix D. The EWs must be bought within the first 48 months a car is purchased and within the first 48,000 miles of a new car's life, and are not transferable to a second owner.

Comment: Other car manufacturers (e.g. GM, Ford, Volkswagen, Chrysler, and Honda) offer a range of EWs.[22] All are available for an assortment of durations and distances varying from 12 to 84 months and from 12,000 to 100,000 miles.

5.6.2 EWs and MSCs for Industrial Products

Case 5.4 (*Computer Servers [Hewlett Packard]*)
Hewlett-Packard Company (commonly referred to as HP) is an American multi-national information technology corporation that provides products, technologies, software, solutions and services to consumers, small- and medium-sized businesses (SMBs) and large enterprises.

The HP service contract depends on the product and in its most generic form contains 19 elements and these are listed in Appendix D.

An interesting feature is the guarantee on service response time. The cost of the EW depends on the level of service offered as illustrated by the two EW options for the HP ProLiant ML 150 servers—"4 years, 4 h, 13 × 5, hardware support at an additional cost of $434.00" and "4 years, 4 h, 24 × 7, hardware support at an additional cost of $690.00".[23]

---

[21] In 2007, Chrysler began to offer non-transferable vehicle lifetime powertrain warranty for the first registered owner or retail lessee in U.S., Puerto Rico and the Virgin Islands. After Chrysler's restructuring, the warranty program was replaced by five-year/100,000 mile transferrable warranty for 2010 or later vehicles.

[22] The GM Vehicle service contracts (VSCs) come in three types:

- Basic Guard: covers just the powertrain
- Value Guard: Basic Guard + coverage for the brakes, air conditioning, steering, and some other components
- Major Guard: Is the comprehensive exclusionary policy.

[23] Quote from Chu and Chintagunta (2009).

Case 5.5 (*Diesel Engines [Wärtsilä]*)
Wärtsilä is Finnish company and a global leader in complete lifecycle power solutions for the marine and energy markets.

Wärtsilä Marine

Is the leading provider of ship machinery, propulsion and manoeuvring solutions. It supplies engines and generating sets, reduction gears, propulsion equipment, control systems and sealing solutions for all types of vessels and offshore applications.

Wärtsilä Power Plants

It is a leading supplier of power plants for the decentralised power generation market. It offers power plants for base-load, peaking and industrial self-generation purposes as well as for the oil and gas industry.

Wärtsilä Services

It supports Wärtsilä customers throughout the lifecycle of their installations. It provides service, maintenance and reconditioning solutions both for ship machinery and power plants.

Wärtsilä offers the following four types of service contracts for its diesel and gas engines used in power generation and marine (ships)

MSC-I: Supply Agreement [Type I in the MSC classification[24]]
MSC-II: Technical Maintenance Agreement [Type II in the MSC classification]
MSC-III: Maintenance Agreement [Type II in the MSC classification]
MSC-IV: Asset Management Agreement [Type III in the MSC classification].

The key elements of each of these are given in Appendix D. Each MSC contracted is a complex document covering items discussed in Sect. 5.5.3

5.7 Infrastructure

In most countries, infrastructures used to be financed by the public sector (PUS), and were constructed, maintained and operated by agencies under the control of national, state or local governments. Over the last few decades there has been a trend towards the involvement of the private sector (PRS) in all stages—finance (capital needed), construction, maintenance and operation and maintenance.[25]

5.7.1 Public Private Partnership

In the context of infrastructures, the term 'public–private partnership' (PPP) was coined to reflect the involvement of the private sector as a partner of the public

[24] The classification is given in Sect. 5.5.5

[25] For more on privatisation in the transport infrastructure see Estache (2001).

sector. There are many different types of PPPs and Hall et al. (2003) group them into five categories as indicated below.

1. Outsourcing
2. PFI [Private financing initiative]
3. Concession
4. BOT [Build, operate, transfer]
5. Lease.

Comment: There are a range of terms used to describe variations of concessions, PFIs and BOTs.

A comparison of the five types is given in Table 5.4 involving the elements—Finance, Construction, Operation (including maintenance) and Ownership. The various symbols used are as follows:

- X: denotes the responsibility of the PRS
- Y: denotes the mode of recovery of the investment
- Z: denotes ownership status.

Variants of PPPs

A PPP can be viewed as a contract and the variants of the different PPPs are as follows[26]:

DBFO (*Design, Build, Finance and Operate*)

A contract made under the principles of the private finance initiative whereby the same supplier undertakes the design and construction of an infrastructure and thereafter maintains it for an extended period, often 25 or 30 years.

DB (*Design and Build*)

A contract where a single supplier is responsible for designing and constructing an infrastructure.

FM (*Facilities Management*)

Management of services relating to the operation of a building involving activities such as maintenance, security, catering and external and internal cleaning.

O&M (*Operation and Maintenance Contract*)

This involves the private sector operating a publicly-owned facility under contract with the Government.[27]

LDO (*Lease Develop Operate*)[28]

This involves a private developer being given a long-term lease to operate and expand an existing facility.

BOOT (*Build Own Operate Transfer*)[29]

---

[26] This section is based on material from Hall et al. (2003).

[27] In this contract, the private sector operator assumes the risks of operating and maintaining the infrastructure, and the government retains the investment risk.

[28] This type of contract is also referred to as a "*concession contract*" or "*franchise*".

[29] This type of contract is similar to a "*concession contract*" or "*franchise*".

Table 5.4 Comparison of different types of PPPs [adapted from Hall et al. (2003)]

		Outsourcing	PFI	Concession	BOT	Lease
Finance	Capital investment		X	X	X	
	Recouped by user charges			Y		Y
	Recouped from government	Y	Y			Y
Construction	By PRS		X	X	X	
Operation	Operation of service	X	X	X	X	X
Ownership	PUS (during and after contract)	Z	Z	Z		Z
	PRS during contract, PUS after			Z	Z	

This involves a private developer financing, building, owning and operating a facility for a specified period. At the expiration of the specified period, the facility is returned to the Government.

BOO *(Build Own Operate)*

This is similar to a BOT, except that the private sector owns the facility in perpetuity.

5.7.2 British Rail

Prior to 1994 British rail (BR) operated the rail system in Great Britain.[30] In 1994 a new government owned company, Railtrack, took ownership and responsibility for maintaining BR's railway infrastructure. BR's other activities were split into more than 100 companies which involved setting up "shadow" companies within BR. The ownership of railway assets was then transferred to the private sector as follows:

- Railtrack was sold in 1996 to the private sector through flotation on the stock market. BR's infrastructure support departments were geographically and functionally divided: seven infrastructure maintenance, seven infrastructure services design, and six track renewal companies. These were then sold by tender.
- BR's passenger rolling stock was sold as three rolling stock leasing companies ("ROSCOs"); these companies lease vehicles to passenger and freight train operators. The ROSCOs combined to buy the company owning the vehicle spare-parts pool. Their vehicles are maintained by seven ex-BR heavy maintenance suppliers.
- BR's freight train operations (including rolling stock) were split into six companies: three geographically-based bulk operations, container operations, non-bulk/international freight and postal contractor. These were then sold by tender to the private sector.

[30] This section is based on material from Kain (1998) and Fig. 5.6 is adapted from it.

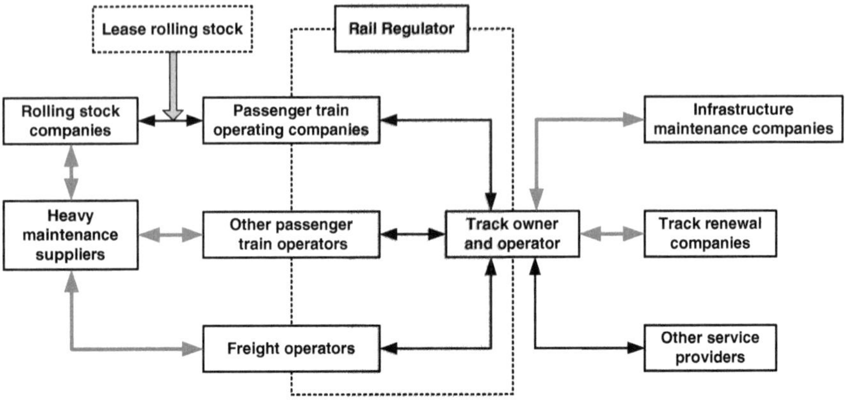

Fig. 5.6 Privatisation of rail infrastructure in UK

- In contrast to freight operations, passenger train operations were not sold; instead, the right to run the ex-BR passenger trains was franchised to 25 private sector train operating companies (TOCs), through the newly created (passenger) Passenger Franchising Director.

The government also set up the Office of Rail Regulator. As a result, several different parties are now involved in the operating and maintenance of the rail system in UK and the relationships between them are shown in Fig. 5.6.

The Regulator allocates the "Network Licence" to Railtrack, permitting Railtrack to be the operator of the network and binding it to the regulatory conditions set out in the Railways Act 1993. The Regulator's interests include:

- Maintaining network advantages: regulation is imposed on the twenty-five TOCs to ensure coordinated action between passenger franchisees.
- Setting and agreeing Railtrack's levels of passenger and freight track access charges.
- Appraising access contract terms and conditions.
- Setting the level of, and arbitrating on, open access.
- Reviewing Railtrack's investment levels and asset disposals.

The 25 franchises are subject to regulations overseen by the Franchising Director (of the Office of passenger rail franchising—OPRAF). OPRAF's activities are centred on drawing up franchise agreements and franchise plans with train operating companies, which set out TOC obligations. The agreements include:

- Given levels of service that franchisees must meet (including service connections).
- Government subsidies to (or premium from) franchisees based on service levels.
- The term of the franchise.
- The fares that are regulated (for example, "Savers" and "Weekly Seasons").
- Provision of data on train operation performance.
- Performance incentives on operational standards.

5.7.3 *Study of Infrastructure MSCs*

A complicating factor in the maintenance of infrastructures is that it needs to take into account the interests of all the stakeholders involved.[31] The government plays a critical role in terms of providing loans to and/or acting as a guarantor for the owner and the regulators are independent authorities responsible for ensuring public safety. The role of maintenance now becomes important in the context of safety and risk.[32]

For PFIs, Concessions and BOT contracts the responsibility for maintenance is with the PRS party involved. In contrast, in the case of outsourcing and leasing it is the responsibility of the PUS parties involved. The maintenance can be either done in-house or outsourced to some third party. This results in many different scenarios for the maintenance of infrastructures. The maintenance contracts are more complex and involve performance guarantees, incentives and penalties. An increasing issue in privatised infrastructures is the appropriate incentives needed to ensure adequate maintenance of the infrastructure as a public resource.

The literature on MSCs for infrastructures is vast. It can be broadly grouped into two categories—(1) general and (2) industry sector specific. We present a small illustrative list of the more recent literature.[33]

General

- Regulation and tendering: Hensher and Stanley (2008).
- Incentive contracting: Kraus (1996).
- Contract negotiations: Kuo and Wilson (2001), Ngee et al. (1997).
- Regulatory contracts: Marques and Berg (2010).

Industry Specific

- Buildings: Lai et al. (2004, 2006), Lai and Yik (2007).
- Highways and Roads: Anastapoulos et al. (2010), Ozbek et al. (2010), Tamin et al. (2011).
- Transport infrastructure: Estache (2001), Vickerman (2004).
- Pavements: Armstrong and Cook (1981).
- Rail: Macbeth and de Opacua (2010), Espling and Olsson (2004), Famurewa et al. (2011), Fearnley et al. (2004), Smith et al. (2010).

---

[31] Depending on the infrastructure one or more of the stakeholders might not be relevant. In some cases two or more of stakeholders might be the same—e.g. owner and operator being the same or service agent and operator being the same if maintenance is done in-house.

[32] The risk issue is discussed further in Chap. 11.

[33] Maintenance of items under a lease contract is discussed in Chap. 10.

References

Albaum G, Wiley J (2010) Consumer perceptions of extended warranties and service providers. J Consum Mark 27:516–523

Anastapoulos PC, Florax RJGM, Labi S, Karlaftis MG (2010) Contracting in highway maintenance and rehabilitation: are spatial effects important? Transp Res Part A 44:136–146

Armstrong RD, Cook WD (1981) The contract formation problem in preventive pavement maintenance: a fixed-charge goal-programming model. Comp Environ Urban Syst 6:147–155

Bertolini M, Bevilacqua M, Braglia M, Frosolini M (2004) An analytical method for maintenance outsourcing service selection. Int J Qual Reliab Manag 21:772–788

Blischke WR, Murthy DNP (1994) Warranty cost analysis. Marcel Dekker, New York

Blischke WR, Murthy DNP (eds) (1996) Product warranty handbook. Marcel Dekker, New York

Bollapragada S, Gupta A, Lawsirat C (2007) Managing a portfolio of long term service agreements. Eur J Oper Res 182:1399–1411

Bolton RN, Myers MB (2003) Price-based global market segmentation for services. J Mark 67:108–128

Bouguerra S, Chelbi A, Rezg N (2012) A decision model for adopting an extended warranty under different maintenance policies. Int J Prod Econ 135:840–849

Bowman RA, Schmee J (2001) Pricing and managing maintenance contract for a fleet of aircraft engines. Simulation 76:69–77

Brennan JR (1994) Warranties. Planning, analysis and implementation. McGraw-Hill, New York

Brito EPZ, Aguilar RLB, Brito LAL (2007) Customer choice of a car maintenance service provider: a model to identify the service attributes that determine choice. Int J Oper Prod Manag 27:464–481

Brooks R, White DW (1996) Pricing of the option to delay the purchase of extended service contracts. J Retailing Consum Serv 3:225–231

Bryant WK, Gerner JL (1982) The demand for service contracts. J Bus 55:345–366

Campbell JD (1995) Outsourcing in maintenance management: a valid alternative to self-provision. J Qual Maintenance Eng 1:18–24

Chen KY, Kaya M, Ozer O (2008) Dual sales channel management with service competition. Manufact Serv Oper Manag 10:654–675

Chen T, Kalra A, Sun B (2009) Why do consumers buy extended service contracts. J Consum Res 36:611–623

Chu J, Chintagunta PK (2009) Quantifying the economic value of warranties in the US server market. Manag Sci 28:99–121

Datta PP, Roy R (2010) Cost modelling techniques for availability type service support contracts: a literature review and empirical study. CIRP J Manufact Sci Technol 3:142–157

Dausch M, Hsu C (2003) Mass-customize service agreements for heavy industrial equipment. In: IEEE international conference on systems, man and cybernetics, vol 5, pp 4809–4814

Day E, Fox RJ (1985) Extended warranties, service contracts and maintenance agreements—a marketing opportunity? J Consum Mark 2:77–86

de Almeida AT (2001) Repair contract decision model through additive utility function. J Qual Maintenance Eng 7:42–48

de Almeida AT (2005) Multicriteria modelling of repair contract based on utility and ELECTRE-1 method with dependability and service quality criteria. Ann Oper Res 128:113–126

de Almeida AT (2007) Multicriteria decision model for outsourcing contracts selection based on utility function and ELECTRE method. Comput Oper Res 34:3569–3574

Desai PS, Padmanabhan V (2004) Durable good, extended warranty and channel coordination. Rev Mark Sci 2:1–23

Djamaludin I, Murthy DNP, Blischke WR (1996) Bibliography on warranties. In: Blischke WR, Murthy DNP (eds) Product warranty handbook. Marcel Dekker, New York

Dunn S (1999) Maintenance outsourcing—critical issues. Available at: www.plant-maintenance.com/maintenance_articles_outsources.html

Embleton PR, Wright PC (1998) A practical guide to successful outsourcing. Empowerment Organ 6:94–106
Espling U, Olsson U (2004) Partnering in railway infrastructure maintenance contract. J Qual Maintenance Eng 10:248–253
Estache A (2001) Privatization and regulation of transport infrastructure in the 1990's. World Bank Res Observer 16:85–107
Famurewa SM, Juntti U, Kumar U (2011) Performance based railway infrastructure maintenance: towards achieving maintenance objectives. In: Proceedings of maintenance performance measurement and management conference 2011
Fearnley N, Bekken JT, Norheim B (2004) Optimal performance-based subsidies in Norwegian intercity rail transport. Int J Transport Manag 2:29–38
Fox RJ, Day E (1998) Enhancing the appeal of service contracts: an empirical investigation of alternative offerings. J Retail 64:335–351
Gandara A, Rich MD (1997) Reliability improvement warranties for military procurement. Report no. R-2264-AF, RAND Corp., Santa Monica, CA
Graves J, Levitin MS (1990) Accounting for extended warranty contracts. J Accountancy 170:101–102
Green PE, Srinivasan V (1978) Conjoint analysis in consumer research: issues and outlook. J Consum Res 5:103–123
Guajardo JA, Cohen MA, Netessine S, Kim SH (2012) Impact of performance based contracting on product reliability: an empirical study. Manag Sci 58:961–979
Hall D, de la Motte R, Davies S (2003) Terminology of public–private partnerships (PPPs). PSIRU research paper (obtained from www.psiro.org)
Hartman JC, Laksana K (2009) Designing and pricing menus of extended warranty contracts. Nav Res Logistics 56:199–214
Hensher DA, Stanley J (2008) Transacting under a performance-based contract: the role of regulation and competitive tendering. Transp Res Part A 42:1142–1151
Hollis A (1999) Extended warranties, adverse selection and aftermarkets. J Risk Insur 66:321–343
Huang HZ, Liu ZJ, Murthy DNP (2007) Optimal reliability, warranty and price for new products. IIE Trans 39:819–827
Huber S, Spinler S (2012) Pricing of full-service contracts. Eur J Oper Res 222:113–131
Huysentruyt M, Read D (2010) How do people value extended warranties?—evidence from two field surveys. J Risk Uncertainty 40:197–218
Jack N, Murthy DNP (2007) A flexible extended warranty and related optimal strategies. J Oper Res Soc 58:1612–1620
Jensen PH, Stonecash RE (2009) Contract type and the cost of provision: evidence from maintenance service contracts. Fiscal Stud 30:279–296
Kain P (1998) The reform of rail transport in Great Britain. J Transp Econ Policy 32:246–247
Karmarkar US, Pitbladdo R (1995) Service markets and competition. J Oper Manag 12:397–411
Khosrowpour M (ed) (1995) Managing information technology investments with outsourcing. Idea Group Publishing, Harrisburg
Kim SH, Cohen MA, Netessine S, Veeraraghavan S (2010) Contracting for infrequent restoration and recovery of mission-critical systems. Manag Sci 56:1551–1567
Kraus S (1996) An overview of incentive contracting. Artif Intell 83:297–346
Kumar R, Kumar U (2004a) Service delivery strategy: trends in mining industries. Int J Surf Min Reclam Environ 18:299–307
Kumar R, Kumar U (2004b) A conceptual framework for the development of a service delivery strategy for industrial systems and products. J Bus Ind Mark 19:310–319
Kumar R, Markeset T, Kumar U (2004) Maintenance of machinery: negotiating service contracts in business-to-business marketing. Int J Serv Ind Manag 15:400–413
Kuo K, Wilson N (2001) The scientific art of contract negotiation. Educause Q 1:32–38
Lai JHK, Yik FWH (2007) Monitoring building operation and maintenance contracts. Facilities 25:238–251

Lai JHK, Yik FWH, Jones P (2004) Disputes arising from vaguely defined contractual responsibilities in building services maintenance contracts. Facilities 22:44–52
Lai JHK, Yik FWH, Jones P (2006) Critical contractual issues of outsourced operation and maintenance service for commercial buildings. Int J Serv Ind Manag 17:320–343
Laksana K, Hartman JC (2010) Planning product design refreshes with service contract and competition considerations. Int J Prod Econ 126:189–203
Lam Y, Lam PKW (2001) An extended warranty policy with options open to consumers. Eur J Oper Res 131:514–529
Levery M (2002) Making maintenance contracts perform. Eng Manag J 12:76–82
Li K, Mallik S, Chahajed D (2012) Design of extended warranties in supply chains under additive demand. Prod Oper Manag 21:730–746
Lutz NA, Padmanabhan V (1994) Income variation and warranty policy. Working Paper, Graduate School of Business, Stanford University, Stanford
Lutz NA, Padmanabhan V (1998) Warranties, extended warranties and product quality. Int J Ind Organ 16:463–493
Macbeth DK, de Opacua AI (2010) Review of service science and possible application in rail maintenance. Eur Manag J 28:1–13
Markeset T, Kumar U (2003a) Design and development of product support and maintenance concepts for industrial systems. Qual Maintenance Eng 9:376–392
Markeset T, Kumar U (2003b) Product support strategy: conventional versus functional products. J Qual Maintenance Eng 11:53–67
Maronick TJ (2007) Consumer perceptions of extended warranties. J Retail Consum Serv 14:224–231
Marques RC, Berg S (2010) Revisiting the strengths and limitations of regulatory contracts in infrastructure industries. J Infrastruct Syst 16:334–343
Martin HH (1997) Contracting out maintenance and a plan for future research. J Qual Maintenance Eng 3:81–90
Murthy DNP, Blischke WR (2000) Strategic warranty management—a life cycle approach. IEEE Trans Eng Manag 47:40–54
Murthy DNP, Blischke WR (2005) Warranty management and product manufacture. Springer, London
Murthy DNP, Djamaludin I (2002) Product warranty—a review. Int J Prod Econ 79:231–260
Murthy DNP, Solem O, Roren T (2004) Product warranty logistics: issues and challenges. Eur J Oper Res 156:110–126
Ng ICL, Nudurupati SS (2010) Outcome-based service contracts in the defence industry—mitigating the challenges. J Serv Manag 21:656–674
Ng ICL, Maull R, Yip N (2009) Outcome-based contracts as a driver for systems thinking and service-dominant logic in service science: evidence from the defence industry. Eur Manag J 27:377–387
Ngee L, Tiong RLK, Alum J (1997) Automated approach to negotiation of BOT contracts. J Comput Civ Eng 11:121–129
Ozbek ME, de la Garza JM, Pinero JC (2010) Implementation of level-of-service component for performance-based road maintenance contracts. Transp Res Rec xx:1–9
Padmanabhan V (1995) Usage heterogeneity and extended warranties. J Econ Manag Strategy 4:33–53
Padmanabhan V (1996) Extended warranties. In: Blischke WR, Murthy DNP (eds) Product warranty handbook. Marcel Dekker, New York
Padmanabhan V, Rao RC (1993) Warranty policy and extended warranties: theory and an application to automobiles. Mark Sci 12:230–247
Panesar SS, Markeset T (2008) Industrial service innovation through improved contractual relationship: a case study in maintenance. J Qual Maintenance Eng 14:290–305
Smith G, Bachman J (2008) The offshoring of airplane care. Bus Week Online, 4 Oct 2008

Smith ASJ, Wheat PE, Nash CA (2010) Exploring the effects of passenger rail franchising in Britain: evidence from the first two rounds of franchising (1997–2008). Res Transp Econ 29:72–79

Stremersch S, Wuyts S, Frambach RT (2001) The purchasing of full-service contracts: an exploratory study within the industrial maintenance market. Ind Market Manag 30:1–12

Sundarraj RP (2004) A web-based AHP approach to standardize the process of managing service contracts. Decis Support Syst 37:343–365

Tamin RZ, Tamin AZ, Marzuki PF (2011) Performance based contract application opportunity and challenges in indonesian national roads management. Proc Eng 14:851–858

Tarakci H, Tang K, Moskowitz H, Plante R (2006) Incentive maintenance contracts for channel coordination. IIE Trans 38:671–684

Tarakci H, Tang K, Teyarachakul S (2009) Learning effects on maintenance outsourcing. Eur J Oper Res 192:138–150

Thomas MU, Rao SS (1999) Warranty economic decision models: a summary and some suggested directions for future research. Oper Res 47:807–820

Tseng FS, Moskowitz H, Plante R (2009) Maintenance outsourcing contracts for new technology adoptions. IEEE Trans Eng Manag 56:203–288

UK Competition Commission (2003) A report into the supply of extended warranties on domestic electrical goods within the UK. Available at www.competition-commission.org.uk/inquiries/completed/2003/warranty/index.htm

Vickerman R (2004) Maintenance incentives under different infrastructure regimes. Utilities Policy 12:315–322

Warranty Week (2006) Extended warranty pricing. Warranty Week, 26 Oct 2006. http://www.warrantyweek.com/archive/ww20061024.html

Warranty Week (2010) Vehicle service contract industry. Warranty Week, 21 Jan 2010. http://www.warrantyweek.com/archive/ww20061024.html

Zhou Z, Li Y, Tang K (2009) Dynamic pricing and warranty policies for products with fixed lifetime. Eur J Oper Res 196:940–948

Chapter 6
EW/MSC Processes

6.1 Introduction

The EW/MSC process can be viewed as a chain involving the following four stages:

- Formulating the terms and conditions of EWs/MSCs
- Marketing (pricing and promotion) and selling EWs/MSCs
- Coordinating claims, service, payments, etc.
- Servicing of claims.

Many different parties (with one or more members in each party) are involved in the chain and include EW/MSC providers (sellers of EWs/MSCs), customers, administrators (responsible for the EWs/MSCs sold), underwriters, insurers, service agents and others (such as regulators and governments). The interactions between the EW/MSC providers and customers define the EW/MSC market. Each member in the chain is faced with several decision problems. The solution to these problems depends on the type of object (product, plant or infrastructure), the type of EW/MSC market and the information available to the different parties involved.

The systems approach and a game-theoretic framework are the most appropriate methods to find the solution to the problems faced by the various parties mentioned above. The systems approach involves several steps (system characterisation, building a mathematical model, analysis and optimisation of the model) as indicated in Chap. 1, and game theory is discussed in Chap. 4. In this chapter, we focus on the systems and game-theoretic characterisation of the EW/MSC process. This will be used in Chap. 8 to build models to find the solutions to the decision problems.

The outline of the chapter is as follows. Section 6.2 deals with the systems approach to study EWs/MSCs from the perspective of the two key parties—customers and EW/MSC providers. Section 6.3 looks at the characterisation of the EW process, the different parties involved and the interactions between them.

D. N. P. Murthy and N. Jack, *Extended Warranties, Maintenance Service and Lease Contracts*, Springer Series in Reliability Engineering,
DOI: 10.1007/978-1-4471-6440-1_6, © Springer-Verlag London 2014

In Sect. 6.4, the focus is on the system characterisation of MSCs for products, plants and infrastructures. We look at the game-theoretic characterisation of EWs and MSCs in Sects. 6.5 and 6.6, respectively.

6.2 Systems Approach to Study EWs/MSCs

As indicated in Chap. 1, the system characterisation is a description of the real world that is relevant and adequate for solving the problem(s) under consideration.

6.2.1 Decision Problems

In this section, we list some of the decision problems for customers and EW/MSC providers.[1]

6.2.1.1 Customer Problems

The two problems for a customer are the following:

1. Deciding on whether to buy an EW/MSC or not.
2. Selecting the best choice when there are two or more options.

6.2.1.2 EW/MSC Provider Problems

In the case of products (consumer, commercial and industrial) and plants, the three different EW/MSC providers are (1) manufacturers, (2) retailers and (3) external providers.[2] There are several decision problems common to all types of provider, and these include the following:

1. The range of EW/MSC offerings
2. The terms and conditions of the different EWs and MSCs
3. The pricing of EWs/MSCs
4. Whether to self-insure or take out an insurance
5. Whether to self-administer or partner with an external administrator.

---

[1] Decision problems for customers and EW providers are discussed in Chap. 8. The decision problems for insurers and underwriters are different and not discussed.

[2] Manufacturers produce products and retailers sell them. As such they deal with the physical product and the post-sale service. In contrast, external EW/MSC providers deal only with the post-sale service.

Table 6.1 EW–MSC market scenarios

		Number of customers		
		One	Few	Many
Number of EW/MSC providers	One	M-11	M-12	M-13
	Few	M-21	M-22	M-23
	Many	M-31	M-32	M-33

In the case of a manufacturer, another decision problem is the following:

1. The channels for sale of EWs/MSCs—direct versus through retailers.

6.2.2 EW/MSC Markets

The EW/MSC market for products (and plants) is the outcome of interactions between customers and EW/MSC providers. The important factor is the number of customers and EW/MSC providers—one, few or many. As a result, there are several different market scenarios as shown in Table 6.1.

The EW/MSC market can be divided into three types based on the number of EW/MSC providers:

1. *Monopolistic* (one EW/MSC provider): Manufacturer is the sole EW/MSC provider.
2. *Oligopolistic* (a few EW/MSC providers): Manufacturer and large retailers or just large retailers.
3. *Competitive* (many EW/MSC providers): Retailers and several external EW/MSC providers.

The correspondence between type of customer and type of product is as indicated below.

Single customer: Complex systems (hydropower plant and mining equipment)
Few Customers: Specialised industrial and commercial products
Many customers: Standard consumer, commercial and industrial products.

6.2.3 System Characterisation

System characterisation involves the following:

- Identification and characterisation of the important variables
- Characterisation of the interaction between the variables
- The degree of detail can vary and depends on several factors—data available for estimation of model parameters and model validation, complexity, tractability, etc. (As an example, the simplest characterisation of consumers would be a homogeneous customer population, whereas a more detailed characterisation would treat

the customer population as heterogeneous with customers grouped into different groups based on some characteristics or attributes such as attitude to risk, income, level of education, information available, usage rate or intensity, and so on.)

6.2.4 Informational Aspect

Information plays a very important role in the decision-making processes for all the parties involved. There are different kinds of information as indicated below.

- *Product Related*: Product/system reliability, past usage and operating environment, maintenance history, cost of repairing different types of failures, etc.
- *Customer Related*: Usage pattern, care and maintenance, attitude to risk, etc.
- *EW/MSC Provider Related*: Competence, ability to provide proper service, financial state, reputation, etc.

Different situations in relation to the information that members of different parties may have are the following:

- Complete or incomplete (partial or no) information
- Asymmetry in information—different members having different information.
- Uncertainty in information.

These lead to issues such as moral hazard and adverse selection which are discussed in Chaps. 4 and 5. Characterisations of the different situations lead to different scenarios and impact on the optimal decisions.

6.3 Characterisation of the EW Process

A simple characterisation of the EW process for products (consumer, commercial and industrial) is given in Fig. 6.1. It contains several elements, and the characterisation of each element involves one or more variables. The characterisation of the interactions can involve variables from several elements as indicated by the directed arcs in the figure.

6.3.1 Characterisation of the Key Elements and Interactions

6.3.1.1 BWs and EWs

A BW is integral to the sale of the product/system, and the terms are usually defined by the manufacturer. The BW plays an important role in providing assurance to consumers. Better warranty terms imply greater assurance and signal

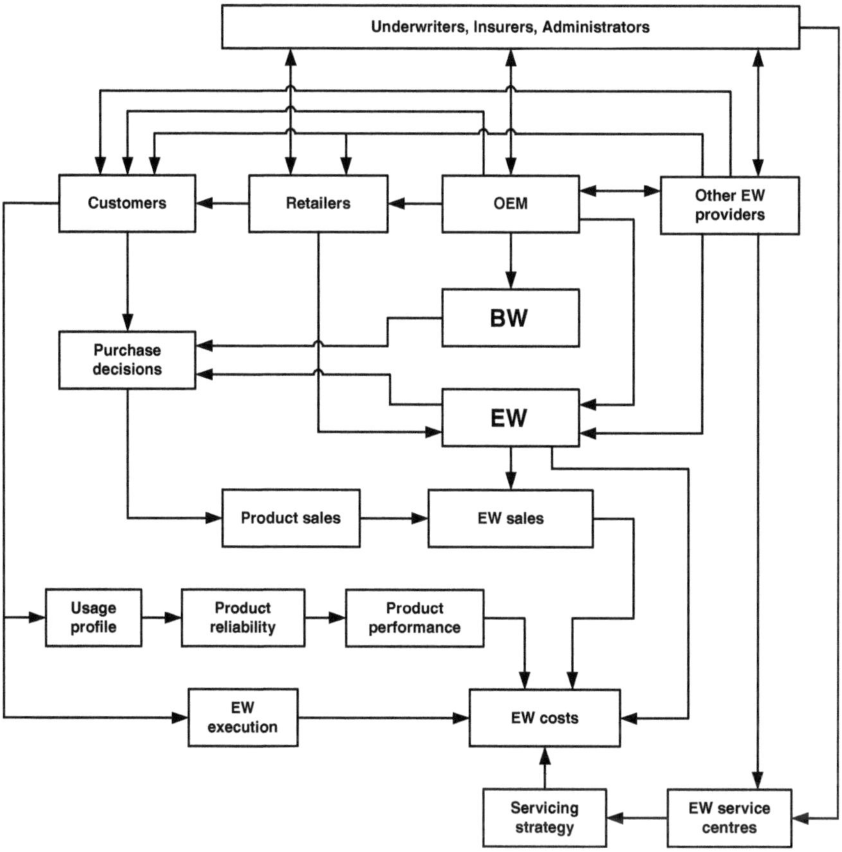

Fig. 6.1 Simple characterisation of the EW process

a more reliable product/system. In some cases, the retailer might increase the BW coverage offered by the manufacturer at no additional cost to the consumer. In this case, the warranty servicing costs for the initial period are borne by the manufacturer and for the latter period by the retailer.

An EW provides extra coverage, and its terms and conditions can be the same as those of a BW or they may differ.

6.3.1.2 EW Providers

EW providers can be one of the following: (1) manufacturers, (2) retailers and (3) other independent providers.

Manufacturers[3]
Historically, manufacturers have stayed out of the EW business for several reasons.

1. Offering an EW would indicate to the customers that their product is inferior.
2. Selling an EW is like selling insurance, and they have an advantage over other EW providers in terms of the information regarding product reliability, risks, etc. As such, it would not be a fair trade practice.
3. Most manufacturers depend upon dealers, distributors and retailers to connect them to customers for the sale of their products. When these parties offer their own EWs, the manufacturers are competing with people who are selling their products.

Two good strategies for manufacturers to follow if they want to enter into the EW market are (1) partnering with their dealers and retailers and (2) developing some kind of "in-the-box" service plan that allows the customer to sign up at any point between the time of sale and the expiry of the BW. The latter is aimed at customers who might not buy the EW at product sale (either being unsure of its worth or being turned off by the high-pressure sales tactics of stores' sales people). When customers take the product home and open the box, they might respond more favourably to an EW offer, especially if it carries the name brand of the manufacturer. Also, manufacturers can wait until the BW is about to expire before making their sales pitch for the EW by phone, email or postcard directly to the customers.

Retailers [Dealers]
Retailers (dealers) view their main activity as selling new products, and as such, most sell EWs rather than offering their own EWs. They earn commission for selling EWs, and this can be as high as 30–50 % of the EW sale price. As such, selling EWs is a very attractive source of income. Some retailers depend on the income they receive from EW sales commissions to keep them profitable. An EW is simply a commodity to be sold with potential profits being the main driver. The retailers who sell their own EWs need to partner with administrators.

Other EW Providers
These include insurance companies, financial institutions (e.g. credit card agencies offering an EW if the product purchase is made by credit card) and others. These are often referred to as *Third-Party EW providers*. Now, there are a small (but growing) number of companies, unaffiliated with either the retailer or the manufacturer who sells EWs direct to customers over the Internet, and these companies compete aggressively with retailers for a share of the EW market. One such EW provider is *GreenUmbrella.com*.[4] Their EW can be purchased by registering the

[3] For more on OEM Extended Warranties, see Warranty Week (2006 and 2009c).

[4] See also, Warranty Week (2008).

product on the GreenUmbrella.com web site within 30 days of purchase. The EW policy is as follows:

> Only two claims are allowed per customer within any 12-month period; In terms of a no-lemon policy, after three repair attempts, GreenUmbrella.com will replace the item, but the replacement unit will not then be covered; The cost of removal, reinstallation, or in the case of a replacement, the disposal of the old unit, is not covered.

The following product types are eligible for coverage: electronics,[5] appliances[6] and computers.[7]

6.3.1.3 EW Administrators

An EW administrator is responsible for all paperwork relating to (1) registering the sale, (2) arranging for repairs and (3) paying claims. There are two different scenarios:

1. The Administrator is either the EW provider or a subsidiary of the EW provider. General Motor's EWs are administered by a subsidiary—*Universal Warranty Corp.*
2. The Administrator is an independent entity.[8] *Assurant Solutions* is an independent EW administrator who acquired the Warranty Management Group unit from GE Consumer and Industrial. As part of that acquisition, Assurant signed

---

[5] Alarm Clocks, Camcorders (Analogue and Digital), Cameras (Analogue and Digital), Car Audio (Radios, Amplifiers, CD Players, CD Changers, Equalizers, Speakers, Subwoofers), Car Videos (DVD Players and Video Monitors), Home Audio Components (Non-Portable: Amplifiers, CD Players, CD Changers, CD Players/Recorders, CD Recorders, Equalizers, Receivers, Tuners), Home Speakers, Home Theatres in a Box, Home Video Products (DVD Players, TV/DVD Combos, TV/VCR Combos, TV/VCR/DVD Combos, Digital Video Recorders, Digital Satellite Systems, HDTV Receivers), MP3 Players, Portable Electronics [PDA's, Satellite Radios, GPS, DVD Players, Telephone (Not Cellular)], Radar Detectors, Boom boxes, Televisions (CRT Projection, CRT Televisions, Front Projectors, LCD Flat Panels, Micro-display Rear, Projection and Plasma).

[6] Air Purifiers, Blenders, Bread Makers, Clothes Steamers, Coffee/Espresso Machines, Cook-tops, Dehumidifiers, Dishwashers, Disposals, Downdrafts, Dryers, Electronic Can Openers, Electronic Tooth Brushes, Electronic Shavers, Floor Cleaners, Food Processors, Freezers, Fryers, Griddles, Grills, Grinders, Hair Dryers, Humidifiers, Ice Machines (Free-standing), Indoor Grills, Irons, Juicers, Microwaves, Mini-Refrigerators/Freezers, Mixers, Ovens, Portable Heaters, Ranges, Range Hoods, Refrigerators, Rice Cookers/Steamers, Rotisseries, Sewing Machines, Slow Cookers (Crock Pots), Steamers, Small Portable Appliances, Space Heaters, Toaster or Toaster Ovens, Trash Compactors, Vacuums, Vacuum Sealers, Waffle Makers, Warming Drawers, Washers, Window Air Conditioners, and Wine Coolers.

[7] Copiers, Desktop Computer Systems, External Electronic Computer Accessories and Electronic Peripheral Devices, Flat Screen Monitors, Laptop Computers, Monitors, Pocket PCs, Printers (Laser, Dot Matrix, or Ink Jet), Printers (Multifunctional), and Paper Shredders.

[8] See Warranty Week (January 10, 2005) for a list of Warranty Administrators in the USA.

a 10-year agreement to market EWs and service contracts on all GE-branded major appliances sold in the USA.

There are many benefits to EW providers to work with EW administrators. The administrators not only manage the claim process but provide expertise on service contract compliance and regulations which the EW provider might lack. In the case of manufacturer-branded EWs, the manufacturer has flexibility in what to outsource to an administrator. The options can include the following:

- Answer the first call from a customer in need of service, while outsourcing other key functions such as compliance and other aspects of claim processing. This allows the manufacturer to have direct contact with customers.
- Outsource everything relating to dealing with customers and servicing of warranty claims. This can include pricing and marketing of EWs.
- Outsource the analysis of warranty data and involve the administrator in the continuous improvement process.[9]

For retailers, the focus is on selling EWs and it is up to the administrator to do all other work. The same applies to EW providers selling EWs on the Internet.

6.3.1.4 Underwriters and Insurers[10]

In most cases, the retailer sells EWs from one or more of the EW providers (the manufacturer or third-party EW providers) and passes the money collected minus the commission to the administrator of the EW (manufacturer, manufacturer's subsidiary or independent administrator) or self-administers the EW. In all cases, the administrator (retailer) takes on the liabilities and the risks associated with the servicing of the EWs. An administrator (retailer, manufacturer or its subsidiary, independent business) can become bankrupt if the cost of servicing EWs exceeds the amount paid by the retailer or the EW provider. Often, this is the result of administrators underpricing the EW in relation to the risk they face, and in some cases, it can be due to poor management of warranty servicing.

Through proper insuring of an EW, the administrator (retailer, dealer, manufacturer or independent) can survive bankruptcy. This can be done either through an insurance company or an underwriter.[11] In the USA, *AIG* is one of the major

[9] Asko Appliances is a Swedish manufacturer of high-end washers, dryers and dishwashers. Service Net is the administrator responsible for handling the claims, doing the warranty data analysis and also help in the product development.

[10] For more on warranty underwriters, see Warranty Week (2010c) and on warranty insurance and insurance companies, see Warranty Week (2009a, b and 2010b).

[11] An underwriter went under because the administrator wrote extended warranty contracts on high-mileage used vehicles, not knowing how costly that would turn out to be in terms of claims.

players in the EW insurance underwriting business that backs up the EW operations of Wal-Mart, Best Buy and others. The EW (and service contract) underwriters in the USA are nestled within the specialty insurance sector of the industry, which gets little attention. These are companies which are parents of warranty administrators—such as *Assurant Inc.* (AIZ), parent of *Assurant Solutions* and *American International Group Inc.* (AIG), parent of *AIG Warranty.*[12]

In the appliance and electronics industries in the USA, there has been a regime in place for decades which ensures that customers never lose their warranty cover. Most of the EWs are sold by retailers who work with third-party administrators, who in turn are either part of or partnered with an established specialty insurance company.

An advantage of insuring is that in the event of an insurance company's collapse, various state insurance guarantee funds have been set up to step in and take over the claims administration. The insurance companies who back EWs and service contracts are graded (positive, negative and stable) based on their financial stability.[13]

Some Comments regarding EW Providers, Administrators, Insurers/Underwriters

1. In some cases, the seller, administrator and underwriter are all one entity. In others, the administrator and underwriter are the same or the underwriters and administrators are different.
2. In the USA for EWs that cover electronics, computers and appliances, the vast majority of the contracts are sold by retailers backed by third-party administrators and insurance underwriters. In Europe, some of the retailers are self-insured and also in some cases act as their own administrators.
3. Some of the EW providers have taken a hybrid approach—managing their own repairs but contracting out the call centre and insurance underwriting functions.
4. Administrators of MSCs are privately held, and several of the insurance underwriters who stand behind them are publicly traded companies who report both revenue and earnings.

---

[12] In the USA most of the EW underwriting business is now handled by one of four insurance companies: Aon Corp., Assurant Inc., the American International Group Inc. (AIG), and the Great American Insurance Group, part of the American Financial Group Inc.

[13] A *positive* outlook indicates that a company is experiencing favourable financial and market trends, relative to its current rating level. If these trends continue, the company has a good possibility of having its rating upgraded.

A *negative* outlook indicates that a company is experiencing unfavourable financial and market trends, relative to its current rating level. If these trends continue, the company has a good possibility of having its rating downgraded.

A *stable* outlook indicates that a company is experiencing stable financial and market trends, and that there is a low likelihood the company's rating will change over an intermediate period. [Warranty Week (2010a)].

6.3.1.5 Customers

Customers for products can be grouped into the following three categories:

1. Individuals/households
2. Businesses
3. Governments.

Individuals can be divided into several groups based on their attributes and characteristics such as (1) attitude to risk, (2) product usage profile, (3) disposable income, (4) level of education and so on.

Businesses can be divided into several groups based on (1) industry sectors (extraction, production, services, etc.), (2) size (small to large based on output volume, total sales, etc.), (3) attitude to risk (small businesses tend to be risk-averse and large ones risk neutral) and so on.

6.3.1.6 Customer Purchase Decisions

The purchasing process involves several stages—need recognition, search for different brands that meet the requirements, evaluation of different brands and the final purchase. The process depends on the product, and several factors play an important role. One needs to differentiate product purchase from EW purchase.

Individuals/households possess limited information regarding technical attributes such as product reliability. The decision process is influenced by advertising, sales people, reputation of manufacturer, price, BW, etc. An EW is important for customers who are not satisfied with the BW and need greater assurance. The willingness to pay extra for the EW depends on attitude to risk. Some wise customers try to find out the details of the EW provider (if not the manufacturer) and the options available should the provider (and/or administrator) go bankrupt.

For businesses buying expensive commercial and industrial products, the decision process involves several persons and so is a group decision. The group members have a better understanding of the technical aspects of products and can get extra information through interactions with sales people. In some cases, the product is very critical (e.g. computers in a bank or travel agency and certain medical equipment in a hospital) so that the BW and the EW play an important role. Performance guarantees and service response options (e.g. 8 a.m.–6 p.m. Monday through Friday or 24 h seven days a week; repair technician on site within some specified interval) play an important role in the final purchase.

The process used by governments for the purchase of consumer products (e.g. computers used in schools or in a large department) can involve a tendering process with the life cycle cost being an important factor in the final decision with EWs playing an important role.

6.3.1.7 Product and EW Sales

Individuals/households often buy one item at a time. Businesses and governments can either buy items individually or in lots. As a result, sales can be viewed as being either continuous or lumpy depending on the type of product. Promotion of a product occurs in two ways—advertising and word of mouth. In the former case, different channels are used for different products (e.g. newspaper, radio, TV, distribution of pamphlets to individual houses in the case of consumer products and trade magazines in the case of commercial and industrial products). For very expensive products (such as aircraft), sales occur at (random) points along the time axis and depend on several factors such the state of economy, interest rate, etc. When there are two or more manufacturers, competition between them has an impact on product sales and the reputation of the manufacturer and also brand names are important variables.

EW sales depend on the sale of products. Some EW providers require that an EW be purchased at the same time as product purchase, and others are more flexible—the customer can buy the EW any time before the expiry of the BW or even after the expiry. As a result, the characterisation of EW sales over time is more complex than that for product sales.

6.3.1.8 Usage Profile

Usage profile can be defined in terms of the following variables:

- Usage intensity (e.g. number of washes per week in the case of a washing machine and km travelled per year in the case of a car)
- Operating load (e.g. load in a washing machine or on a truck)
- Operating environment (e.g. trucks being driven on dirt tracks in a mine operation versus those being driven on highways in a transport operation).

In the simplest characterisations, all customers are viewed as being identical in terms of usage intensity, operating load and operating environment. A more detailed characterisation would involve dividing customers into several groups—such as those with low, medium and high usage intensities. A still more detailed characterisation would involve modelling the variation in usage as a continuous random variable.

6.3.1.9 Product Reliability and Performance

In Chap. 2, we defined several notions of product reliability. If customers are assumed to be homogeneous (in terms of usage), and use the product in the manner for which it was designed, then the product reliability is the inherent reliability. If the customers are heterogeneous (in terms of their usage), then it is the field reliability (which takes into account the effect of usage intensity, load and/or operating environment).

6.3.1.10 EW Service Centres

The service centres for servicing EW claims are often the same as those that service BW claims. The service centres can be owned by

1. The manufacturer (for specialised products).
2. The retailer (e.g. automobile dealers).
3. A third-party independent business.

Some EW providers might partner with an administrator but require that the EW servicing be done at their own centres. Others might decide to outsource the EW (and BW) servicing based on economic considerations.

6.3.1.11 Servicing Strategy

For a non-repairable component, the service centre has the option of replacing a failed item by a new or used unit. For some EWs, the contract requires using only parts from vendors (component suppliers) recommended by the manufacturer. For others, the service centre has the freedom to choose a cheaper brand or a used component in the replacement process. For a repairable item, the service centre has the option of either repairing or replacing. This leads to an interesting issue—repair versus replace strategies—and this is discussed further in Chap. 7 since repair actions have implications for EW costs.

6.3.1.12 EW Costs

EW costs are the costs associated with servicing of claims under an EW. As with BW costs (discussed in Sect. 5.2.3), there are several different notions of EW costs. These include the following:

- Cost per unit (EW) sold
- Life cycle costs
- Costs per unit time.

The costs from the EW provider perspective are different from the customer perspective. Also, the costs are uncertain since they depend on product reliability and the servicing strategy used. These costs play an important role in the pricing of EWs, and they are discussed further in Chap. 7.

6.3.2 Detailed EW Characterisation

The EW process outlined in Fig. 6.1 covers the main elements of the EW process. A more detailed characterisation would involve many other elements—such as component manufacturers, contracts between different parties, information available to different players, etc. Figure 6.2 is a detailed characterisation which includes two sources for components (in-house and those obtained from an external vendor), three types of service centres (owned by the manufacturer, by the retailer and those which are independently owned), three types of EW (EW-1: manufacturer warranty sold direct, EW-2: manufacturer warranty sold through retailer; and EW-3: retailer warranty) and customers grouped into three categories (based on the type of EW purchased).

6.4 System Characterisation of the MSC Process

MSCs are similar to EWs in some ways but differ in others. As such, a simplified characterisation of the MSC process is very similar to the EW process. The process for products (mainly industrial and commercial but in some cases they can be consumer—e.g. air conditioners in big buildings and in homes) is different from that for complex systems (built using equipment from more than one manufacturer) and infrastructures.

6.4.1 MSC Process for Products

The key elements of the MSC process and the interactions between them are shown in Fig. 6.3. Since the key elements are the same as in the EW process, we omit any further discussion on their characterisation. One important item to note is that an MSC can be either a standard contract with no flexibility or one which allows for some customisation to meet the needs of different customers. In the latter case, a few of the terms can differ from contract to contract. The terms and pricing of an MSC depend on the age and condition of the product at the start of the contract.[14]

6.4.2 MSC Process for Complex Systems and Infrastructures

From the owner's perspective, a complex system or infrastructure can be viewed as an asset, and we will use this term in the remainder of the section to denote both types of item. The key elements of the MSC process and the interactions between the elements are shown in Fig. 6.4. We give a brief characterisation of some of the elements contained in the figure.

---

[14] For more on service contract underwriters, see Warranty Week (2010a).

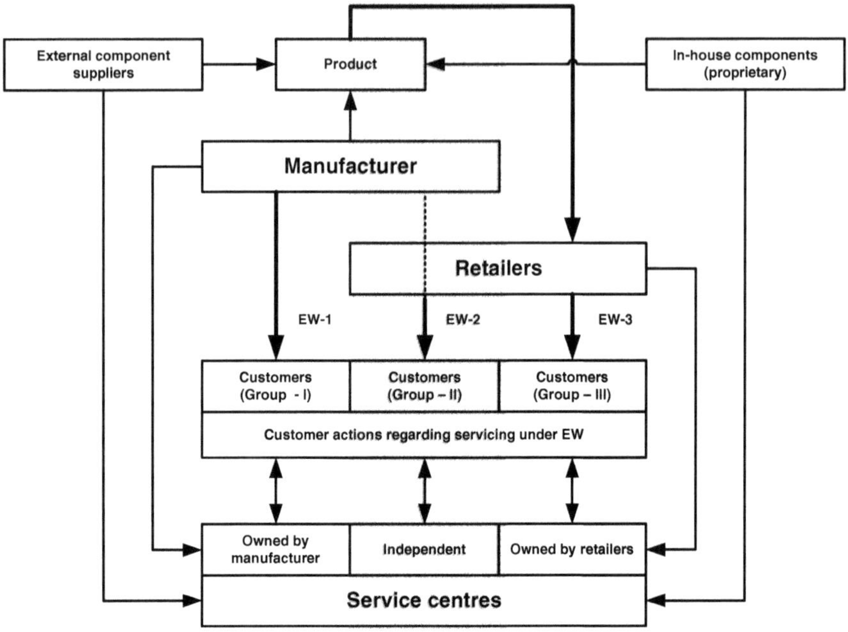

Fig. 6.2 Detailed EW process

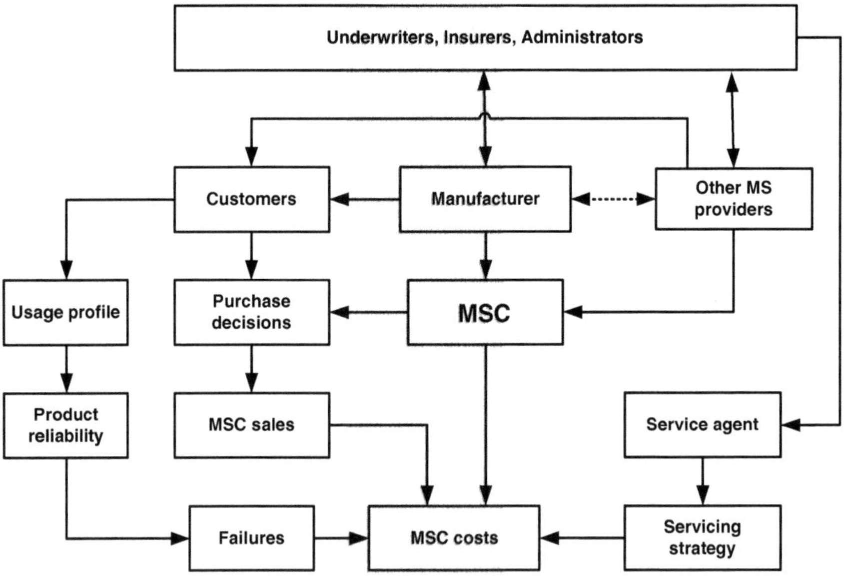

Fig. 6.3 MSC process for products

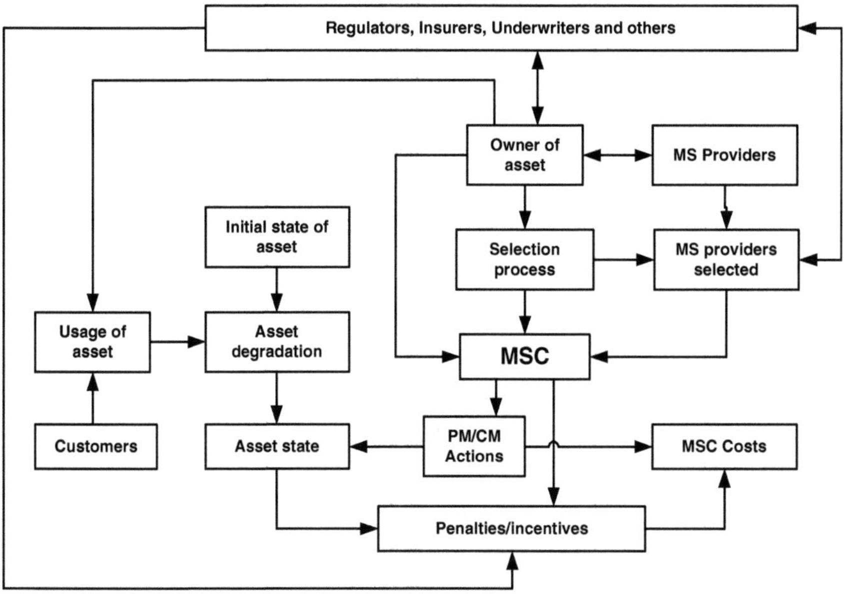

Fig. 6.4 MSC process for complex systems and infrastructures

6.4.2.1 Owners

The owner is either an agency in the public sector (PUS) or a business in the private sector (PRS). Both need to take into account the interests of other stakeholders as discussed in Chap. 5.

6.4.2.2 Asset State

This indicates the condition of the asset as a function of time. The state can be characterised in terms of discrete levels (ranging from good to bad with several intermediate levels) or as a continuous variable with a lower value implying greater degradation. The asset state during the contract period is a function of the state at the start of contract (initial state) and on the usage of the asset.

6.4.2.3 Initial State of Asset

The initial state depends on the past history of the asset—usage, failures, maintenance, etc. This information regarding the initial state depends on keeping a proper record of usage and maintenance. When this is not done, the owner might not know the true initial state of the asset, and this leads to uncertainty. The owner might or might not reveal all the information which leads to information asymmetry.

6.4.2.4 Usage of Asset

In the case of a complex system used in processing or manufacturing operations, usage can be characterised through the production rate—quantity of ore mined per unit (hour, day, week, etc.) in a mine operation, number of items produced per unit in a manufacturing process and so on. In the case of an infrastructure, it is the flow rate—the volume of fluid pumped per unit in water, gas or sewerage network; the number of cars passing over a section of a road network; the volume of goods moved over a rail network; and so on.

6.4.2.5 MSC

The duration of a MSC can vary from a few years to several tens of years. A contract is usually very complex and is customised for each customer. Often (depending on the initial asset state), it can involve a degree of upgrade before the MSC comes into operation. An MSC can include performance guarantees with penalties and incentives depending on the efforts of the MSC provider.

6.4.2.6 MSC Providers

The number of MSC providers can vary from one to several, and this depends on the asset and its location. An important issue is the competency of MSC provider. The owner of asset can get a feel for this through information such as the following:

- Reputation of the MSC provider
- Previous experience in maintaining similar assets
- Financial status of the MSC provider
- Feedback from previous clients of the MSC provider

Another factor is the networking ability of MSC provider so that specialist tasks can be outsourced to competent third-party subcontractors.

6.4.2.7 MSC Selection Process

For PUS assets, the selection of a MSC provider can be complicated. It starts with a public notice calling for bids from interested providers. The bids received are evaluated, and all but a few are rejected. Then, a detailed process starts to select the MSC provider. The evaluation is based on lots of factors—cost, competence, reputation, risks, etc.[15]

---

[15] The risk issues are discussed in Chap. 11.

6.4.2.8 MSC Costs

The cost of a MSC needs to be studied from two different perspectives—owner of the asset and the MS provider. This is discussed further in Chap. 7.

6.5 Game-Theoretic Characterisation of EW Decision-Making

Game theory (GT) provides the most appropriate framework for finding the optimal solutions to decision problems for both the customer (owner) and the EW/MSC provider. The characterisation depends on the decision-makers (players) in the market for EWs and the power structure between the players.

6.5.1 Characterisation of the EW Market

The most general characterisation of an EW market is as shown in Fig. 6.5, and this involves several interacting elements. We discuss each of these briefly and use the following terminology.[16]

Parties: These are distinct groups (EW providers, retailers, service agents and customers) or parties in the market. An example is the case where the manufacturer is the sole EW provider, sells the EWs directly and services the EW through manufacturer-owned service centres. In this case, there are only two parties—manufacturer and customers.

Players: There can be one or more players making up each party. In the above-mentioned example, there is only one manufacturer, but there can be several customer groups with different characteristics.

EW Providers: The EW providers can be divided into three distinct groups as indicated below:

- Manufacturers
- Retailers
- Third parties

In some EW markets, one or more of these groups of providers may not be present. Also, there may be one or more players in each group. For example, in a monopolistic market, there is only one EW provider, whereas in an oligopolistic EW market, there are two or more EW providers.

[16] EW-1–EW-3 are the same as those in Fig. 6.2.

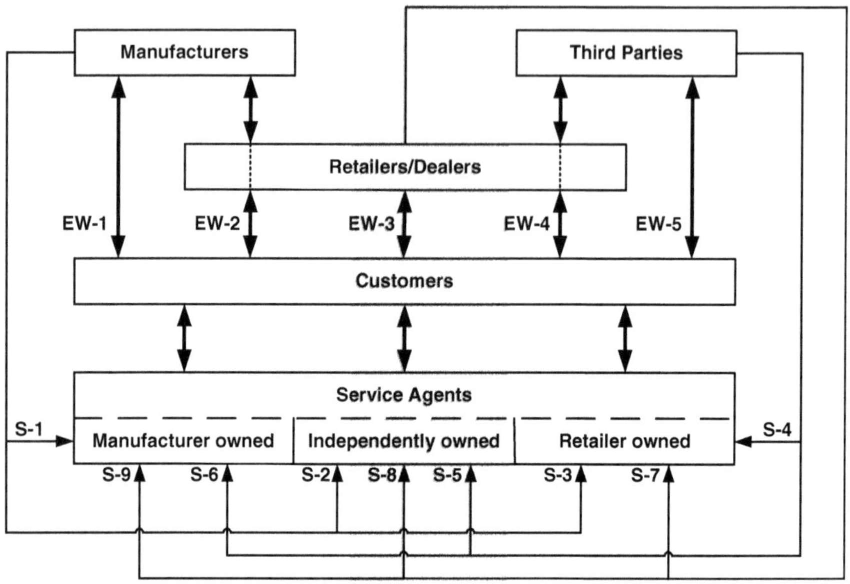

Fig. 6.5 Key elements of the EW market

Retailers: There can be one or several retailers in the EW market. The retailers sell both the product as well as the EWs.

EWs: The EWs can be divided into five distinct groups based on the type of EW provider and whether the EWs are marketed directly by EW providers or through retailers. The different groupings are as indicated below.

- EW-1: Manufacturer EW sold directly to customers
- EW-2: Manufacturer EW sold through retailer to customers
- EW-3: Retailer EW sold directly to customers
- EW-4: Third-party EW sold through retailer to customers
- EW-5: Third-party EW sold directly to customers

Within each group, an EW provider may offer one or more types of EW which differ in their terms and price.

Customers: There can be one customer or several customers, and then, the customer population can be either homogeneous or heterogeneous. In the latter case, the population can be divided into several groups based on characteristics such as attitude to risk, usage intensity, etc.

Service agents: The service agents (SAs) can be divided into three groups as indicated below.

- Manufacturer owned
- Retailer owned
- Independently owned

Each EW provider can choose one or more of the SAs to service their EWs. This leads to the following nine different servicing channels.

- S-1: EW-1 and EW-2 serviced by manufacturer-owned SA
- S-2: EW-1 and EW-2 serviced by independently owned SA
- S-3: EW-1 and EW-2 serviced by retailer-owned SA
- S-4: EW-4 and EW-5 serviced by retailer-owned SA
- S-5: EW-4 and EW-5 serviced by independently owned SA
- S-6: EW-4 and EW-5 serviced by manufacturer-owned SA
- S-7: EW-3 serviced by retailer-owned SA
- S-8: EW-3 serviced by independently owned SA
- S-9: EW-3 serviced by manufacturer-owned SA.

6.5.2 Illustrative GT Scenarios for EW Decision-Making

There are many possible scenarios based on different combinations of parties/groups (EW providers, retailers, customers and service agents) in the market, the number of players in each of the groups and the power structure between the parties/players. As discussed in Sect. 4.5, there are two kinds of power structure between any two players—dominance (which we denote by $\rightarrow$ in our schematic representations) and equal or no dominance (which we denote by $\leftrightarrow$ in our schematic representations). In the case where there is a dominance relationship between two players, the follower's decisions depend on the decisions of the leader. In the equal or no dominance case, the players' decisions are assumed to be made simultaneously. A player's response function to the decisions made by another player is indicated by a broken arrow.

It is not possible to discuss all possible scenarios. Instead, we look at a few, and some of these will be discussed further in Chap. 8. Note that we only show the EW group (EW-1–EW-5). This is only one of the decision variables involved. In general, there are other decision variables, and these are discussed further in Chap. 8.

Scenario 1: Two Parties
The two parties are EW providers (who directly market and service their EWs) and customers. Depending on the number of customers, the EW market (see Table 6.1) is either M-11 (single customer), M-12 (few customers) or M-13 (many customers). The EW providers are the dominant players (leaders), and the customers are followers.

Scenario 1 (a) [Monopolistic EW Market]
The manufacturer's decision variables are (1) the number of different EWs to offer, (2) the terms (e.g. duration) and conditions (e.g. exclusions) of each EW and (3) the price of each EW. For a given set of EWs, the customer's (s') decision variables are (1) whether to purchase EW or not and (2) the EW to select if there are two or more EWs to choose from.

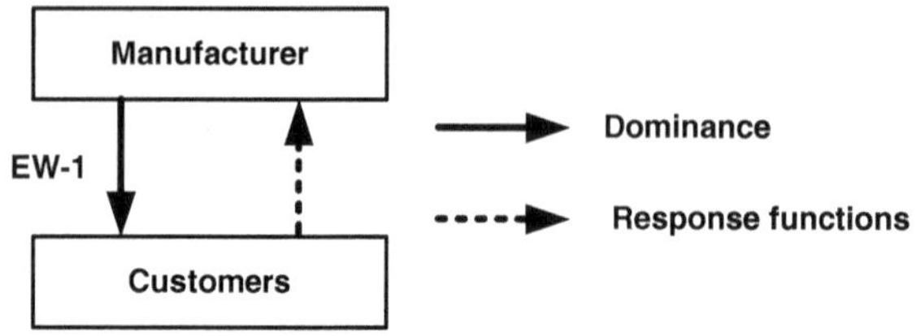

Fig. 6.6 Scenario 1(a) (Single EW provider)

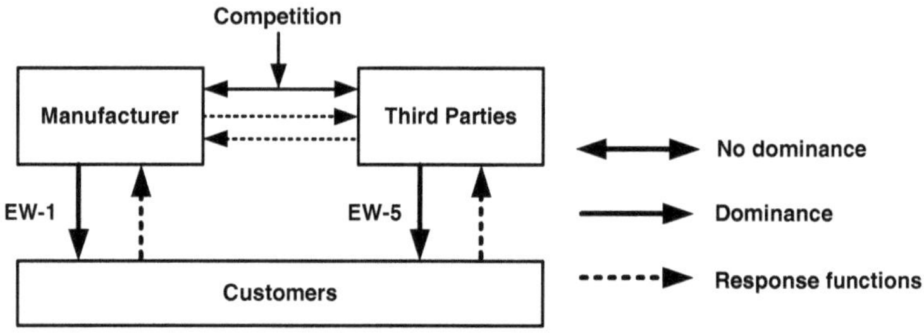

Fig. 6.7 Scenario 1(b) (Multiple EW Providers)

For a given set of EW options (price, duration, etc.), the customers choose the best option (discussed further in Chap. 8), and this defines their response function. The manufacturer then makes the optimal decision taking into account the response function. This is shown schematically in Fig. 6.6 and is a Stackelberg game.

Scenario 1 (b) [Oligopolistic or Competitive EW Market]
Here, there are two or more EW providers with one being the manufacturer. If the number of EW providers is few, the EW market is oligopolistic (M-21 or M-23 depending on the number of customers) or competitive (M-31 or M-33 depending on the number of customers). The response functions (see, Fig. 6.7) in the vertical direction define the optimal decisions of the customers based on the "leader–follower" Stackelberg formulation. The EW providers then optimise their decisions taking into account these response functions. The final optimisation for the EW providers can be viewed as a Nash game with horizontal response functions for the players.

Scenario 2: Three Parties
The three parties are (1) EW providers, (2) retailers and (3) customers. We look at two special cases.

Scenario 2 (a) [Monopolistic EW Market]
The manufacturer is the sole EW provider, and the retailer/dealer is the seller of both the product and the EWs. The decision variables for the manufacturer and customer are the same as in Scenario 6.1 (a). There are two EW prices—the

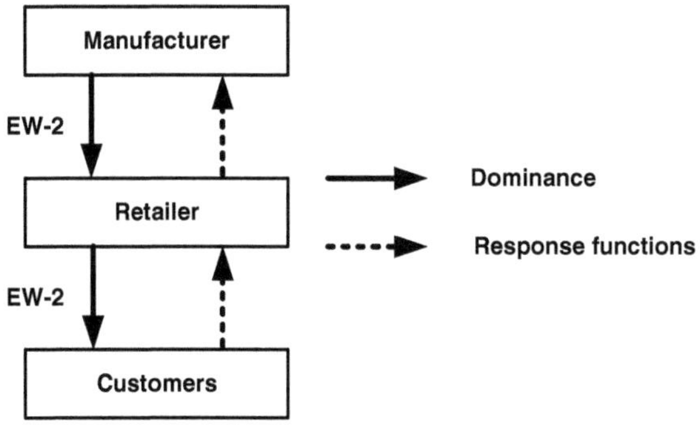

Fig. 6.8 Scenario 2(a) (Single EW Provider)

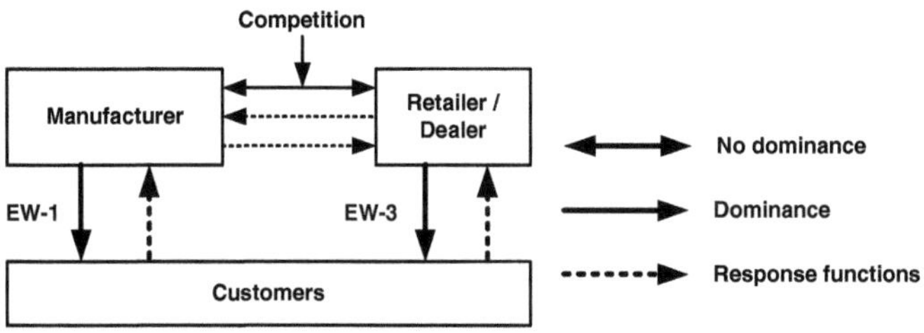

Fig. 6.9 Scenario 2(b) (Two EW Providers)

"wholesale" price (price that the retailer/dealer pays to the manufacturer) for each EW sold and the "retail" price (the price charged for the EW to customers). The difference between the two is the markup on price—a decision variable for the retailer/dealer. As a result, we have a three-stage Stackelberg game with two separate vertical response functions as shown in Fig. 6.8. The retailer's optimal decision is obtained as the solution of the lower-level game taking into account the EW wholesale price charged by the manufacturer and the response function of customers. The manufacturer's optimal decision is obtained as the solution of the higher-level game which takes into account the response function of the retailer.

Scenario 2 (b) [Oligopolistic EW Market]
Both manufacturer and retailer sell their own brands of EW directly to customers. In this case, we have two EW providers competing, and Fig. 6.9 gives the game-theoretic characterisation. This scenario is identical to Scenario 1(b) discussed earlier.

Scenario 2 (c) [Oligopolistic EW Market]
Here, the EW providers are the manufacturer and a small number of retailers. The manufacturer uses two channels for the marketing of EWs as indicated in Fig. 6.5. The game-theoretic characterisation is as shown in Fig. 6.10. Note that in this case, we have the manufacturer and retailers cooperating [as in Scenario 1(a)] and also competing [as in Scenario 2(b)].

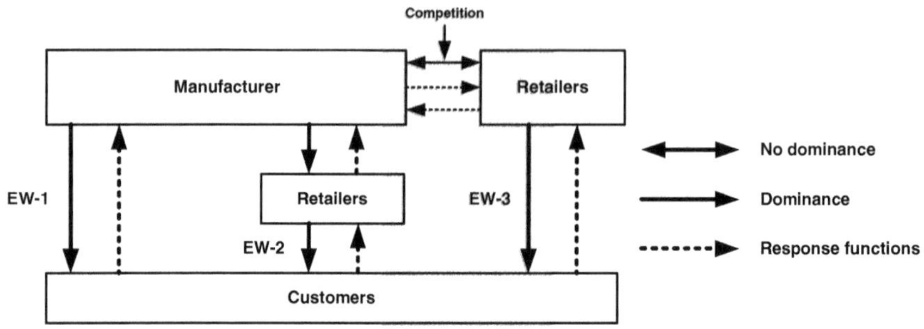

Fig. 6.10 Scenario 2(c) (Two EW providers—two channels for manufacturer)

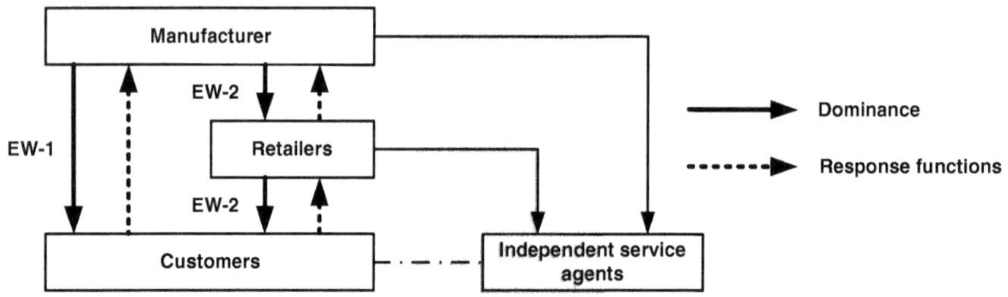

Fig. 6.11 Scenario 3 (Multiple EW providers and independent service centres)

Scenario 3: Four Parties
This is an extension of Scenario 2 (c) with all EW servicing being carried out by independently owned service centres. The contract between the EW providers and the service centres introduces new decision variables (charging for different kinds of repair), and the game-theoretic characterisation which is shown in Fig. 6.11 is more complex. Note here that the independent service agents are followers in the game.

6.6 Game-Theoretic Characterisation of MSC Decision-Making

Several different scenarios can be considered (depending on the type of system—product, plant or infrastructure; the number of parties involved, etc.). We will restrict our discussion to the two-party scenario.

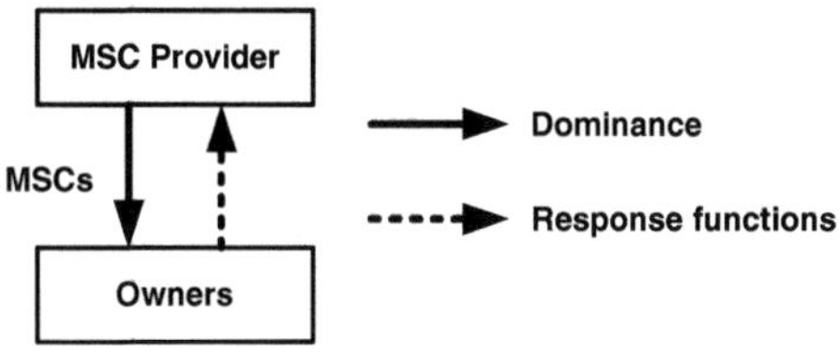

Fig. 6.12 Scenario 4(a) (Single MSC provider and several customers)

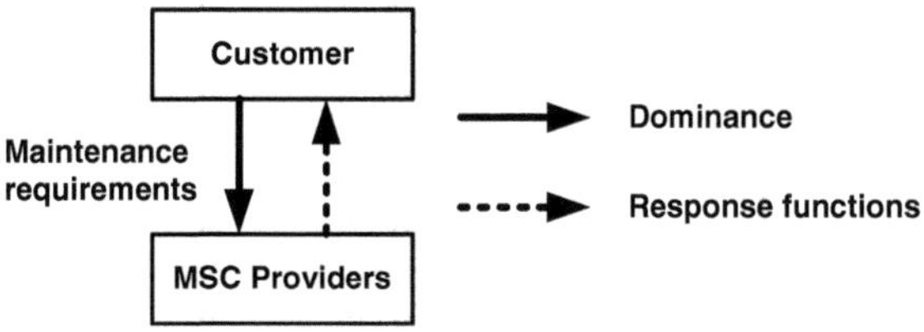

Fig. 6.13 Scenario 4(b) (Single customer and several MSC providers)

Fig. 6.14 Scenario 4(b) (Single customer and single MSC provider)

Scenario 4: Two Parties

The two parties are (1) MSC providers and (2) customers (system owners) for MS.We look at three special cases which characterise the maintenance for most real-world systems (products, plants and infrastructures).

Scenario 4 (a): Single MSC Provider and Several Customers

There is a single MSC provider (manufacturer, retailer or some third party) and several owners. An example of this is a retailer (of very specialised equipment used in hospitals) providing a MSC to several owners (hospitals). In this case, the MSC provider is the leader and the customers are the followers so that the game theoretic characterisation is as shown in Fig. 6.12.

Note that this is very similar to Scenario 1(a) for EW decision-making.

Scenario 4 (b): Single Customer and Several MSC Providers

We look at the simplest case where there is a single owner (e.g. a transport company owning several trucks or owner of a complex plant or infrastructure) and several MSC providers who can service the trucks. In this case, the customer is the leader and the MSC providers are the followers. The game-theoretic characterisation is shown in Fig. 6.13.

Scenario 4 (c): Single Customer and Single MSC Provider
This scenario corresponds to a plant situated in a remote location where the maintenance requires highly specialised equipment which is provided by only one MSC provider. The game-theoretic characterisation is shown in Fig. 6.14.

References

http://www.warrantyweek.com/archive/ww20061024.html
Warranty Week (2005) Extended warranty administrators. Warranty Week, Jan 10
Warranty Week (2006) Manufacturer's extended warranties. Warranty Week, Feb 24
Warranty Week (2008) Flat rate extended warranties. Warranty Week, Aug 7
Warranty Week (2009a) Extended warranty insurance companies. Warranty Week, May 14
Warranty Week (2009b) Warranty insurance. Warranty Week, May 21
Warranty Week (2009c) OEM extended warranties. Warranty Week, May 28
Warranty Week (2010a) Service contract underwriters. Warranty Week, Feb 11
Warranty Week (2010b) Extended warranty insurance companies. Warranty Week, Aug 5
Warranty Week (2010c) Vehicle service contract underwriters. Part 1—Warranty Week, Aug 26; Part 2—Warranty Week, Sep 2

Chapter 7
EW and MSC Cost Analysis

7.1 Introduction

A BW is provided by a manufacturer and is included with the sale of a product. An EW may be obtained from the manufacturer, retailer or an independent provider and is an optional purchase by the customer. An EW lasts for a specific period beyond that of the BW and its terms may be identical to those of the BW or they may include additional features for the rectification of product failure such as cost sharing, parts exclusions, cost limits and cost deductibles. The customer may be able to choose a specific type of EW from a set of options being offered by the service provider and then may purchase the EW at the time of the product sale or when the BW expires.

A MSC is similar to an EW in that the maintenance of a product (plant or infrastructure) is carried out by an external service provider. The period over which the maintenance actions are to be carried out and the payment to be made by the customer (system owner) to the service provider are specified in the contract. The terms of the contract may also specify the different types of maintenance action to be carried out, parts exclusions, cost limits, cost deductibles, product or service performance guarantees and incentives/penalties if these performance levels are achieved/not achieved. The contract may be either a standard one offered by the service provider, a customised version designed to meet a customer's specific needs or one which is initiated and dictated by the customer.

An EW/MSC provider must service all claims (failures that require CM actions) over a contract period. A schedule for PM actions may also be specified in the contract. The cost associated with servicing each claim and each PM action is either borne by the service provider or it may be shared between the service provider and the customer, depending on the terms of the contract. Thus, it is necessary to look at the costs of servicing EWs and MSCs from the perspective of the service provider and the customer. Servicing costs cannot be predicted with certainty since they depend on the frequency of occurrence of claims which are influenced by operating environment, usage intensity and PM actions. The actual costs to rectify claims may also vary significantly.

D. N. P. Murthy and N. Jack, *Extended Warranties, Maintenance Service and Lease Contracts*, Springer Series in Reliability Engineering,
DOI: 10.1007/978-1-4471-6440-1_7,

Two types of servicing costs which are important to study are cost per unit sale and cost per unit time.[1] In each case, a proper framework is needed to build models in order to estimate the relevant costs. The focus of this chapter is on modelling for cost estimation. The results are then used in Chaps. 8 and 10 for optimal decision-making by service providers and customers.

The outline of the chapter is as follows. Section 7.2 looks at the system characterisations needed to model the two servicing cost notions which were mentioned previously. In Sect. 7.3, the modelling assumptions are specified and details are given on the modelling of sales of products and contracts. The cost analysis of 1-D and 2-D BWs is dealt with in Sect. 7.4. The results from this section are then used to provide the corresponding cost analysis of EWs in Sect. 7.5. In Sect. 7.6, MSC cost analysis is discussed and some examples are given. Finally, we look at few decision models in maintenance outsourcing.

7.2 System Characterisation for Cost Analysis

The basic cost of providing an EW or MSC consists of the costs associated with the servicing of claims due to item failures (CM actions) plus the servicing costs for any PM actions which may be part of the contract. The key cost elements for each type of maintenance action are the material costs for replacement parts, the labour costs which depend on the time taken to perform the maintenance action, the transport costs incurred whether the servicing is carried out on site or if the failed item needs to be shipped to a repair facility plus other costs for administration, inventory, etc.

Cost per unit sale refers to the cost of a single contact purchased by a customer. EW/MSC providers normally sell contracts to a collection of customers at different points in time. The *cost per unit time* measure refers to the aggregation of the servicing costs from all of these contracts. A contract provider needs to estimate cost per unit sale in order to determine the proper contract price so that a profit can be made on each sale. A customer needs the cost information to assess whether purchasing the contract is worthwhile compared to other options that may be available to service the item. Cost per unit time together with the price of each contract sold gives the provider an estimated profile of profits earned over time. Both of these cost notions are now described in detail.

[1] Another type of cost is the life cycle cost (LCC). This is discussed in Blischke and Murthy (1994).

7.2.1 Cost per Unit Sale

The cost per unit sale is the servicing cost associated with providing a single contract (which may be an EW, a MSC or even a BW). The system characterisation for modelling this type of cost consists of several interlinked elements and is indicated in Fig. 7.1. Some of these elements are under the control of the contract provider while others are influenced by the actions of the customers.

The number of claims made under a contract is influenced by the inherent product reliability, the product usage during the contract by the customer, and the servicing strategy used by the provider. CM costs are incurred by rectifying product failures and the cost of servicing the contract may also include the cost of performing PM. In the case of a MSC, the past usage and maintenance history affects the product reliability during the contract period.

7.2.2 Cost per Unit Time

The second servicing cost notion uses information on sales of contracts over time. Cost per unit time is based on an aggregation of costs from all contracts sold and still in force at any given time. The extra elements needed to complete the system characterisation for modelling this type of cost (in addition to those already given in Fig. 7.1) are indicated in Fig. 7.2.

7.3 Modelling for Cost Analysis

7.3.1 Assumptions

In order to simplify the building and analysis of the models, we make the following assumptions:

1. All customers are alike in terms of their usage. One can relax this assumption by dividing the customers into two or more groups based on usage intensity.[2]
2. All items are statistically similar. One can relax this assumption by including two types of items (conforming and non-conforming) to take into account quality variations in manufacturing.[3]
3. Each failure that occurs under a contract results in an immediate claim by a customer. Relaxing this assumption involves modelling the delay time between failure and claim.

---

[2] Cost analysis with heterogeneous usage intensity is discussed in Kim et al. (2001).

[3] For more on this, see Blischke and Murthy (1994, 1996) and Murthy and Djamaludin (2001).

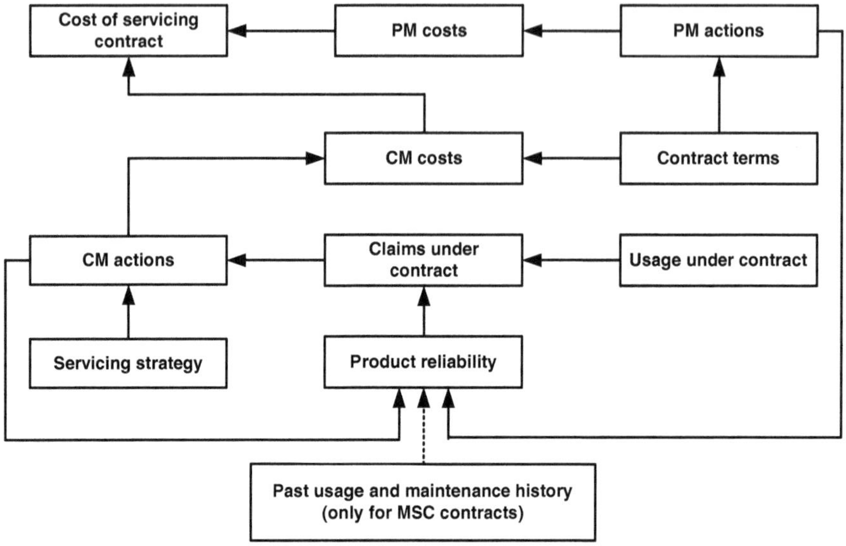

Fig. 7.1 Key elements for estimating cost per unit sold

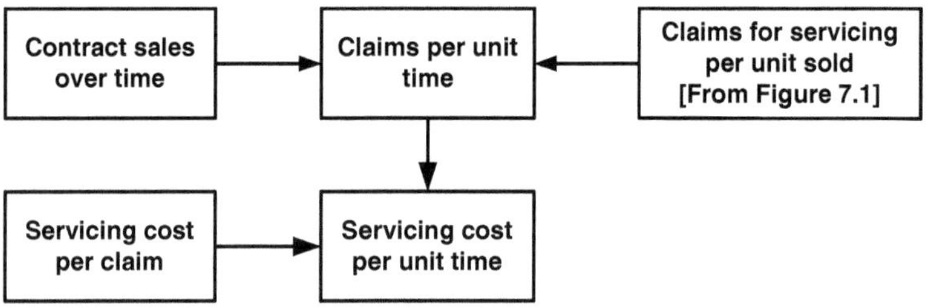

Fig. 7.2 Key elements for estimating cost per unit time

4. All claims are valid. This can be relaxed by assuming that a fraction of the claims are invalid, either because the item was used in a mode not covered by the contract or because it was a bogus claim.
5. The time to rectify a failure (either through repair or replacement) is sufficiently small in relation to the mean time between failures, so it can be assumed to be zero. Although these rectification times are ignored in order to model successive failures over time, they can be treated as being non-zero for the purposes of calculating downtime penalty costs that may specified in the contract.
6. The service provider has the logistic support (spares and facilities) needed to carry out the necessary rectification actions without any delays.
7. There are no PM actions specified in the service contract, so the servicing costs only refer to the cost of claims requiring CM actions.

7.3.2 Modelling Sales

As mentioned in Chap. 3, every product has a finite life cycle. Let L denote the length of the time period over which new product sales (with BWs) take place and $S(t)$ denote the total sales that occur over the interval [0, t). The product sales rate $s(t) = \mathrm{d}S(t)/\mathrm{d}t,\ 0 \le t \le L$, can be modelled in many different ways. One well-known model is the Bass diffusion model given by

$$s(t) = (a + bS(t))(S_m - S(t)), \quad S(0) = 0, \tag{7.1}$$

with $a > 0$ denoting the effect of advertising, $b > 0$ denoting the word-of-mouth effect[4] and

$$S_m = S(L) = \int_0^L s(t)\mathrm{d}t \tag{7.2}$$

the total number of first purchase sales over the life cycle.

As mentioned previously, customers can either purchase an EW either at the time of product purchase or at the expiry of the BW. Let ϕ_1 and ϕ_2 denote the fractions of customers who buy the EW at the instant of product purchase and at the expiry of the BW, respectively. Note that $\phi_1, \phi_2 \ge 0$ and $\phi_1 + \phi_2 \le 1$. As a result, the sales rate for EWs is given by

$$s_e(t) = \begin{cases} \phi_1 s(t), & 0 \le t < W \\ \phi_1 s(t) + \phi_2 s(t - W), & W \le t < L \\ \phi_2 s(t - W), & L \le t < L + W \end{cases} \tag{7.3}$$

Figure 7.3 shows the EW sales rate in the case of a 1-D warranty.

7.4 Cost Analysis of BWs

In this section, we consider the cost analysis of BWs. The results are then used in the Sect. 7.5 to model the costs of EWs. We confine our attention to non-renewing FRW policies and look at the servicing cost analysis of a 1-D BW with warranty period W and a 2-D BW with a rectangular warranty region given by $[0, W) \times [0, U)$.[5] In each case, we use the conditional approach to derive the results for

[4] This is the simple diffusion model first proposed in Bass (1969). Since then, the basic model has been extended to take into account other factors, e.g. advertising effort, negative and positive word-of-mouth effects. Details of these can be found in Mahajan and Wind (1986).

[5] Cost analysis of several different types of 1-D and 2-D warranties can be found in Blischke and Murthy (1994, 1996).

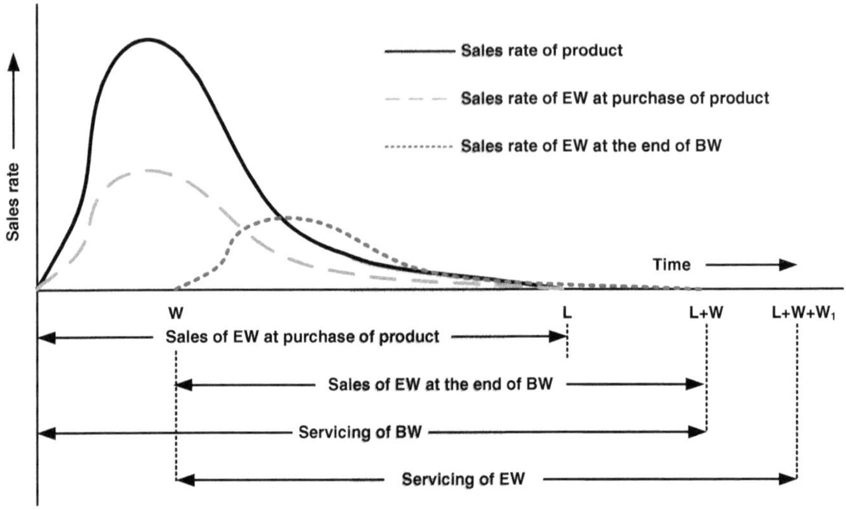

Fig. 7.3 Sales rate for new products and EWs

moments of warranty servicing costs.[6] In the 1-D case, both cost measures are discussed for replacement by new items and minimal repair, respectively. In the 2-D case, only details of the cost per unit sale are provided.

7.4.1 1-D BWs

7.4.1.1 Cost Per Unit Sale

Let $N(t)$ denote the number of claims over the interval $[0, t)$ with $\tilde{C}_i$ the cost of servicing the ith claim. $\{N(t), t \geq 0\}$ is a 1-D point process (for details, see Sect. 3.5). Each individual servicing cost is a random variable, and we assume that these variables are *iid* with distribution function $F_C(c)$. The total servicing cost for the interval $[0, t)$ is given by

$$C(t) = \sum_{i=1}^{N(t)} \tilde{C}_i. \tag{7.4}$$

A complete probabilistic characterisation of $C(t)$ is extremely difficult, even for the most simple cases. It is therefore necessary to look only at the first and second moments. Using the conditional approach (see Appendix B), it is easy to show that the expected total servicing cost is given by

[6] The conditional approach is discussed in Appendix A.

$$E[C(t)] = E[\tilde{C}_i]E[N(t)]. \tag{7.5}$$

The second moment of total servicing cost is also easily obtained using the conditional approach. Note that

$$\begin{aligned} E\left[\{C(t)\}^2 | N(t) = n\right] &= E\left[\{\tilde{C}_1 + \tilde{C}_2 + \cdots + \tilde{C}_n\}^2\right] \\ &= nE\left[\tilde{C}_i^2\right] + n(n-1)\{E[\tilde{C}_i]\}^2 \end{aligned} \tag{7.6}$$

and

$$E\left[\tilde{C}_i^2\right] = \mathrm{Var}[\tilde{C}_i] + \{E[\tilde{C}_i]\}^2. \tag{7.7}$$

Using (7.7) in (7.6) we have

$$E\left[\{C(t)\}^2 | N(t) = n\right] = n\mathrm{Var}[\tilde{C}_i] + n^2\{E[\tilde{C}_i]\}^2 \tag{7.8}$$

and then removing the conditioning gives

$$E\left[\{C(t)\}^2\right] = \mathrm{Var}[\tilde{C}_i]E[N(t)] + \{E[\tilde{C}_i]\}^2 E\left[\{N(t)\}^2\right]. \tag{7.9}$$

Always replace by new

Each failed item is replaced by a new item, so $\{N(t), t \geq 0\}$ is an ordinary renewal process with *iid* inter-failure times having distribution function $F(t)$. For this ordinary renewal process, the expected number of claims (failures) over the interval $[0, t)$ is given by

$$E[N(t)] = M(t) \tag{7.10}$$

with $M(t)$ the solution of the integral equation specified in (3.24). The cost of each new item is C_f, a constant, so using (7.5), we have

$$E[C(t)] = C_f M(t). \tag{7.11}$$

For an ordinary renewal process (see Appendix B), we have

$$E\left[\{N(t)\}^2\right] = M_2(t) = \sum_{n=1}^{\infty} (2n-1)F^{[n]}(t), \tag{7.12}$$

where $F^{[n]}(t)$ is the n-fold convolution of the distribution function $F(t)$ with itself. Thus, using (7.9), we have

$$E\left[\{C(t)\}^2\right] = C_f^2 M_2(t). \tag{7.13}$$

The first and second moments for the total servicing cost over the BW period are given by (7.11) and (7.13), respectively, with $t = W$.

Always do minimal repair

The failure of an item is often due to one or few components failing and this number is very small in relation to the total number of components contained in the item. A minimal repair usually involves simply replacing only failed components, so that after the repair is completed, the item is basically as it was at the time of failure since all non-replaced components have the same usage and age as they had previously. The counting process $\{N(t), t \geq 0\}$ for the number of claims (failures) is a non-homogeneous Poisson process (NHPP) with intensity function $\lambda(t) = h(t)$ [the hazard function associated with $F(t)$], as discussed in Sect. 3.5. The cumulative intensity function for the NHPP is given by $\Lambda(t) = \int_0^t \lambda(x)\mathrm{d}x = H(t)$.

The expected number of claims over the interval $[0, t)$ is given by

$$E[N(t)] = \Lambda(t) = H(t) \tag{7.14}$$

and

$$E\left[\{N(t)\}^2\right] = \Lambda(t) + \{\Lambda(t)\}^2 = H(t) + \{H(t)\}^2. \tag{7.15}$$

We assume that the variability in minimal repair costs is small and we denote the mean by $C_r = E[\tilde{C}_i]$. Thus, each repair can be assumed to cost C_r. Using (7.14) and (7.15) in (7.5) and (7.9) gives the following expressions for the first two cost moments

$$E[C(t)] = C_r H(t), \tag{7.16}$$

and

$$E\left[\{C(t)\}^2\right] = C_r^2\left[H(t) + \{H(t)\}^2\right]. \tag{7.17}$$

The first and second moments for the total servicing cost over the BW period are given by (7.16) and (7.17) respectively with $t = W$.

7.4.1.2 Cost Per Unit Time

Here, the focus is on using the pattern of item sales over the product life cycle to estimate the total servicing costs as a function of time. The cost incurred in the small time interval $[t, t + \delta t]$ is due to servicing claims from items that were sold during the previous period $[\psi, t)$, where ψ is given by

$$\psi = \max\{0, t - W\} \tag{7.18}$$

Let $\rho(t)\delta t$ denote the expected number of claims in the interval $[t, t + \delta t)$. $\rho(t)$ is called the claims rate, and we derive an expression for this rate and then multiply this by the expected cost of servicing each claim to give the expected servicing cost per unit time.

Always replace by new

For an item sold at time x, the expected number of claims in $[t, t + \delta t)$ is given by $m(t - x)\delta t$ where $m(t)$ is the renewal density function. It is the derivative of the renewal function and is defined by the integral equation

$$m(t) = f(t) + \int_0^t m(t - x)f(x)\,\mathrm{d}x \tag{7.19}$$

Since the sales rate at time x is given by $s(x)$, integrating the expected claims from sales over the period $[\psi, t)$ yields the following expression for the claims rate:

$$\rho_1(t) = \int_\psi^t s(x)\,m(t - x)\,\mathrm{d}x \tag{7.20}$$

for $0 \le t \le L + W$. The expected servicing cost per unit time is then $C_f\rho_1(t)$.

Always do minimal repair

The approach is very similar to the previous case. For an item sold at time x, the expected number of claims in $[t, t + \delta t)$ is given by $\lambda(t - x)\delta t$ where $\lambda(t) = h(t)$, the hazard function associated with $F(t)$. As a result, the claims rate is now given by

$$\rho_2(t) = \int_\psi^t s(x)\lambda(t - x)\,\mathrm{d}x \tag{7.21}$$

for $0 \le t \le L + W$. The expected servicing cost per unit time is $C_r\rho_2(t)$.

Repair versus Replace

In the case of a repairable item that fails under warranty, the service agent (SA) has the choice either to repair or replaced the item. The optimal choice depends on the relative costs, the age of the failed item and the duration of the remainder of the warranty. There are many strategies that utilise average repair cost in making repair versus replace decisions. In all of these, the warranty period is divided into distinct intervals for repair and replacement. Nguyen and Murthy (1989) discuss a strategy where the warranty period is split into a replacement interval followed by a repair interval. Jack and Murthy (2001) propose a strategy under which the warranty period is divided into three distinct intervals—$[0, x)$, $[x, y]$ and $(y, \text{WP}]$. The first failure in the middle interval is remedied by replacement and all other failures are minimally repaired.[7]

7.4.2 2-D BWs

Claims (resulting from failures) under a 2-D warranty are random points in a 2-D warranty region. In Sect. 3.8, we discussed different approaches to modelling the occurrence of claims. We use Approach 1 to build models for the cost analysis of a 2-D BW with a rectangular warranty region given by $[0, W) \times [0, U)$. Let $C(W, U)$ denote the expected warranty servicing cost per unit sale.

The 2-D problem is effectively reduced to a 1-D problem by treating usage as a random function of age. In addition, it is assumed that the usage rate for a customer is constant over the warranty period but varies across the customer population. As a result, the usage rate Z is a random variable which can be either discrete (with for example, categories low, medium and high users) or continuous with density function $g(z)$ as shown in Fig. 7.4. Modelling of item failures under warranty is done using 1-D models obtained by conditioning on the usage rate. The bulk of the 2-D warranty literature assumes a linear relationship between usage and age.[8] The warranty period (for a non-renewing FRW) depends on the usage rate and, conditional on $Z = z$, is given by $\text{WP} = \min\{W, (U/z)\}$.

The time to first item failure conditioned on the usage rate $Z = z$ has distribution function $F_z(t)$. This is related to $F(t)$ by the relationship indicated in (3.41) so that

$$F_z(t) = F(\tilde{z}^{\nu} t) \tag{7.22}$$

where $\tilde{z} = z/z_0$ and $\nu > 1$. z_0 is the nominal usage rate for the item so that $F_z(t) = F(t)$ when $z = z_0$.

As we did with a 1-D BW, we look at the two cases—(1) always replace by new and (2) always do minimal repair. We confine our attention to the expected servicing cost per unit sale.

---

[7] These are suboptimal strategies. The characterisation of the optimal strategy is more complex. Jack and van der Duyn Schouten (2000) conjectured the form of the optimal strategy and Jiang et al. (2006) proved that the conjecture was true.

[8] See, for example, Blischke and Murthy (1994), Lawless et al. (1995) and Gertsbakh and Kordonsky (1998). Iskandar and Blischke (2003) deal with motorcycle data. See Lawless et al. (1995) and Yang and Zaghati (2002) for automobile warranty data analyses based on this approach.

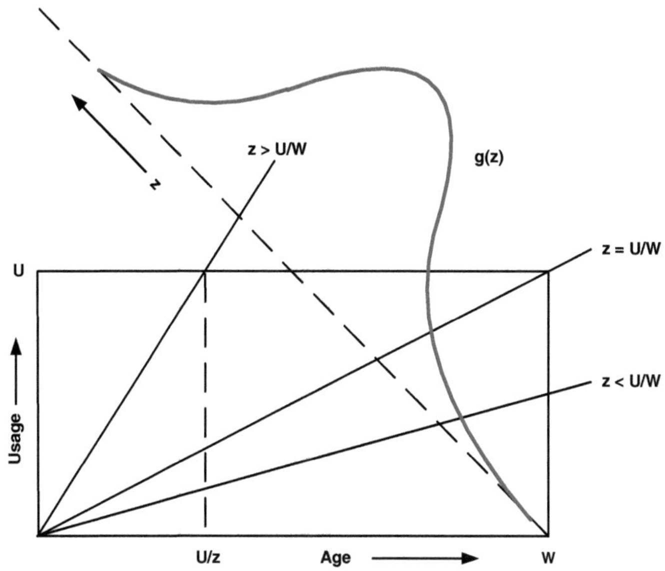

Fig. 7.4 WP for different usage rates

Always replace by new

Since failed items are replaced by new ones, conditional on $Z = z$, claims over the warranty period occur according to the ordinary renewal process associated with $F_z(t)$. As a result, the conditional expected servicing cost per unit sale is given by

$$E[C(W, U|Z = z)] = \begin{cases} c_r M_z(U/z), & \text{if } z > \gamma \\ c_r M_z(W), & \text{if } z \leq \gamma \end{cases} \tag{7.23}$$

where $\gamma = U/W$ and $M_z(t)$ is given by the integral equation

$$M_z(t) = F_z(t) + \int_0^t M_z(t - t')\, \mathrm{d}F_z(t'). \tag{7.24}$$

By unconditioning, the expected warranty servicing cost per unit sale is given by

$$E[C(W, U)] = c_r \left[\int_0^{\gamma} M_z(W)\, g(z)\, \mathrm{d}z + \int_{\gamma}^{\infty} M_z(U/z)\, g(z)\, \mathrm{d}z \right]. \tag{7.25}$$

Always do minimal repair
Since failed items are all minimally repaired, conditional on $Z = z$, claims over the warranty period occur according to an NHPP with conditional intensity function given by $\lambda_z(t) = h_z(t) = f_z(t)/[1 - F_z(t)]$. Defining

$$\Lambda_z(t) = \int_0^t \lambda_z(t)\,\mathrm{d}t, \tag{7.26}$$

the conditional expected servicing cost per unit sale is given by

$$E[C(W, U|Z = z)] = \begin{cases} c_r\Lambda_z(U/z), & \text{if } z > \gamma \\ c_r\Lambda_z(W), & \text{if } z \leq \gamma \end{cases}. \tag{7.27}$$

By unconditioning, the expected warranty servicing cost per unit sale is given by

$$E[C(W, U)] = c_r\left[\int_0^{\gamma} \Lambda_z(W)\,g(z)\,\mathrm{d}z + \int_{\gamma}^{\infty} \Lambda_z(U/z)\,g(z)\,\mathrm{d}z\right]. \tag{7.28}$$

Repair versus Replace
Similar to the 1-D case, many strategies that involve a choice between repair and replacement based on average repair cost have been proposed. These typically involve dividing the warranty region into several distinct subregions. Iskandar and Murthy (2003) and Iskandar et al. (2005) study two such strategies that involve rectangular subregions. Jack et al. (2009) proposed a strategy which involves the complex shape indicated in Fig. 7.5. Here, Γ denotes the region enclosed by the curve. The servicing strategy is as follows: replace with a new item at the first failure occurring in the region Γ and minimally repair all other failures. The shape of the curve is selected to minimise the expected warranty servicing cost per unit sale.

7.5 Cost Analysis of EWs

An EW starts once the BW on an item expires. It can be purchased either when the item is purchased or at the end of the BW period. The EW terms may be the same or they may be different from those of the BW, and this leads to several possible cost scenarios. We again consider non-renewing FRW policies, and we look at the servicing cost analysis of a 1-D EW with warranty period W_1 and then a 2-D EW with a warranty region lying outside the rectangular BW region $[0, W) \times [0, U)$.

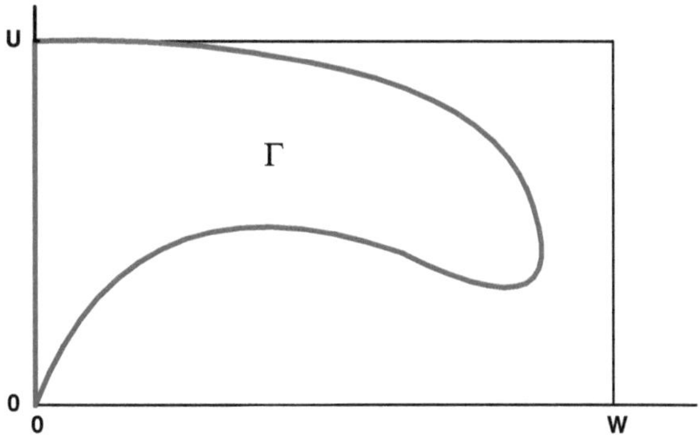

Fig. 7.5 Warranty servicing strategy

The exact location and shape of the 2-D EW region depends on when the EW is purchased. In each case, we confine our attention to expected warranty servicing costs per unit sale. In the 1-D case, we deal with both replacing failed items by new items and with minimal repair at each failure, whereas in the 2-D case, we only consider minimal repair and the EW given by a rectangle with time limit U_1 and usage limit W_1.

7.5.1 Identical EW and BW Terms

7.5.1.1 1-D EWs

For a 1-D EW which begins at time W after the product sale and which ends at time $W + W_1$, the time of the EW purchase makes no difference to the method of cost analysis. The warranty servicing cost $C_E(W_1; W)$ is simply the difference between the servicing costs for two BWs—one with warranty period W and the other with warranty period $W + W_1$. Thus, the expected EW servicing cost per unit sale is given by

$$E[C_E(W_1; W)] = E[C(W + W_1)] - E[C(W)] \tag{7.29}$$

Always replace by new
Using (7.29) and (7.11), it follows that the expected servicing cost for the EW period is given by

$$E[C_E(W_1; W)] = C_f[M(W + W_1) - M(W)]. \tag{7.30}$$

Always do minimal repair

In this case, substituting (7.16) into (7.29) gives the expected servicing cost for the EW period as

$$E[C_E(W_1; W)] = C_r[H(W + W_1) - H(W)]. \tag{7.31}$$

7.5.1.2 2-D EWs: I (EW Bought at Time of Item Purchase)

At the time of item purchase, the EW specifies additional limits for age and usage beyond the BW limits to be W_1 and U_1, respectively. Thus, the EW region is given by the shaded area in Fig. 5.3 lying outside the BW region $[0,W) \times [0,U)$.

Let $C_E(W_1, U_1; W, U)$ denote the expected EW servicing cost per unit sale with $\gamma_0 = (U + U_1)/(W + W_1)$. We assume that the item is always minimally repaired on failure, and we use the conditional approach to determine the expected servicing cost. The item usage rate across the population of customers who purchase an EW is given by the random variable Z with density function $g_1(z)$ [if this is the same as that for those people who bought the item, then it follows that $g_1(z) = g(z)$].

The conditional expected EW servicing cost per unit sale (conditioned on usage rate) is given by

$$E[C_E(W_1, U_1; W, U|Z = z)] = \begin{cases} c_r[\Lambda_z(W + W_1) - \Lambda_z(W)], & \text{if } z \le \gamma_0 \\ c_r[\Lambda_z((U + U_1)/z) - \Lambda_z(U/z)], & \text{if } z > \gamma_0 \end{cases}. \tag{7.32}$$

On removing the conditioning, we have the expected EW servicing cost as

$$E[C_E(W_1, U_1; W, U)] = c_r\left\{\int_0^{\gamma_0} [\Lambda_z(W + W_1) - \Lambda_z(W)]g_1(z)\,dz + \int_{\gamma_0}^{\infty} [\Lambda_z((U + U_1)/z) - \Lambda_z(U/z)]g_1(z)\,dz\right\}, \tag{7.33}$$

where $\Lambda_z(t)$ is given by (7.26).

7.5.1.3 2-D EWs: II (EW Bought when the BW Expires)

The BW expires either at time W when the initial age limit W is reached (and the usage is zW) or at time U/z when the initial usage limit U is reached. The additional limits on age and usage which are specified in the EW terms are W_1 and U_1, respectively, and these are now measured from the particular point at which the BW

Table 7.1 EW starting times and finishing times as z varies for Case (1)

Usage rate	Age of item	
	Start of EW	Expiry of EW
$0 < z \le \gamma_2$	W	$W + W_1$
$\gamma_2 < z \le \gamma_1$	W	$W + U_1/z$
$\gamma_1 < z < \infty$	U/z	$(U + U_1)/z$

expires. The two possible locations for the rectangular EW region are shown in Fig. 5.4 and are given by $[U/z, U/z + W_1) \times [U, U + U_1)$ and $[W, W + W_1) \times [zW, zW + U_1)$, respectively. $C_E(W_1, U_1; W, U)$ denotes the expected EW servicing cost per unit sale. Again, we assume that the item is always minimally repaired on failure and we use the conditional approach to determine the expected servicing cost.

Let the usage limit to age limit ratios for the BW and EW regions be $\gamma_1 = U/W$ and $\gamma_2 = U_1/W_1$, respectively. The two cases (1) $\gamma_2 \le \gamma_1$ and (2) $\gamma_2 > \gamma_1$ need to be analysed separately. Case (2) is shown in Fig. 5.4. In both situations, the EW can expire due to the age or usage limit being exceeded.

Case (1) $\gamma_2 \le \gamma_1$

The age of the item at the beginning and end of the EW depends on the value of the usage rate and the three possible situations are summarised in Table 7.1.

Conditional on the usage rate, we have

$$E[C_E(W_1, U_1; W, U|Z = z)] = \begin{cases} c_r[\Lambda_z(W + W_1) - \Lambda_z(W)] & \text{for } 0 < z \le \gamma_2 \\ c_r[\Lambda_z(W + U_1/z) - \Lambda_z(W)] & \text{for } \gamma_2 < z \le \gamma_1 \\ c_r[\Lambda_z((U + U_1)/z) - \Lambda_z(U/z)] & \text{for } \gamma_1 < z < \infty \end{cases}. \tag{7.34}$$

On removing the conditioning, the expected EW servicing cost per unit sold is given by

$$\begin{aligned} E[C_E(W_1, U_1; W, U)] = c_r \Bigg\{ & \int_0^{\gamma_2} [\Lambda_z(W + W_1) - \Lambda_z(W)] g_1(z)\,\mathrm{d}z \\ & + \int_{\gamma_2}^{\gamma_1} [\Lambda_z(W + U_1/z) - \Lambda_z(W)] g_1(z)\,\mathrm{d}z \\ & + \int_{\gamma_1}^{\infty} [\Lambda_z((U + U_1)/z) - \Lambda_z(U/z)] g_1(z)\,\mathrm{d}z \Bigg\}. \end{aligned} \tag{7.35}$$

Case (2) $\gamma_2 > \gamma_1$

The three possible situations for the age of the item at the beginning and end of the EW are now given in Table 7.2.

Table 7.2 EW starting times and finishing times as z varies for Case (2)

Usage rate	Age of item	
	Start of EW	Expiry of EW
$0 < z \le \gamma_1$	W	$W + W_1$
$\gamma_1 < z \le \gamma_2$	U/z	$U/z + W_1$
$\gamma_2 < z < \infty$	U/z	$(U + U_1)/z$

Using a similar approach to Case (1), the expected EW servicing cost per unit sold for Case (2) is given by

$$
\begin{aligned}
E[C_E(W_1, U_1; W, U)] = c_r \Bigg[& \int_0^{\gamma_1} \{\Lambda_z(W + W_1) - \Lambda_z(W)\} g_1(z)\,\mathrm{d}z \\
& + \int_{\gamma_1}^{\gamma_2} \{\Lambda_z(U/z + W_1) - \Lambda_z(U/z)\} g_1(z)\,\mathrm{d}z \\
& + \int_{\gamma_2}^{\infty} \{\Lambda_z((U + U_1)/z) - \Lambda_z(U/z)\} g_1(z) \Bigg] \mathrm{d}z
\end{aligned}
\tag{7.36}
$$

7.5.2 Non-identical EW and BW Terms

We confine our attention to some of the cost sharing and cost limit 1-D EWs that were defined in Sect. 5.3.3. In each case, we assume minimal repair is performed at each item failure and, as before, let $\tilde{C}_i$ denote the random cost of rectifying any failure, with mean $C_r = E[\tilde{C}_i]$. We focus only on the expected EW servicing cost per unit sold. The approach needed for the cost analysis of these EWs is similar to that used at the beginning of Sect. 7.6.1, so we only indicate the changes needed to derive the results.

7.5.2.1 Policy 6(a): Specific Parts Excluded (SPE)

Let SS_1 denote the subsystem consisting of the set of components which are covered under the EW, and SS_2 is the subsystem consisting of the set of components which are not covered. All failures from subsystem SS_1 are rectified by the EW provider at no cost to the customer. The cost of rectifying failures from subsystem SS_2 is borne by the customer. The probability that a failure comes from subsystem SS_1 (and so there is a warranty claim) is p and the probability the failure is from subsystem is $1 - p$.

The expected servicing cost for the EW provider is given by

$$E[C_E^p(W_1; W)] = pC_r[\Lambda(W + W_1) - \Lambda(W)], \tag{7.37}$$

whereas the customer's expected servicing cost for the EW period is

$$E\left[C_E^c(W_1; W)\right] = (1 - p)C_r[\Lambda(W + W_1) - \Lambda(W)]. \tag{7.38}$$

7.5.2.2 Policy 6(b): Lump Sum Cost Sharing (LCS)

Let ϕ denote the proportion of the cost borne by the customer for servicing the ith claim under the EW. The customer's servicing cost for the ith claim is $\phi\tilde{C}_i$ and the cost to the EW provider is $(1 - \phi)\tilde{C}_i$. Note that the cost proportion may or may not vary over the EW period (see Fig. 5.2 for the case where ϕ changes once under the EW).

The expected servicing costs for the EW provider and the customer for the EW period are given by (7.37) and (7.38) with p replaced by $1 - \phi$.

7.5.2.3 Policy 6(c): Material or Labour Cost Sharing (MLCS)

The customer pays for the material needed to repair a failure, and the EW provider pays for the labour. ϕ is the ratio of the material cost to the material + labour cost for the servicing of the ith claim under the EW. This may be constant or may vary with each claim. In the former case, the expressions for the expected servicing costs for the two parties for the EW period are the same as in Policy 6(b).

7.5.2.4 Policy 7(a): Limit on Individual Cost (LIC)

The EW provider rectifies a failure at no cost to the customer if the cost of the rectification action is below a specified limit c_I. If the rectification cost exceeds c_I, then the customer pays the excess cost. Let $\tilde{C}_i$, the random cost of rectifying any failure, have distribution function $F_C(c)$, survivor function $\bar{F}_C(c)$ and density function $f_C(c)$. The customer's servicing cost for the ith claim is $\max\{0, \tilde{C}_i - c_I\}$ and the cost to the EW provider is $\min\{\tilde{C}_i, c_I\}$.

The expected servicing cost for the EW provider is given by

$$E[C_E^p(W_1; W)] = C_r^p[\Lambda(W + W_1) - \Lambda(W)], \tag{7.39}$$

and the customer's expected servicing cost for the EW period is

$$E\left[C_E^c(W_1;W)\right] = C_r^c[\Lambda(W+W_1) - \Lambda(W)], \tag{7.40}$$

with $C_r^p = \int_0^{c_I} cf_C(c)\,\mathrm{d}c + c_I\bar{F}_C(c_I)$ and $C_r^c = \int_{c_I}^{\infty} (c - c_I)f_C(c)\,\mathrm{d}c$.

7.5.2.5 Policy 7(b): Individual Cost Deductible (ICD)

The customer pays the EW provider the amount (deductible) c_d to service each claim. The cost to the EW provider to service the ith claim under the EW is $\tilde{C}_i - c_d$. Note that the EW makes a profit on the ith claim if $\tilde{C}_i < c_d$.

The expected servicing cost for the EW provider is given by

$$E[C_E^p(W_1;W)] = (C_r - c_d)[\Lambda(W+W_1) - \Lambda(W)], \tag{7.41}$$

and the customer's expected servicing cost for the EW period is

$$E\left[C_E^c(W_1;W)\right] = c_d[\Lambda(W+W_1) - \Lambda(W)]. \tag{7.42}$$

7.6 Cost Analysis of MSCs

A MSC is similar to an EW, but there are differences as discussed in Sect. 7.1. The maintenance of a product (plant or infrastructure) is carried out by an external SA for a specified time period. A MSC can include penalties which are caused by (1) inadequate maintenance effort from the SA (resulting in the number of failures exceeding some limit, item availability falling below an agreed value, etc.) and (2) by the customer violating the terms of operation (usage intensity, mode, etc.) during the contract period. These penalties determine the eventual cost to the customer and the SA. The key elements involved in the cost analysis of a MSC are shown in Fig. 7.6.

Let A denote the age of the item at the start of the MSC and T the duration of the contract period. Failures are assumed to occur according to a point process with intensity function $\lambda(t)$ where t denotes calendar clock (based on the time since the owner first purchased the item). CM costs over the contract period are uncertain and these can also be affected by performing PM as part of the MSC. The effect of any scheduled PM actions can be modelled by suitably modifying the failure intensity function and so PM costs can also be assessed. In the following, we shall confine our attention to the analysis of CM costs only and consider cost per unit sale. Any cost sharing between the customer and the SA (if this is part of the contract) and any resulting penalties are determined by the contract terms.

The condition of the item at the beginning of the MSC period and the usage intensity by the customer during the period both affect the rate of occurrence of

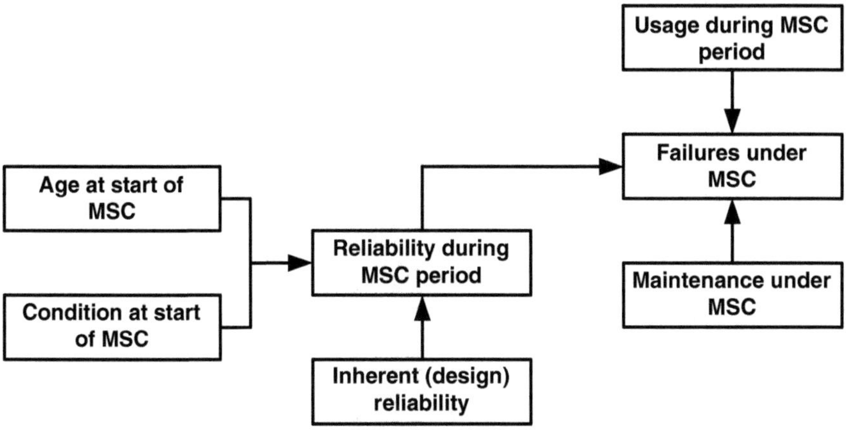

Fig. 7.6 Key elements for cost analysis of a MSC

failures and hence CM costs. Let ζ denote the item's condition at age A, s_0 the stress on the item under the manufacturer's recommended operating conditions and s the actual stress that the item experiences during the MSC period. One possible form for the failure intensity function during this period is

$$\lambda(t) = \left[\zeta + \left(\frac{s}{s_0}\right)^{\gamma}\right]\lambda_0(t), \quad A \le t < A + T, \tag{7.43}$$

where the parameter $\gamma > 0$ and $\lambda_0(t)$ is the item's inherent (baseline) failure intensity.

This type of multiplicative scaling of the intensity function allows several scenarios to be modelled. For example, the item may

1. be operated and maintained as per the manufacturer's recommendations prior to and during the MSC period ($\zeta = 0$ and $s = s_0$),
2. not have been maintained well, and/or the usage intensity has exceeded the manufacturer's recommended limit prior to the start of the MSC period, but normal usage occurs during the MSC period ($\zeta > 0$ and $s = s_0$),
3. not have been maintained well and/or the usage intensity exceeds the manufacturer's recommended limit prior to and during the MSC period ($\zeta > 0$ and $s > s_0$).

These three cases are indicated in Fig. 7.7.

The value of s is a deterministic quantity and its value is specified in the terms of the MSC. The amount of information available regarding the item's initial condition also needs to be considered. The true value of ζ may or may not be able to be assessed before the MSC is signed. When there is uncertainty in this factor, ζ becomes a random variable. There may also be asymmetry of information about ζ, with the true value being known to the customer but not to the SA.

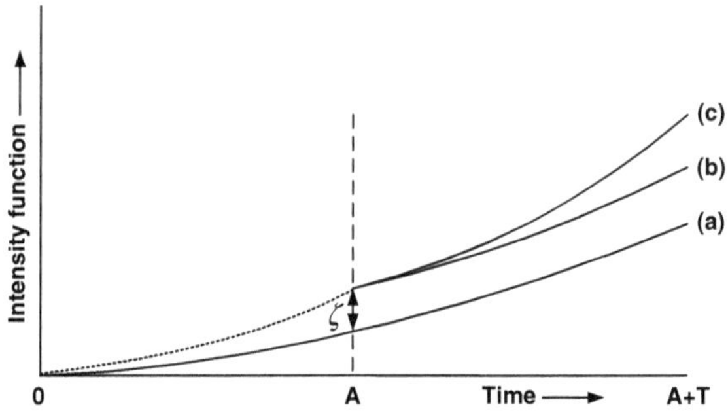

Fig. 7.7 Intensity function for failures over the MSC period

7.6.1 No Uncertainty in Initial Condition

We assume that the item is repaired at every failure at cost c_r and ζ is known with certainty before the MSC is signed. Using (7.43), the expected CM costs for the MSC period are given by

$$\begin{aligned} E[C_M(T;A,\zeta,s)] &= \left[\zeta + \left(\frac{s}{s_0}\right)^{\gamma}\right] c_r \int_A^{A+T} \lambda_0(t)\,\mathrm{d}t \\ &= \left[\zeta + \left(\frac{s}{s_0}\right)^{\gamma}\right] c_r[\Lambda_0(A+T) - \Lambda_0(A)], \end{aligned} \tag{7.44}$$

where $\Lambda_0(u) = \int_0^u \lambda_0(t)\,\mathrm{d}t$.

The MSC terms may include the provision for a major upgrade (overhaul) of the item to be performed before the contract period begins. The effect of such an upgrade will be to reduce the value of ζ and hence the expected CM costs.

7.6.2 Uncertainty in Initial Condition

We now assume that ζ is a random variable with density function $f_\zeta(x), 0 \le x \le l$, so the minimum and maximum possible values are 0 and l, respectively. Expected CM costs for the MSC period are obtained by conditioning on the value of ζ and then removing the conditioning. This gives

$$E[C_M(T;A,\zeta,s)] = \left\{\int_0^l \left[x + \left(\frac{s}{s_0}\right)^{\gamma}\right] f_\zeta(x)\,\mathrm{d}x\right\} c_r[\Lambda_0(A+T) - \Lambda_0(A)]. \tag{7.45}$$

7.6.3 Some Examples

Example 7.1 (Photocopier)
The item is a photocopier used in an office environment that has been designed for a nominal usage rate of $s_0 = 750$ pages copied per day.[9] Under this nominal usage rate ($s = 750$) and with minimal repairs performed at each failure, photocopier failures have been found to occur according to a power-law (Weibull) process with intensity function $\lambda_0(t) = (\beta/\alpha)(t/\alpha)^{\beta-1}$ where $\alpha = 157.5$ days and $\beta = 1.55$. The photocopier is maintained by a SA under a MSC.

Using (7.44), the expected CM costs for a MSC of length $T = 1$ year (365 days) when the photocopier is operated under nominal usage, is of age A years (365A days) and has initial condition $\zeta = 0$ are given by

$$c_r[\Lambda_0(365(A+1)) - \Lambda_0(365A)] = c_r\left[\left(\frac{365(A+1)}{157.5}\right)^{1.55} - \left(\frac{365A}{157.5}\right)^{1.55}\right]. \tag{7.46}$$

Table 7.3 gives the values of these expected costs from when the photocopier is new in steps of 1 year until it 4 years old.

Thus, the age of the photocopier is an important factor for the SA to consider when pricing the MSC, as is the usage rate. Table 7.4 shows the effect on expected CM costs of increasing the usage rate s when the value of the stress scaling parameter in (7.44) is $\gamma = 1.1$.

For example, we can see from the table that the expected CM costs for the next year for a machine which is now two years old and which produces 1,000 copies per day is [(12.9 – 3.7)/3.7] 100 % (almost 250 %) more than that for a new photocopier operating at the nominal usage rate. □

Example 7.2 (Street lights in an urban region)
A city council has decided to outsource the maintenance of street lights within areas controlled by the city and it has called for tenders from several SAs. The maintenance service contract requires replacing all existing lights in an area and then replacing each light that fails during the contract period by a new light immediately a failure occurs. Each SA needs to submit a detailed cost estimate and a contract price so that the city council can make the final decision regarding the awarding of the contract.

There are several light manufacturers and the reliability characteristics and costs/per bulb are different. We focus on one particular SA who can buy lights from one or more of these suppliers. Since the cost of a CM replacement is much higher than that for a PM replacement, the SA wants to use a block replacement

[9] This example is adapted from Bulmer and Eccleston (2003).

Table 7.3 Expected CM costs for a 1-year MSC as a function of photocopier age A years

Age A (years)	01	1	2	3	4
Expected CM costs for MSC	3.7 c_r	7.1 c_r	9.4 c_r	11.3 c_r	13.0 c_r

Table 7.4 Expected CM costs for the photocopier as its age A years and usage rate s increase

Usage rate (s) (copies per day)	Age A (years)				
	0	1	2	3	4
750	3.7 c_r	7.1 c_r	9.4 c_r	11.3 c_r	13.0 c_r
800	4.0 c_r	7.6 c_r	10.1 c_r	12.1 c_r	14.0 c_r
1,000	5.1 c_r	9.7 c_r	12.9 c_r	15.5 c_r	17.8 c_r
1,200	6.2 c_r	11.9 c_r	15.8 c_r	19.0 c_r	21.8 c_r
1,500	7.9 c_r	15.2 c_r	20.1 c_r	24.2 c_r	27.9 c_r

policy for the maintenance of the lights. The block policy has parameters which need to be selected optimally. Here, we look at the case where the SA purchases from a single supplier and derive the expected cost and the optimal block policy for maintenance. We confine our attention to the analysis for one supplier. Similar analyses for bulbs bought from other suppliers will then allow the SA to decide on the best supplier to select.

The SA has obtained information from the supplier that the time to failure of the lights has distribution function $F(t)$. The contract is for L years, and there are n lights in the area under the control of the city council. A block policy with interval v is used to maintain the lights during the contract period. Let $K = \mathrm{int}(L/v)$, denote the largest integer less than L/v, then all the lights are replaced by new lights (under PM action) at set times $t = jv, j = 0, 1, 2, \ldots K$, and any light that fails in between these times is replaced by a new light (under CM action). The cost of each CM action (involving the replacement of a single failed light) is c_f and a PM action costs c_p (where all n lights are replaced at the same time) and $c_p < nc_f$ so that performing PM is worthwhile.

Using (3.36), the expected maintenance cost for the SA is given by

$$J(L; v, K) = (K + 1)c_p + nc_f[KM(v) + M(L - Kv)], \tag{7.47}$$

where $M(t)$ is the renewal function associated with $F(t)$ and which is given by (3.24).

The SA needs to determine the optimal values of v and K by minimising the objective function given in (7.47). This is done using a two-stage process as indicated below.

Stage 1: Fix the value of K and let $v^*(K)$ denote the value of v which minimises $J(L; v, K)$. $v^*(K)$ can be obtained from the first-order necessary condition $\partial J(L; v, K)/\partial v = 0$. As a result, it is obtained by solving the following equation:

$$nc_f K[m(v) - m(L - Kv)] = 0, \tag{7.48}$$

where $m(t) = \mathrm{d}M(t)/\mathrm{d}t$ is the renewal density function. The solution of (7.48) is

$$v^*(K) = \frac{L}{K+1}. \tag{7.49}$$

Thus, the optimal v is an integer divisor of the contract period L.

Stage 2: Using (7.49) in (7.47) gives

$$\tilde{J}(L;K) = (K+1)\left[c_p + nc_f M\left(\frac{L}{K+1}\right)\right]. \tag{7.50}$$

Let K^* denote the value of K which minimises $\tilde{J}(L;K)$. K^* can be found through an exhaustive search by evaluating $\tilde{J}(L;K)$ for $K = 0, 1, 2, \ldots$, and then identifying the K^* which yields the smallest value for $\tilde{J}(L;K)$.

Finally, when K^* is found, it follows that $v^* = v^*(K^*) = L/(K^*+1)$. If $K^* = 0$, this implies that the optimal policy for the SA is to perform only one PM action at the beginning of the contract period.

Comment: If L is very long ($\gg$the mean time to light failure), then the expected maintenance cost over the contract period can be approximated by $c_p + [L \times J_\infty(v)]$ where $J_\infty(v)$ is the asymptotic cost per unit time under the block policy which is given by

$$J_\infty(v) = \frac{c_p + nc_f M(v)}{v}. \tag{7.51}$$

In this case, if we find that $v^* \geq L$, then this implies that the SA should only perform the initial PM action when the contract period begins and the expected maintenance costs are then given by $c_p + nc_f\, M(L)$.

There are 5,000 lights in one of the city areas, and the time to failure of each of the lights has a two-parameter Weibull distribution with $F(t) = 1 - e^{-(t/\alpha)^\beta}$ where $\alpha = 4$ years and $\beta = 2$. This implies that the mean time to failure per light is 3.55 years. The SA has estimated that c_f, the cost to replace each light under CM action, will be \$30 and c_p, the cost of a PM action where all 5,000 lights are replaced, will be \$10,000. Table 7.5 shows the optimal number of PM actions that the SA should use during the contract period, the optimal interval between these PM actions and the expected maintenance costs for contract periods of length $L = 5$, 10, 15 and 20 years, respectively.

Comment: The value of v that minimises the asymptotic cost per unit time for the block policy given in (7.51) is $v^* = 1.07$. □

Table 7.5 Optimal block policy variables for varying contract lengths

L	v^*	K^*	$J(L; v^*, K^*)$
5	1.00	4	\$95,918.75
10	1.11	8	\$191,356.65
15	1.07	13	\$286,766.90
20	1.05	18	\$381,973.15

Example 7.3 (Hydraulic pumps)
In open cut mines, coal and overburden are transported using excavators and dump trucks. The excavator is a complex machine with a hydraulic system as one of the important elements. The hydraulic system consists of several hydraulic pumps and a pump is considered to have failed if it cannot provide the required flow rate at the required pressure. The current maintenance policy for the pumps is based on the age policy where a pump is subjected to a PM action when it reaches an age v or to a CM action should it fail earlier. PM and CM actions both involve evaluating the condition of a pump to decide whether it should be scrapped or reconditioned. A reconditioned pump can be considered to be nearly as good as new.

A mining company owns 4 excavators and 17 dump trucks. The excavators are operated continuously except when down for either CM or PM. The maintenance is outsourced by the company so that all maintenance actions are carried out by an external SA who uses a PM age of $v = 12{,}000$ h for the pumps in the hydraulic systems of the excavators.

The cost to the mining company is \$30,000 for the reconditioning of an item (under PM or CM) and \$50,000 to purchase a new item. The mining company wants to revaluate its maintenance policy based on the maintenance data (failure data + censored data due to PM actions) collected in the past.

The data collected indicate that a two-parameter Weibull distribution with $F(t) = 1 - \exp\{-(t/\alpha)^{\beta}\}$ where $\alpha = 15{,}000$ h and $\beta = 2.2$ is adequate to model the time to pump failure. Since a mine has a long operating life (around 30–40 years), the mining company is interested in determining if $v = 12{,}000$ is optimal for the PM of a pump based on expected maintenance costs per unit time over an infinite time horizon of operation. If this age for carrying out PM actions is not optimal, then the company wants to determine the true optimal value v^* and then renegotiate the MSC based on this new value.

The objective function for the optimisation is the asymptotic expected maintenance cost per unit time under the age PM policy, and this is given by

$$J_{\infty}(v) = \frac{c_f F(v) + c_p \bar{F}(v)}{\int_0^v \bar{F}(t)\,dt}. \tag{7.52}$$

The cost of each CM action (c_f) and each PM action (c_p) is the sum of the material and labour costs and the loss in production due to downtime costs. The downtime costs for a CM action are significantly higher than those for a PM action.

Table 7.6 Optimal age PM policies as η varies

η	v^*	$J_\infty(v^*)$
1.8	16,500	\$13.14
1.9	15,500	\$13.73
2.0	14,650	\$14.28

Based on the scrap/replacement decision and the time to carry out each type of action, the mining company has estimated that $c_p = \$100,000$ and $c_f = \eta c_p$ with η lying in the interval 1.8–2.0.

Table 7.6 shows the optimal ages for performing a PM action and the minimum expected maintenance costs per hour for the three values of η. The results show that the mining company can safely renegotiate the MSC with the SA based on an increased value for the PM age of about 15,000 h. □

7.7 Maintenance Outsourcing Decision Models

Maintenance outsourcing decision models can be grouped into two categories—game-theoretic and non-game-theoretic. This section deals with three models belonging to the latter category and those belonging to the former group are the focus of the next chapter. In Models 7.1 and 7.2, the duration of the MSC is very long so that it can be treated as being infinite. In Model 7.3, the duration is finite.

Model 7.1 (Tarakci et al. 2006a)
Here, the object is a manufacturing plant used in continuous production. All failures are minimally repaired (CM actions), and the plant is also subjected to periodic major overhauls (PM actions) which restore it to as-good-as-new condition. A PM action is initiated when the plant has operated for a time period v subsequent to the last PM action. The plant is not producing any output when it is undergoing a PM or CM action. Both PM and CM actions are outsourced to an external SA.

Notation:

$\lambda(t)$	Failure intensity function of plant ($t = 0$ corresponds to the plant becoming operational after a PM action)
$\Lambda(t)$	Cumulative failure intensity ($\Lambda(t) = \int_0^t \lambda(t')\,dt'$)
t_p	Average time to carry a PM action
t_r	Average time to carry out a CM action
c_p	Average cost of a PM action
c_r	Average cost of a CM action
R	Revenue generated per unit time when the plant is operational
P	Per unit time payment to the service agent to carry out the maintenance

v	Decision variable for determining the timing of a PM action
$U(v)$	Asymptotic availability of the plant
$\pi(v)$	Asymptotic maintenance cost per unit time incurred by the service agent
$J_{\mathrm{M}}(v)$	Asymptotic expected profit per unit time for the manufacturer
$J_{\mathrm{A}}(v)$	Asymptotic expected profit per unit time for the service agent.

We now present the results of the model analysis. Note that every PM action is a renewal point. The interval between two successive PM completion times defines the renewal cycle. One can then derive the expressions given below using the Renewal Reward Theorem (see Appendix B).

The expected number of failures (CM actions) over an operational period of length v is given by $\Lambda(v)$. As a result, the expected cycle length is given by

$$\mathrm{ECL} = v + \Lambda(v)t_r + t_p. \tag{7.53}$$

The asymptotic availability is given by

$$U(v) = \frac{v}{\mathrm{ECL}} = \frac{v}{v + \Lambda(v)t_r + t_p}. \tag{7.54}$$

The asymptotic expected maintenance cost per unit time is given by

$$\pi(v) = \frac{c_r\Lambda(v) + c_p}{\mathrm{ECL}} = \frac{c_r\Lambda(v) + c_p}{v + \Lambda(v)t_r + t_p}. \tag{7.55}$$

The asymptotic expected profits per unit time for the manufacturer and the SA are given by

$$J_{\mathrm{M}} = R\,U(v) - P \quad \text{and} \quad J_{\mathrm{A}}(v) = P - \pi(v), \tag{7.56}$$

respectively.

The manufacturer and the SA have to decide on the optimal v. The manufacturer wants to maximise $J_{\mathrm{M}}(v)$ whereas the SA wants to maximise $J_{\mathrm{A}}(v)$. This is a vector optimisation problem where the objective function is a 2-D vector and the decision variable is a scalar. In such cases, one can only achieve Pareto optimality.

The model formulation given above is slightly different to that given by Tarakci et al. (2006a). In their formulation, v is the time interval between the start of two PM actions. As such, it includes uptime (when the plant is operational) and the downtimes to carry out CM actions over the renewal cycle. Since the uptime is $\leq v$, the expected number of failures (and CM actions) over a cycle is $<\Lambda(v)$. However, they assume that the expected number of failures over a cycle is $\Lambda(v)$. The cycle length is $\mathrm{ECL} = v + t_p$, and the asymptotic availability is given by $(v - \Lambda(v)\,t_r)/(v + t_p)$.

Since both parties have to agree on v, Tarakci et al. (2006a) propose the following three coordination mechanisms to achieve this, resulting in three different types of MSC.

1. Cost subsidisation (CS)
2. Uptime bonus (UB)
3. Combination of uptime target and bonus (UTB).

The CS Contract

Here, the manufacturer pays the SA an amount Δc_p each time the agent performs a PM action. Tarakci et al. (2006a) derive conditions which result in both parties agreeing on the optimal v. They also look at the case where a similar subsidisation for each repair results in a similar outcome.

The UB Contract

Here, the SA receives a bonus proportional to the availability level achieved by performing the PM actions. In this case, the optimal strategy for the SA is to reduce v to a value which leaves zero profit for the manufacturer.

The UTB Contract

Here, the bonus is given only when the availability level exceeds some target level τ. The bonus is B so that the SA's asymptotic expected profit per unit time is given by

$$\tilde{J}_{\mathrm{A}}(v) = P - \pi(v) + B[U(v) - \tau]^{+}, \tag{7.57}$$

where $[x]^{+} = \max\{x, 0\}$. Tarakci et al. (2006a) derive conditions which result in both parties agreeing to the optimal v.

They also look at the case where the manufacturer has the option to choose a SA from n different SAs with each agent having a different expected cost and expected time to carry out a PM action.

Model 7.2 (Tarakci et al. 2006b)

This is an extension of the model in Tarakci et al. (2006a). Here, the manufacturing plant consists of three subsystems. Subsystems 1 and 2 produce components which are fed into subsystem 3 where the components are assembled and finally sold to customers.

The manufacturer outsources the maintenance of subsystems 1 and 2 to two independent SAs under a maintenance contact similar to that in their earlier model. The modelling of the objective functions for the two service contractors is identical to that of their earlier model so that the agents' asymptotic expected costs per unit time and expected profits per unit time are similar to that given by (7.55) and the second part of (7.56) (with subscripts $i = 1, 2$ used to differentiate the two SAs). The problem is to find the optimal PM intervals for the two agents to maximise

$$J_M(v_1, v_2) = R \min\{U_1(v_1), U(v_2)\} - \pi_1(v_1) - \pi_2(v_2). \tag{7.58}$$

Tarakci et al. (2006b) look at alternative contracts to arrive at the optimal PM intervals that the manufacturer and the two SAs will agree to.

Model 7.3 (Tarakci et al. 2009)

Here, the service contract period L is finite. Over this period, the SA carries out N (a decision variable) PM actions, and the effect of each PM action is to renew the plant to as-good-as-new condition. Any failures between PM actions are minimally repaired and the failure intensity function for the plant is the same as in Model 7.1.

The new features in this model are the learning effects in PM actions. Tarakci et al. (2009) define two kinds of learning, each of which impact on the time and cost to do a PM action.

Natural learning

Here, the expected time and expected cost of a PM action decreases with the number of previous PM actions carried out. The expected time and the expected cost of the ith PM action are given by $t_p i^{-a}$ and $c_p i^{-a}$, respectively, with a the learning parameter ($a \geq 0$) representing the effect of learning.[10] The higher the value of a, the faster is the learning process.

Learning through effort

Here, the learning takes place through a costly training programme which the SA carries out at the start of the service contract, and this determines the value of the learning parameter a. The cost of the programme $K(a)$ is an increasing function with $K(0) = 0$, $K'(a) > 0$ and $K''(a) > 0$.

Tarakci et al. (2009) derive several interesting results.

References

Bass FW (1969) A new product growth model for consumer durables. Manage Sci 15:215–227

Blischke WR, Murthy DNP (1994) Warranty cost analysis. Marcel Dekker, New York

Blischke WR, Murthy DNP (eds) (1996) Product warranty handbook. Marcel Dekker, New York

Bulmer M, Eccleston J (2003) Photocopier reliability modelling using evolutionary algorithms. In: Blischke WR, Murthy DNP (eds) Case studies in reliability and maintenance. Wiley, New York

Gertsbakh IB, Kordonsky HB (1998) Parallel time scales and two-dimensional manufacturer and individual customer warranties. IIE Trans 30:1181–1189

Iskandar BP, Blischke WR (2003) Reliability and warranty analysis of a motorcycle based on claims data. In: Blischke WR, Murthy DNP (eds) Case studies in reliability and maintenance. Wiley, New York

[10] If the log of the expected times and costs are plotted against I, it is a straight line with negative slope.

Iskandar BP, Murthy DNP (2003) Repair-replace strategies for two-dimensional warranty policies. Math Comput Modell 38:1233–1241

Iskandar BP, Murthy DNP, Jack N (2005) A new repair-replace strategy for items sold with a two-dimensional warranty. Comput Oper Res 32:669–682

Jack N, Murthy DNP (2001) A servicing strategy for items sold under warranty. J Oper Res 2:1284–1288

Jack N, van der Schouten Duyn F (2000) Optimal repair-replace strategies for a warranted product. Int J Prod Econ 67:95–100

Jack N, Iskandar BP, Murthy DNP (2009) A repair-replace strategy based on usage rate for items sold with a two-dimensional warranty. Reliab Eng Syst Saf 94:611–617

Jiang X, Jardine AKS, Lugitigheid D (2006) On a conjecture of optimal repair-replacement strategies for warranted products. Math Comput Model 44:963–972

Kim CS, Djamaludin I, Murthy DNP (2001) Warranty cost analysis with heterogeneous usage intensity. Int Trans Oper Res 8:337–347

Lawless JF, Hu J, Cao J (1995) Methods for the estimation of failure distributions and rates from automobile warranty data. Lifetime Data Anal 1:227–240

Mahajan V, Wind Y (1986) Innovation diffusion models of new product acceptance. Ballinger Publishing Company, Cambridge

Murthy DNP, Djamaludin I (2001) Warranty and quality. In: Rahim MA, Ben-Daya M (eds) Integrated optimal modelling in PIQM: production planning, inventory, quality and maintenance. Kluwer Academic Publishers, New York

Nguyen DG, Murthy DNP (1989) Optimal replace-repair strategy for servicing items sold with warranty. Eur J Oper Res 39:206–212

Tarakci H, Tang K, Moskowitz H, Plante R (2006a) Incentive maintenance contracts for channel coordination. IIE Trans 38:671–684

Tarakci H, Tang K, Moskowitz H, Plante R (2006b) Maintenance outsourcing of a multi-process manufacturing system with multiple contractors. IIE Trans 38:67–78

Tarakci H, Tang K, Teyarachakul S (2009) Learning effects on maintenance outsourcing. Eur J Oper Res 192:138–150

Yang G, Zaghati Z (2002) Two-dimensional reliability modeling from warranty data. Proceedings of annual Reliability and Maintainability Symposium, pp 272–278

Chapter 8
Game-Theoretic Models for EW/MSC Decision-Making

8.1 Introduction

As mentioned in Chap. 6, game theory (GT) is the appropriate framework to use to study the decision problems for the different parties involved and obtain their optimal decisions. The GT approach to decision-making is described in Chap. 4. In this chapter, we focus on EW/MSC decision-making, in the context of products and plants (but excluding infrastructure), from the perspectives of the different parties, and we discuss the GT models proposed in the literature. The outline of this chapter is as follows. Section 8.2 deals with the framework for building GT models for EW/MSC decision-making and discusses the key issues, different scenarios and model formulations. The GT models can be broadly categorised into two different groups—static and dynamic. In Sect. 8.3, details of the model formulations and analyses of a number of static EW decision models are given. Dynamic EW decision modelling is covered in Sect. 8.4. Finally, Sect. 8.5 deals with GT models for MSC decision-making.

8.2 Framework for GT Modelling

EW and MSC processes and markets involve several interacting elements as shown in Figs. 6.1 and 6.2, respectively. The characterisation and modelling of each element can be done in several ways leading to many different scenarios and to the several different GT models discussed in later sections of the chapter.

8.2.1 Key Elements and Their Characterisations

Parties: The distinct parties (or groups) are the EW/MSC providers (manufacturers, retailers and other independent providers), the service agents and the customers. The service agents may be manufacturer owned, retailer owned or independently owned,

D. N. P. Murthy and N. Jack, *Extended Warranties, Maintenance Service and Lease Contracts*, Springer Series in Reliability Engineering,
DOI: 10.1007/978-1-4471-6440-1_8, © Springer-Verlag London 2014

and the customers may be either homogeneous or heterogeneous. In the latter case, the customer heterogeneity may be due to differences in usage, risk attitude, income and information. Each party in an EW/MSC market may consist of one or more *players,* and this leads to different market structures as illustrated through examples in Sects. 6.5 and 6.6.

Product/Plant: As discussed in Chap. 2, every product/plant is unreliable. The reliability changes with age which is a dynamic characterisation. Many GT models (especially in the economic and marketing literature) model reliability in a static sense—either the item fails or does not fail over the interval of interest (EW or MSC period). Often terms such as durability and quality are sometimes used instead of reliability.

Demand for EWs/MSCs: This can be considered to be either exogenous (treated as an external variable) or endogenous (so that it is a function of other variables in the formulation—such as price and duration of the EW/MSC).

BWs/EWs/MSCs: These are characterised by variables such as price, duration and terms (such as exclusions and deductibles).

Maintenance (PM and CM): The maintenance requirements on the part of the customers and service agents are defined by the terms and conditions of the EW/MSC.

Power Structure: As discussed in Sect. 6.5, the two possible types of power structure that can occur between any two players A and B in an EW/MSC market are *dominance* which we indicate by A $\rightarrow$ B and *no dominance* (equal power) indicated by A $\leftrightarrow$ B. In the former case, the dominant player A's decisions are known to and influence the decisions made by the dominated player B. A is known as the *leader* and B the *follower* in this type of power structure. In the latter case, the two players are assumed to make their decisions simultaneously or at least are unaware of each others' decisions.

Decision Problems: The decision problem for each player is different. It is characterised by an objective function which may be expected cost, expected revenue, expected profits, expected utility, sales, etc. The decision variables can be the choice between two or more alternatives (for customers), price and duration of contract (for EW/MSC providers), actions such as repair versus replace and type of repair (for service agents).

Information: There are three types of information—(1) product related (reliability, quality, durability, etc.), (2) customer related (homogeneous or heterogeneous, attitude to risk, income, etc.) and (3) service related (terms of the EW/MSC, service delivery guarantees, etc.). Other issues include symmetry versus asymmetry in information between players, perfect (complete) or imperfect (incomplete and uncertain).

8.2.2 *Different Scenarios*

The possible different scenarios to consider arise from the various combinations of the elements discussed earlier and from their characterisation. A multilevel classification can be used to characterise these scenarios. At Level 1, the different scenarios are based on the characterisation of customers and EW/MSC providers as indicated below.

		Number of EW/MSC providers	
		1	≥2
Customers	Homogeneous	A	C
	Heterogeneous	B	D

For each of the four scenarios at Level 1, we have a Level 2 classification based on the characterisation of customers' attitude to risk and type of information available.

		Information			
		Symmetric		Asymmetric	
		Perfect	Imperfect	Perfect	Imperfect
Customers	Risk neutral	a	c	e	g
	Risk averse	b	d	f	h

Thus, combining all the possibilities from these two levels produces a total of thirty-two possible scenarios. It is possible to add further levels by considering other customer characteristics (such as usage and income), and this will result in the number of scenarios increasing still further.

8.2.3 *Model Formulations*

Model formulations can be either static or dynamic. Most of the GT models reported in the economics and marketing literature are static and single period. The items under consideration either function properly or do not function properly under the EW/MSC. A few models deal with multiperiod problems with the formulation in each period again being static. Realistic stochastic failure models require the possibility of multiple failures over time and the use of dynamic formulations. The operational research and reliability literature contain GT models which are dynamic in nature.

8.3 Static GT Models for EW Decision-Making

In these "economic-type" GT models, there is a fixed EW period (the BW period is normally ignored), and the product either works properly or does not work properly during the EW period. The EW acts as insurance for the customer who obtains a refund from the EW provider if the product fails.

For each model, we state the assumptions, characterise the key elements and the decision variables for the different parties, derive expressions for their objective functions and then obtain the optimal decisions.

8.3.1 Single EW Provider (Scenarios A and B)

Customers purchase a product directly from the manufacturer, and we begin by assuming that the manufacturer is the sole EW provider (EWP). We focus on the decisions made by the two different parties in the market (manufacturer and customers) per unit EW sale. In Models 8.1–8.3, the customers are assumed to be homogeneous in all attributes (risk attitude, usage intensity, etc.) so this is market scenario A. Model 8.4 deals with market scenario B, where there is heterogeneity in customer usage. Scenarios A and B are also considered in Models 8.5–8.9.

Model 8.1 (Stackelberg game)
Assumptions: A monopolist manufacturer sells a product directly to customers and also offers each customer the option to buy an EW. The manufacturer (EWP) and the customers are all risk neutral, and each party has complete information about product reliability and costs. The customers are assumed to be homogeneous in attributes such as risk attitude and usage intensity. In the power structure between the manufacturer and each customer, the manufacturer is assumed to be the leader and the customer is the follower, so the dominance is indicated by Fig. 6.6.

Key elements and decision variables: The manufacturing cost per unit of product is c_m, and the selling price to the customer is p_p^r. The customer earns a monetary benefit of m using the product during the EW period but incurs a monetary loss of k if the product fails. The probability that the product does not fail is π. Under the terms of the EW offered to the customer, the level of protection is $s(0 \le s \le 1)$, which means that the EWP refunds the amount sk to the customer should the product fail during the EW period, and the price of the EW is $p_e^r(s)$.

The set of decision variables for the EWP is given by $y \equiv \left\{p_p^r, s, p_e^r(s)\right\}$, whereas the customer has the single decision variable

$$x = \begin{cases} 0 & \text{if the product is not purchased,} \\ 1 & \text{if the product is purchased but not the EW,} \\ 2 & \text{if both the product and the EW are purchased.} \end{cases}$$

Objective functions: For a given $y \equiv \left\{p_p^r, s, p_e^r(s)\right\}$ chosen by the EWP, the customer's expected profit $J_C(x; y)$ is given by

$$J_C(x;y) = \begin{cases} 0, & \text{if } x = 0, \\ m - p_p^r - (1-\pi)k, & \text{if } x = 1, \\ m - p_p^r - p_e^r(s) - (1-\pi)(1-s)k, & \text{if } x = 2. \end{cases} \quad (8.1)$$

For a given purchase decision x made by the customer, the EWP's expected profit $J_M\left(p_p^r, s, p_e^r(s); x\right)$ is given by

$$J_M\left(p_p^r, s, p_e^r(s); x\right) = \begin{cases} 0, & \text{if } x = 0, \\ p_p^r - c_m, & \text{if } x = 1, \\ p_p^r - c_m + p_e^r(s) - (1-\pi)sk, & \text{if } x = 2. \end{cases} \quad (8.2)$$

Customer's optimal strategy: The customer's decision between buying the product (without the EW) or not buying the product depends on whether $J_C(1;y) > J_C(0;y)$ or $J_C(1;y) < J_C(0;y)$, and the customer is indifferent between the two options if $J_C(1;y) = J_C(0;y)$. Thus, the customer will buy the product only if

$$p_p^r \leq m - (1-\pi)k. \quad (8.3)$$

The terms on the *rhs* of (8.3) represent the expected profit the customer will earn using the product during the EW period. The customer's decision between buying the EW or not depends on whether $J_C(2;y) > J_C(1;y)$ or $J_C(2;y) < J_C(1;y)$, and the customer is indifferent between the two options if $J_C(2;y) = J_C(1;y)$. Thus, the customer will be indifferent when

$$p_e^r(s) = (1-\pi)ks, \quad (8.4)$$

which is the expected refund the customer will receive from the EWP, an increasing linear function of s.

Figure 8.1 shows this indifference line in the $p_e^r(s)$ versus s diagram. Also shown are the customer's optimal decisions.

EWP's optimal strategy: The EWP will sell the product to the customer only if a positive profit is made on the sale which implies that

$$p_p^r > c_m. \quad (8.5)$$

The EWP will offer the EW to the customer only if the expected profit generated by the EW sale is positive. The $(s, p_e^r(s))$ combination chosen by the EWP must be such that the customer is willing to buy the EW, and at the same time, this choice must maximise the EWP's expected profit. This implies that the optimal

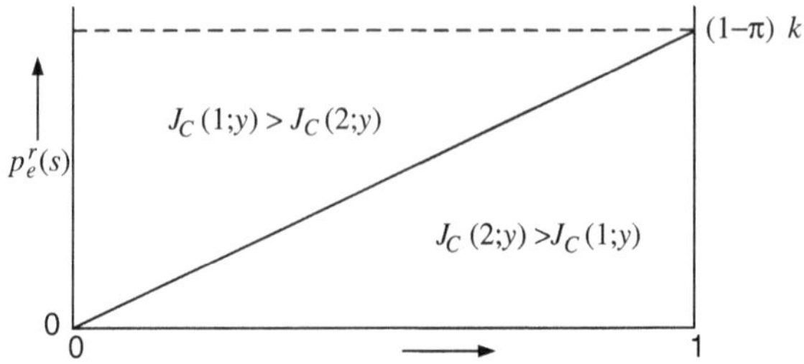

Fig. 8.1 Customer's optimal decisions in the $p_e^r(s)$ versus s diagram

$(s, p_e^r(s))$ combination must satisfy $J_c(2; y) = J_c(1; y)$ and must also maximise $J_M\left(p_p^r, s, p_e^r(s); x^*\right)$.

We do the optimisation in three stages. In the first stage, we derive the optimal $p_e^r(s)^*$; in the second stage, we obtain the optimal s^* which maximises $J_M\left(p_p^r, s, p_e^r(s)^*; x^*\right)$; and in the third stage, we obtain the optimal p_p^{r*} which maximises $J_M\left(p_p^r, s^*, p_e^r(s)^*; x^*\right)$.

Stage 1: For a given s, the EWP's optimal pricing strategy for the EW is given by

$$p_e^r(s)^* = (1 - \pi)ks. \tag{8.6}$$

Note that this corresponds to the straight line in Fig. 8.1, and this choice for $p_e^r(s)^*$ gives

$$J_M(p_p^r, s, p_e^r(s)^*; 2) = J_M(p_p^r, s, p_e^r(s)^*; 1) = p_p^r - c_m. \tag{8.7}$$

Stage 2: Since the EWP's expected profit does not depend on s explicitly, this implies that the EWP can choose any s^* in the interval $(0 \leq s \leq 1)$ and the optimal EW price depends on the value of s^* that is selected.

Stage 3: From (8.7), it can be seen that the EWP's expected profit is an increasing function of p_p^r. Constraint (8.3) implies that

$$p_p^{r*} = m - (1 - \pi)k. \tag{8.8}$$

In summary, the EWP's optimal strategy is to set the price of the product at $p_p^{r*} = m - (1 - \pi)k$, select the level of protection for the EW to any value $s^* \in [0, 1]$ and the price as $p_e^r(s)^* = (1 - \pi)ks^*$, the fair actuarial premium. The EWP's optimal expected profit is then given by

$$J_M(p_p^{r*}, s^*, p_e^r(s)^*; x^*) = m - (1 - \pi)k - c_m. \tag{8.9}$$

The customer's optimal strategy is to purchase the product, but then the customer is indifferent between purchasing and not purchasing the EW, so $x^* = 1$ or 2. The customer's expected net profit is given by

$$J_C(1; y^*) = J_C(2; y^*) = 0. \tag{8.10}$$

Thus, the EWP is able to extract all the consumer surplus from the customer, leaving the customer with zero profit.

Model 8.2 (Stackelberg game)
We now consider the effect of risk attitude on customer decision-making.

Assumptions: The manufacturer remains risk neutral, but the customers are now risk averse with utility function

$$U(V) = \frac{1}{\gamma}\left(1 - e^{-\gamma V}\right), \quad \gamma > 0 \tag{8.11}$$

The other assumptions are the same as in Model 8.1.

Key elements and decision variables: These are the same as in Model 8.1.

Objective functions: For a given $y \equiv \left\{p_p^r, s, p_e^r(s)\right\}$ chosen by the EWP, the customer's expected utility $J_C(x; y)$ is given by

$$J_C(x; y) = \begin{cases} 0, & \text{if } x = 0, \\ \pi U(m - p_p^r) + (1 - \pi)U(m - p_p^r - k), & \text{if } x = 1, \\ \pi U(m - p_p^r - p_e^r(s)) + (1 - \pi)U(m - p_p^r - p_e^r(s) - (1 - s)k), & \text{if } x = 2. \end{cases} \tag{8.12}$$

Using (8.11) and some simple manipulation, this reduces to

$$J_C(x; y) = \begin{cases} 0, & \text{if } x = 0, \\ \frac{1}{\gamma}\left\{1 - e^{-\gamma\left(m - p_p^r\right)}\left[\pi + (1 - \pi)e^{\gamma k}\right]\right\}, & \text{if } x = 1, \\ \frac{1}{\gamma}\left\{1 - e^{-\gamma\left(m - p_p^r - p_e^r(s)\right)}\left[\pi + (1 - \pi)e^{\gamma(1-s)k}\right]\right\}, & \text{if } x = 2. \end{cases} \tag{8.13}$$

For a given purchase decision x made by the customer, the EWP's expected profit $J_M\left(p_p^r, s, p_e^r(s); x\right)$ is again given by

$$J_M(p_p^r, s, p_e^r(s); x) = \begin{cases} 0, & \text{if } x = 0, \\ p_p^r - c_m, & \text{if } x = 1, \\ p_p^r - c_m + p_e^r(s) - (1 - \pi)sk, & \text{if } x = 2. \end{cases} \tag{8.14}$$

Customer's optimal strategy: As in Model 8.1, the customer will decide to buy the product if $J_C(1; y) > J_c(0; y)$ and will be indifferent between buying and not

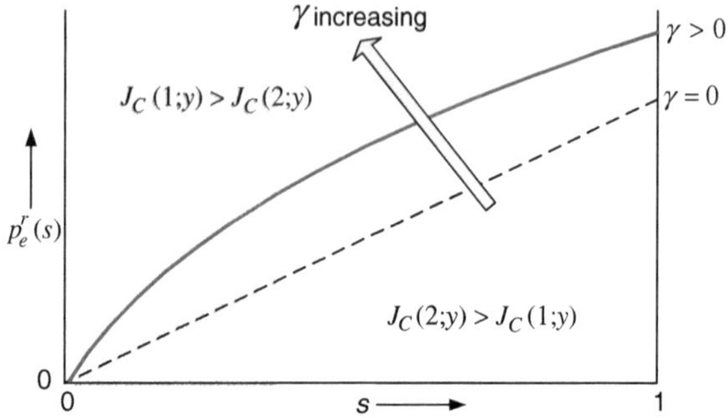

Fig. 8.2 Customer's optimal decisions in the $p_e^r(s)$ versus s diagram

buying if $J_C(1;y) = J_c(0;y)$. Using (8.13), it is easily shown (through simple analysis) that the customer will purchase the product only if

$$p_p^r \leq m - \frac{1}{\gamma}\log\left[\pi + (1-\pi)e^{\gamma k}\right]. \tag{8.15}$$

Note: This reduces to $p_p^r \leq m - (1-\pi)k$ when $\gamma \to 0$ as is to be expected.

For a given $y \equiv \left\{p_p^r, s, p_e^r(s)\right\}$, the customer chooses between $x = 1$ and $x = 2$ by comparing $J_c(1;y)$ and $J_c(2;y)$. By equating these two functions and after some simple manipulation, the indifference curve in the $p_e^r(s)$ versus s diagram is given by

$$p_e^r(s) = \frac{1}{\gamma}\log\left[\frac{\pi + (1-\pi)e^{\gamma k}}{\pi + (1-\pi)e^{\gamma(1-s)k}}\right]. \tag{8.16}$$

Note: This reduces to $p_e^r(s) = (1-\pi)ks$ when $\gamma \to 0$ as is to be expected

It is easily shown that $p_e^r(0) = 0,\ p_e^r(1) = \log\left[\pi + (1-\pi)e^{\gamma k}\right]/\gamma$ and $\mathrm{d}p_e^r(s)/\mathrm{d}s \geq 0$. The customer's optimal decisions are shown in Fig. 8.2.

EWP's optimal strategy: We use the three-stage approach to obtain the optimal $p_e^r(s)^*, s^*$ and p_p^{r*}. For a given s, the optimal $p_e^r(s)^*$ must be on the curve indicated in Fig. 8.2, so

$$p_e^r(s)^* = \frac{1}{\gamma}\log\left[\frac{\pi + (1-\pi)e^{\gamma k}}{\pi + (1-\pi)e^{\gamma(1-s)k}}\right]. \tag{8.17}$$

Using this optimal price, the EWP's objective function when $x = 2$ is given by

$$J_M(p_p^r, s, p_e^r(s)^*; 2) = p_p^r - c_m + \frac{1}{\gamma}\log\left[\frac{\pi + (1-\pi)e^{\gamma k}}{\pi + (1-\pi)e^{\gamma(1-s)k}}\right] - (1-\pi)sk. \tag{8.18}$$

The optimal level of protection s^* is obtained by maximising this expected profit.

Now, $\frac{\partial J_M(p_p^r, s, p_e^r(s)^*; 2)}{\partial s} = (1-\pi)k\left[\frac{e^{\gamma(1-s)k}}{\pi+(1-\pi)e^{\gamma(1-s)k}} - 1\right]$, and the usual first-order condition $\partial J_M(p_p^r, s, p_e^r(s)^*; 2)/\partial s = 0$ yields $s^* = 1$ which implies that the EWP should offer full protection to the customer at a price of

$$p_e^r(s^*)^* = \frac{1}{\gamma}\log\left[\pi + (1-\pi)e^{\gamma k}\right]. \tag{8.19}$$

The EWP's objective function is now given by

$$J_M(p_p^r, s^*, p_e^r(s^*)^*; 2) = p_p^r - c_m + \frac{1}{\gamma}\log\left[\pi + (1-\pi)e^{\gamma k}\right] - (1-\pi)k, \tag{8.20}$$

which is an increasing function of p_p^r. Thus, constraint (8.15) implies that the optimal product price set by the EWP is

$$p_p^{r*} = m - \frac{1}{\gamma}\log\left[\pi + (1-\pi)e^{\gamma k}\right]. \tag{8.21}$$

As in Model 8.1, the customer's optimal strategy is to purchase the product but then be indifferent between purchasing and not purchasing the EW, so $x^* = 1$ or 2. The customer's expected utility is given by

$$J_C(1; y^*) = J_C(2; y^*) = 0. \tag{8.22}$$

Thus, the EWP is able to extract all the consumer surplus from the customer, leaving the customer with zero utility.

Model 8.3 (Stackelberg game)
We now consider a product that requires maintenance effort from customers. The effect of carrying out this maintenance is to improve the reliability of the product.

Assumptions: For a given level of maintenance effort e, the probability that the product does not fail during the EW period is $\pi(e)$ with $\pi'(e) = \mathrm{d}\pi(e)/\mathrm{d}e > 0$. The other assumptions are the same as in Model 8.1.

Key elements and decision variables: There are lower and upper limits on a customer's maintenance effort, so $\underline{e} \leq e \leq \bar{e}$. The probability of no failure is given by the linear function

$$\pi(e) = a + be, \tag{8.23}$$

with $a, b \geq 0$ and $a + b\bar{e} < 1$. The cost of the maintenance effort to the customer is given by the quadratic function

$$\psi(e) = \alpha e^2, \quad \alpha > 0. \tag{8.24}$$

The set of decision variables for the EWP is given by $y \equiv \left\{p_p^r, s, p_e^r(s)\right\}$, whereas the customer has decision variables $x(x = 0, 1, 2)$ and $e(\underline{e} \leq e \leq \bar{e})$.

Objective functions: For a given $y \equiv \left\{p_p^r, s, p_e^r(s)\right\}$ chosen by the EWP, the customer's expected profit $J_C(x, e; y)$ is given by

$$J_C(x, e; y) = \begin{cases} 0, & \text{if } x = 0, \\ m - p_p^r - [1 - \pi(e)]k - \psi(e), & \text{if } x = 1, \\ m - p_p^r - p_e^r(s) - [1 - \pi(e)](1 - s)k - \psi(e), & \text{if } x = 2. \end{cases} \tag{8.25}$$

For a given purchase decision x and maintenance effort level e from the customer, the EWP's expected profit $J_M(p_p^r, s, p_e^r(s); x, e)$ is given by

$$J_M(p_p^r, s, p_e^r(s); x, e) = \begin{cases} 0, & \text{if } x = 0, \\ p_p^r - c_m, & \text{if } x = 1, \\ p_p^r - c_m + p_e^r(s) - [1 - \pi(e)]sk, & \text{if } x = 2. \end{cases} \tag{8.26}$$

Customer's optimal strategy: The customer will buy the product only if

$$p_p^r \leq m - [1 - \pi(e)]k - \psi(e). \tag{8.27}$$

We use a two-stage optimisation procedure to find the customer's optimal maintenance efforts and purchase decisions. Let e_1^* and $e_2^*(s)$ denote the optimal maintenance efforts for $x = 1$ and 2, respectively. These are obtained from the usual first-order conditions

$$\frac{\mathrm{d}J_C(1, e; y)}{\mathrm{d}e} = bk - 2\alpha e = 0 \tag{8.28}$$

and

$$\frac{\mathrm{d}J_C(2, e; y)}{\mathrm{d}e} = b(1 - s)k - 2\alpha e = 0, \tag{8.29}$$

giving

$$e_1^* = bk/2\alpha \quad \text{and} \quad e_2^*(s) = b(1 - s)k/2\alpha, \tag{8.30}$$

respectively.

Note: If the customer purchases an EW, the optimal level of maintenance decreases as s increases. (This agrees with the conventional wisdom that customers expend less effort on maintenance under an EW).

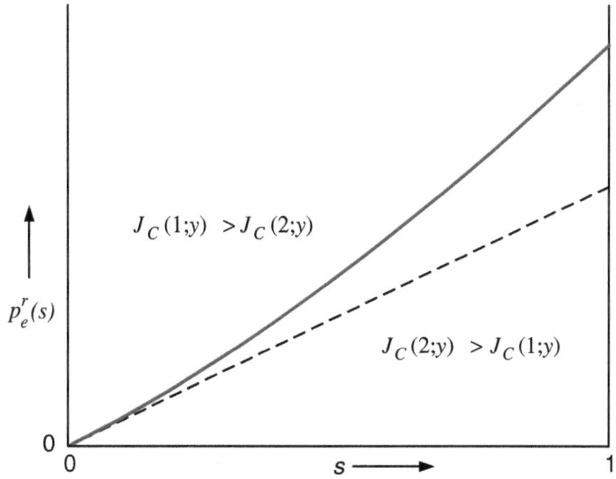

Fig. 8.3 Customer's optimal decisions in the $p_e^r(s)$ versus s diagram (with maintenance)

The customer will buy the product only if $J_C(1, e_1^*; y) \geq J_C(0; y)$ which implies that

$$p_p^r \leq m - [1 - \pi(e_1^*)]k - \psi(e_1^*). \tag{8.31}$$

The analysis of the customer's decision between buying the EW or not involves finding a curve in the $p_e^r(s)$ versus s diagram that separates the two actions. This curve is obtained from the condition $J_C(1, e_1^*; y) = J_C(2, e_2^*(s); y)$ which on simplifying gives

$$p_e^r(s) = \frac{b^2k^2}{4\alpha}s^2 + \left(1 - a - \frac{b^2k}{2\alpha}\right)ks. \tag{8.32}$$

If $2\alpha(1 - a) - b^2k > 0$, then the curve is as shown in Fig. 8.3.

EWP's optimal strategy: We use the three-stage approach to obtain the optimal $p_e^r(s)^*$, s^* and p_p^{r*}. For a given s, the optimal $p_e^r(s)^*$ must be on the curve indicated in Fig. 8.2, so

$$p_e^r(s)^* = \frac{b^2k^2}{4\alpha}s^2 + \left(1 - a - \frac{b^2k}{2\alpha}\right)ks. \tag{8.33}$$

This choice for $p_e^r(s)^*$ gives

$$J_M(p_p^r, s, p_e^r(s)^*; 2, e_2^*(s)) = J_M(p_p^r, s, p_e^r(s)^*; 1, e_1^*) = p_p^r - c_m. \tag{8.34}$$

Fig. 8.4 Customer's optimal decisions in the $p_e^r(s)$ versus s diagram (light and heavy users)

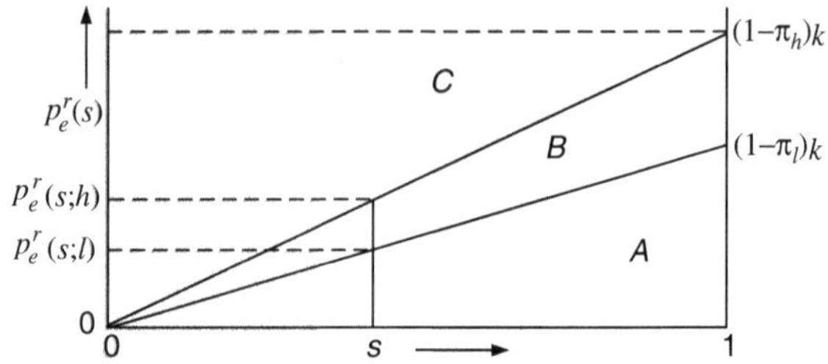

Thus, the EWP can choose any s^* in the interval $(0 \le s \le 1)$, and the optimal EW price depends on the value of s^* that is selected. From (8.34), the EWP's expected profit is an increasing function of p_p^r, so constraint (8.27) implies that

$$p_p^{r*} = m - [1 - \pi(e)]k - \psi(e). \tag{8.35}$$

If the manufacturer (EWP) designs the product so that the required maintenance effort from the customer as part of the EW policy is $\bar{e}$ and the customer does not cheat (no moral hazard), then the optimal product price, EW price and level of protection are given by the results of Model 8.1 with $\pi = \pi(\bar{e})$. However, if the EWP cannot observe the maintenance effort from the customer and so the customer can cheat then the optimal prices and level of protection are given by the results from Model 8.4. Note that the optimal values for the EWP are higher in Model 8.3 than in Model 8.1.

Model 8.4 (Stackelberg game)
The EWP (manufacturer) is now faced with a heterogeneous population of customers. The customers differ in terms of their usage intensity, so π, the probability of no failure during the EW period, varies across the customer population. The approach used can easily be extended to deal with varying risk attitudes, where it is the risk-aversion parameter γ that varies.

Assumptions: There are two types of customer who are either light users of the product (with the probability of no failure during the EW period being π_l) or heavy users (with the probability of no failure being $\pi_h(\pi_h < \pi_l)$). Note that this type of formulation can easily be extended to more than two customer types or, alternatively, customer usage may be assumed to be a continuous random variable with density function $g(u)$, $\underline{u} \le u \le \bar{u}$. Each customer knows exactly what their usage is going to be during the EW period, but this is unknown to EWP. Thus, there is asymmetry of information between the two parties. The EWP offers only one pricing structure and level of EW protection to customers.

The model formulation and analysis is similar to that for Model 8.1. A customer's expected profit is given by (8.1) with $\pi = \pi_l$ for a light user and $\pi = \pi_h$ for a heavy user. The EWP's expected profit is given by (8.2), again with the two possible values of π, depending on the type of customer.

Customer's optimal strategy: A customer who is a light user will buy the product only if

$$p_p^r \leq m - (1 - \pi_l)k, \tag{8.36}$$

with π_h replacing π_l in the case of a heavy user.

For a given s, there are two indifference lines (one for each type of customer) as shown in Fig. 8.4 (with equations $p_e^r(s;l) = (1 - \pi_l)ks$ and $p_e^r(s;h) = (1 - \pi_h)ks$, respectively). These lines divide the diagram into three different regions.

If $p_e^r(s) \leq p_e^r(s;l)$, then both light and heavy users will buy the EW (Region A). If $p_e^r(s,l) < p_e^r(s) \leq p_e^r(s;h)$, then light users will not buy the EW and heavy users will buy the EW (Region B), and if $p_e^r(s) > p_e^r(s;h)$, then neither light nor heavy users will buy the EW (Region C).

These three regions characterise the optimal EW purchase decisions for a customer and the optimal EW price and refund that the EWP should offer.

EWP's optimal strategy: The optimal s^* can be anywhere in the interval $[0, 1]$ and the optimal price is given by

$$p_e^{r*}(s) = p_e^r(s;h) = (1 - \pi_h)ks. \tag{8.37}$$

The above results imply that there will be an automatic separation of customers. Only those customers who are heavy users of the product will purchase the EW.

Comment: Variations in attitude to risk (modelled in terms of low and high risk) follow along similar lines, and we have two curves (each similar to that in Fig. 8.2) lying one above the other. The higher curve is the indifference curve for customers with greater risk aversion. The optimal EW price (for a given s) is a point on this curve, and then, the optimal s is determined. Note that again we have a separated solution—only the more risk-averse consumers buying the EW.

Model 8.5 (Desai and Padmanabhan)
Desai and Padmanabhan (2004) consider a manufacturer who sells a product to a retailer who in turn sells it to customers. Thus, there are three parties in the EW market, and the manufacturer is the sole EW provider. Two channel options for selling the EW to customers are considered, and these are (1) selling through the retailer and (2) selling direct to customers.

Assumptions: The manufacturer and the retailer are both risk neutral. The customers are assumed to be heterogeneous in their attitude to risk. Their utility function for a monetary outcome V is assumed to be given by $U(V) = -e^{-\gamma V}$, where the risk-aversion parameter γ varies across the customer population. The customers make their optimal decisions by maximising the mean–variance approximation to their certainty equivalent for a random pay-off [see (4.3)].

Key elements and decision variables: The manufacturing cost per unit of product (c_m), monetary benefit to a customer from using the product (m) and level of protection (s) under the EW are the same as in Model 8.1. The level of EW protection is assumed to be a fixed parameter and so is no longer a decision variable for the EWP in this model. The monetary loss δ experienced by a customer during the EW period is now a random variable with mean $\bar{\delta}$ and variance σ^2.

As in all previous models, a customer's decision variable is

$$x = \begin{cases} 0 & \text{if the product is not purchased,} \\ 1 & \text{if the product is purchased but not the EW,} \\ 2 & \text{if both the product and the EW are purchased.} \end{cases}$$

The decision variables for the manufacturer and the retailer depend on the channel used to sell the EW. If both the product and the EW are sold through the retailer, then the manufacturer's set of decision variables is $y \equiv \{p_p^w, p_e^w\}$, which comprises the wholesale prices for the two items. The retailer's decision variables are the two corresponding retail prices that are given by the set $v \equiv \{p_p^r, p_e^r\}$. This case is discussed in Model 8.5-(1). Alternatively, the manufacturer might decide to bypass the retailer and sell the EW directly to customers. The sets of decision variables for the manufacturer and retailer are then $y \equiv \{p_p^w, p_e^r\}$ and $v \equiv \{p_p^r\}$, respectively. This case is discussed in Model 8.5-(2).

Objective Functions: A customer's objective function is $J_C(x; y, v)$, the mean–variance approximation to their certainty equivalent, whereas the objective functions of the manufacturer and retailer are their expected profits $J_M(y; x, v)$and $J_R(v; x, y)$, respectively.

Customer's optimal strategy: The customer's objective functions are given by

$$J_C(1; y, v) = m - p_p^r - \bar{\delta} - \frac{\gamma_1}{2}\sigma^2, \tag{8.38}$$

$$J_C(2; y, v) = m - p_p^r - p_e^r - (1 - s)\bar{\delta} - \frac{\gamma_1}{2}(1 - s)^2\sigma^2, \tag{8.39}$$

and $J_C(0; y, v) = 0$. For given retail prices from the sets y and v chosen by the manufacturer and/or retailer, a customer will be indifferent between purchasing only the product $(x = 1)$ or purchasing the product with the EW $(x = 2)$ if $J_C(1; y, v) = J_C(2; y, v)$, which yields

$$\gamma_1 = \frac{2(p_e^r - s\bar{\delta})}{\sigma^2\left[1 - (1 - s)^2\right]}, \tag{8.40}$$

so those customers who have risk-aversion parameter $\gamma < \gamma_1$ will find it optimal to purchase the product but not the EW.

A customer will be indifferent between not purchasing the product ($x = 0$) or purchasing the product with the EW ($x = 2$) if $J_C(0; y, v) = 0 = J_C(2; y, v)$, which implies that $m - p_p^r - p_e^r - (1-s)\bar{\delta} - \gamma_2(1-s)^2\sigma^2/2 = 0$. This yields

$$\gamma_2 = \frac{2\left[m - p_p^r - p_e^r - (1-s)\bar{\delta}\right]}{\sigma^2(1-s)^2}, \tag{8.41}$$

so those customers who have risk-aversion parameter $\gamma > \gamma_2$ will find it optimal not to purchase the product.

Thus, the demand for the product (with and without the EW) is given by

$$D \equiv D(y, v) = \gamma_2 = k_1[a_1 - p_p^r - p_e^r] \tag{8.42}$$

where $k_1 = 2/\sigma^2(1-s)^2$ and $a_1 = m - (1-s)\bar{\delta}$. The demand for the product without the EW is given by $D_p \equiv D_p(y, v) = \gamma_1$ so the demand for the EW is given by

$$D_e \equiv D_e(y, v) = \gamma_2 - \gamma_1 = D - k_2[p_e^r - a_2] \tag{8.43}$$

where $k_2 = 2/\sigma^2\{1-(1-s)^2\}$ and $a_2 = s\bar{\delta}$.

Thus, the two demand functions given in (8.42) and (8.43) are both linear functions of the retail prices for the product and the EW. For a given pair of retail prices belonging to the sets y and v, a customer's optimal decisions are given by

$$x^*(y, v) = \begin{cases} 0 & \text{if } \gamma > \gamma_2, \\ 1 & \text{if } \gamma < \gamma_1, \\ 2 & \text{if } \gamma_1 \le \gamma \le \gamma_2. \end{cases} \tag{8.44}$$

Model 8.5-(1)

The manufacturer sells both the product and the EW through the independent retailer. The manufacturer and retailer are both assumed to be risk neutral so they select values for their decision variables $y = \{p_p^w, p_e^w\}$ and $v = \{p_p^r, p_e^r\}$ in order to maximise their expected profits. A three-stage Stackelberg game takes place with the manufacturer as the dominant player choosing values for y in Stage 1 followed by the retailer choosing v in Stage 2 and then the customer choosing x in Stage 3. The power structure for this type of game is shown in Fig. 4.5i and the particular scenario to be discussed is shown in Fig. 6.8.

Objective functions: The manufacturer's expected profit is given by

$$J_M(y; v, x^*(y, v)) = (p_p^w - c_m)D + (p_e^w - c_{se})D_e \tag{8.45}$$

where $c_{se} = s\bar{\delta}$ is the expected servicing cost per unit EW sold. The retailer's expected profit is given by

$$J_R(v; y, x^*(y, v)) = \left(p_p^r - p_p^w - c_p^m\right)D + \left(p_e^r - p_e^w - c_e^m\right)D_e \tag{8.46}$$

where c_p^m and c_e^m are the marketing costs for selling the product and the EW, respectively.

Retailer's optimal strategy: The retailers optimal prices for the product and the EW based on the manufacturer's prices y and the customer's response $x^*(y, v)$ are given by

$$p_p^{r*}(y, x^*(y, v)) = \frac{a_1 - a_2 + c_p^m + p_p^w}{2} \quad \text{and} \quad p_e^{r*}(y, x^*(y, v)) = \frac{a_2 + c_e^m + p_e^w}{2}. \tag{8.47}$$

These optimal values define v^* for the retailer.

EWP's (manufacturer's) optimal strategy: The manufacturer's optimal prices (taking into account v^* and $x^*(y, v^*)$) are given by

$$p_p^{w*} = \frac{a_1 - a_2 + c_m - c_p^m}{2} \quad \text{and} \quad p_e^{w*} = \frac{a_2 + c_{se} - c_e^m}{2}. \tag{8.48}$$

These optimal values define y^* for the manufacturer.

Model 8.5-(2)

The manufacturer sells the product through the retailer and sells the EW directly to the customers. The structure of the three-stage game is different from that in the previous model. In Stage 1, the manufacturer chooses p_p^w the wholesale price of the product. In Stage 2, the retailer chooses p_p^r the product's retail price, and the manufacturer chooses p_e^r the retail price of the EW with these choices being made simultaneously (Nash game). Finally, in Stage 3, the customer chooses x so decides whether to purchase the product and, if so, whether to purchase the EW.

The manufacturer and retailer are again both assumed to be risk neutral so they select values for their decision variables $y = \left\{p_p^w, p_e^r\right\}$ and $v = \left\{p_p^r\right\}$ in order to maximise their expected profits.

Objective functions: The manufacturer has a different objective function for each of the first two stages of the game. We denote these two functions by J_{M1} and J_{M2}, respectively. Also, since the optimal values for the manufacturer's decision variables are chosen in these two separate stages, it is easier to use actual prices as the arguments of the objective functions rather than the sets y and v.

Customer's optimal strategy: In Stage 3, the customer's optimal decision $x^*\left(p_e^r, p_p^r\right)$ is still given by (8.44).

EWP's (manufacturer's) and retailer's optimal strategies: In Stage 2, the retailer needs to find the value of $p_p^r\left(p_p^w\right)$ that maximises their expected profit given by

$$J_R\left(p_p^r; p_p^w\right) = \left(p_p^r - p_p^w - c_p^m\right)D, \tag{8.49}$$

whereas the EWP needs to find the value of $p_e^r\left(p_p^w\right)$ that maximises their expected profit given by

$$J_{M2}\left(p_e^r; p_p^r\right) = (p_e^r - c_{se} - c_e^m)D_e + \left(p_p^w - c_m\right)D. \tag{8.50}$$

These optimal retail prices (which are both functions of p_p^w) are given by

$$\begin{aligned} p_p^{r**}(p_p^w) &= \frac{a_1 + c_p^m - p_e^r + p_p^w}{2}, \text{ and} \\ p_e^{r**}(p_p^w) &= \frac{\left(a_1 + c_{se} + c_m - p_p^r + c_e^m - p_p^w\right)k_1 + \left(a_2 + c_{se} + c_e^m\right)k_2}{2}. \end{aligned} \tag{8.51}$$

In Stage 1, the manufacturer needs to find the value of p_p^w which maximises their expected profit given by

$$J_{M1}\left(p_p^w; p_p^{r**}(p_p^w), p_e^{r**}(p_p^w)\right) = (p_e^r - c_{se} - c_e^m)D_e + \left(p_p^w - c_m\right)D. \tag{8.52}$$

The optimal wholesale price for the product is given by

$$p_p^{w**} = \frac{\left(8a_1 - 9a_2 + c_{se} + 10c_m + c_e^m - 8c_p^m\right)k_1 + \left(8a_1 - 8a_2 + 8c_m - 8c_p^m\right)k_2}{(18k_1 + 16k_2)}. \tag{8.53}$$

Model 8.6 (Li et al.)
Li et al. (2012) also consider a manufacturer who produces a single product and sells it exclusively through a retailer to customers. The retailer is the EW provider in Model 8.6-(1), whereas in Model 8.6-(2), it is the manufacturer who provides the EW. Model 8.6-(3) is used for comparison purposes, and here, there is no retailer so the manufacturer sells both the product and the EW directly to customers. In Model 8.6-(4), a third-party EW provider is also present. The retailer buys the EW from this provider and then resells it to the customers.

Assumptions: In each model, all parties in the market are assumed to be risk neutral, and there is no information asymmetry among the parties.

Key elements and decision variables: In addition to the price of the EW, its length w_e is also a decision variable. The EW servicing cost is $c_{se}(w_e) = cw_e^2$. The demand for the product and the demand for the EW are given by

$$D \equiv D(p_p^r) = 1 - bp_p^r, \tag{8.54}$$

and

$$D_e \equiv D_e(p_p^r, p_e^r, w_e) = \begin{cases} (1 - bp_p^r) - dp_e^r/w_e, & \text{if } w_e > 0 \\ 0, & \text{if } w_e = 0 \end{cases}, \tag{8.55}$$

respectively. b is the product price sensitivity for customers, and d measures their EW demand sensitivity to the ratio p_e^r/w_e.

The remaining decision variables are specified in each model. We first give the expressions for the objective functions for Models 8.6-(1)—(3) and then the optimal strategies.

Model 8.6-(1)

The retailer is the EW provider, and this is called Model R in Li et al. (2012).

Decision variables: The manufacturer has the single decision variable $y \equiv \{p_p^w\}$, and the set of decision variables for the retailer is $v \equiv \left\{p_p^r, p_e^r, w_e\right\}$.

Objective functions: The manufacturer's and retailer's objective functions (expected profits) are given by

$$J_M(y; v) = (1 - bp_p^r)p_p^w \tag{8.56}$$

and

$$J_R(v; y) = (p_p^p - p_p^w)(1 - bp_p^r) + \left(p_e^r - cw_e^2\right)(1 - b_p^r - d\frac{p_e^r}{w_e}\right), \tag{8.57}$$

respectively.

The manufacturer is the dominant player (leader) in the Stackelberg game and so is able to look ahead and anticipate the decisions made by the retailer and the customers. In Stage 1, the manufacturer chooses the wholesale product price that maximises (8.56). In Stage 2, given the wholesale price set in Stage 1, the retailer chooses the retail product price and the retail price and length of the EW which maximise (8.57). Finally, the customers make their decision.

Model 8.6-(2)

The manufacturer is the EW provider and this is called Model M in Li et al. (2012).

Decision variables: The set of decision variables for the manufacturer is $y \equiv \left\{p_p^w, p_e^r, w_e\right\}$ and the retailer has the single decision variable $v \equiv \{p_p^r\}$.

Objective functions: The manufacturer's and retailer's objective functions (expected profits) are given by

$$J_M(y;v) = p_p^w\left(1 - bp_p^r\right) + \left(p_e^r - cw_e^2\right)\left(1 - b_p^r - d\frac{p_e^r}{w_e}\right) \tag{8.58}$$

and

$$J_R(v;y) = \left(p_p^r - p_p^w\right)\left(1 - bp_p^r\right), \tag{8.59}$$

respectively.

The manufacturer is again the dominant player in the game. In Stage 1, the manufacturer chooses the wholesale product price and the EW terms which maximise (8.58). In Stage 2, given the manufacturer's decisions in Stage 1, the retailer chooses the retail product price which maximises (8.59). Finally, the customers make their decision.

Model 8.6-(3)

There is no retailer in this model which is called Model C in Li et al. (2012). The manufacturer is the EW provider and sells the product and the EW directly to the customers.

Decision variables: The set of decision variables for the manufacturer is $y \equiv \left\{p_p^r, p_e^r, w_e\right\}$.

Objective function: The manufacturer's objective function (expected profit) is given by

$$J_M(y) = p_p^r\left(1 - bp_p^r\right) + \left(p_e^r - cw_e^2\right)\left(1 - bp_p^r - d\frac{p_e^r}{w_e}\right). \tag{8.60}$$

As in the previous cases, this model also has a game element—because customers choose their option (from not purchasing the product, purchasing the product but not the EW, or purchasing both items) after the manufacturer's decisions are made. The customers' choices are implied by the demand functions given in (8.54) and (8.55).

The optimal values of the decision variables for the manufacturer and the retailer in Models 8.6-(1)–8.6-(3) are shown in Table 8.1.

Model 8.6-(4) (Four parties)

The manufacturer only sells the product to the retailer. There is a third-party EW provider who sells the EW through the retailer, and this provider also services the

Table 8.1 Optimal decision variable values for manufacturer and retailer

	Model 8.6-(1)	Model 8.6-(2)	Model 8.6-(3)
p_p^{r*}	$\frac{b - d\left[6cd - \sqrt{3c(12cd^2 - b)}\right]}{b^2}$	$\frac{b - 3d\left[6cd - \sqrt{c(36cd^2 - b)}\right]}{b^2}$	$\frac{b - 3d\left[3cd - \sqrt{c(9cd^2 - b)}\right]}{b^2}$
p_p^{w*}	$\frac{2b - 2d\left[6cd - \sqrt{3c(12cd^2 - b)}\right]}{3b^2}$	$\frac{b - 6d\left[6cd - \sqrt{c(36cd^2 - b)}\right]}{b^2}$	N/A
p_e^{r*}	$\frac{8d\left[6cd - \sqrt{3c(12cd^2 - b)}\right] - 2b}{3b^2}$	$\frac{24d\left[6cd - \sqrt{c(36cd^2 - b)}\right] - 2b}{b^2}$	$\frac{12d\left[3cd - \sqrt{c(9cd^2 - b)}\right] - 2b}{b^2}$
w_e^*	$\frac{6cd - \sqrt{3c(12cd^2 - b)}}{3bc}$	$\frac{6cd - \sqrt{c(36cd^2 - b)}}{bc}$	$\frac{3cd - \sqrt{c(9cd^2 - b)}}{bc}$

EW. The retailer sells the product and the EW to the customers. This is called Model 3R in Li et al. (2012).

Decision variables: The manufacturer has the single decision variable $y \equiv \{p_p^w\}$. The sets of decision variables for the third-party EW provider and the retailer are $z \equiv \{w_e, p_e^w\}$ and $v \equiv \{p_p^r, p_e^r\}$, respectively.

Objective functions: The objective functions (expected profits) for the manufacturer, third-party EW provider and retailer are given by

$$J_M(y; v, z) = p_p^w\left(1 - bp_p^r\right), \tag{8.61}$$

$$J_{TP}(z; y, v) = \left(p_e^w - cw_e^2\right)\left[\left(1 - bp_p^r\right) - d\frac{p_e^r}{w_e}\right] \tag{8.62}$$

and

$$J_R(v; y, z) = (p_p^r - p_r^w)(1 - bp_p^r) + (p_e^r - p_e^w)\left[\left(1 - bp_p^r\right) - d\frac{p_e^r}{w_e}\right]. \tag{8.63}$$

In Stage 1 of the game, the manufacturer chooses the wholesale product price which maximises (8.61) and so takes into account the subsequent actions of retailer in setting the retail price and the behaviour of the customers with regards to purchasing the product at this price. In Stage 2, the third-party provider chooses the length and wholesale price of the EW to maximise (8.62), taking into account the demand for the EW when it is sold through the retailer. In Stage 3, the retailer chooses the retail prices for the product and the EW by maximising (8.63), taking into account the previous choices made by the manufacturer and third-party EW provider. Thus, in this Stackelberg game, there is dominance between the manufacturer and the retailer and also between the third-party EW provider and the retailer. The game is solved by working backwards starting from the retailer's problem in Stage 3.

The optimal values of the decision variables for the manufacturer, third-party EW provider and the retailer are shown in Table 8.2.

As shown in the table, the expressions for the optimal values of the decision variables are rather complicated. The only closed form expression given is that for p_p^{w*} where

$$\Phi = \sqrt[3]{\begin{array}{l}\left\{8b^3 - 378b^2cd^2 - 3888bc^2d^4 + 4330989c^3d^6\right\} + \\ \left\{18d\sqrt{3c}\left(b - 90cd^2\right)\sqrt{8b^3 - 945b^2cd^2 + 71928bc^2d^4 - 793152c^3d^6}\right\}\end{array}}.$$

The optimal values for the other variables are expressed in terms of p_p^{w*}.

Table 8.2 Optimal decision variable values for manufacturer, third-party EW provider and retailer

	Model 8.6-(4)
p_p^{r*}	$\frac{6d\sqrt{c\left(b^2p_p^{w*}+9cd^2-b\right)}+\left(b^2p_p^{w*}+3b-18cd^2\right)}{4b^2}$
p_p^{w*}	$\frac{4b^2-288bcd^2+(171cd^2)^2+\Phi(\Phi+8b+117cd^2)}{12b^2\Phi}$
p_e^{r*}	$\frac{\left(b^2p_p^{w*}-b-24cd^2\right)\sqrt{b^2p_p^{w*}+9cd^2-b}+\left(b^2p_p^{w*}-b+72cd^2\right)d\sqrt{c}}{4b^2d\sqrt{c}}$
p_e^{w*}	$\frac{\left(b^2p_p^{w*}-b-12cd^2\right)\sqrt{b^2p_p^{w*}+9cd^2-b}-\left(b^2p_p^{w*}-b-36cd^2\right)d\sqrt{c}}{4b^2d\sqrt{c}}$
w_e^*	$\frac{3d\sqrt{c}-\sqrt{b^2p_p^{w*}+9cd^2-b}}{b\sqrt{c}}$

Model 8.7 (Kurata and Nam)
A manufacturer produces a single product and sells it exclusively through a retailer. Kurata and Nam (2010) investigate competition between the manufacturer and the retailer in the "after-sales service" offered to customers. The manufacturer provides a base after-sales service, the cost of which is included in the retail price of the product. This can interpreted as the BW for the product. The retailer offers an optional after-sales service for an extra payment to supplement the basic level provided by the manufacturer. This can be interpreted as an EW.

Assumptions: The wholesale price of the product p_p^w and the retail price p_p^r are fixed parameters (exogenous variables) in the model. There are two groups of customers. Those who use only the base after-sales service offered by the manufacturer (Segment 0), and those who pay for the optional after-sales service offered by the retailer in addition to the base service (Segment 1).

Decision variables: The manufacturer's decision variable is y, the basic after-sales service level (length of the BW) and the retailer's decision variable is v, the optional after-sales service level (length of the EW).

Objective functions: The Segment 0 demand (for the product with only the base level of service) is given by

$$D_p \equiv D_p(y) = a_0 + b_0 y(2\bar{y} - y), \tag{8.64}$$

and the Segment 1 demand (for the optional level of service) is given by

$$D_e \equiv D_e(y, v) = a_1 + b_1(y + v)(2\bar{y} + 2\bar{v} - y - v). \tag{8.65}$$

In these two demand functions, the optimal levels for the two types of service are $\bar{y}$ and $\bar{v}$ which can be estimated by conducting consumer behaviour surveys. a_0 and a_1 are the minimum market sizes of the two groups of customers and b_0 and b_1 are the two demand sensitivities to changes in service level. There is no interaction effect between the two service plans.

The objective functions (expected profits) for the manufacturer and retailer are given by

$$J_M(y; v) = (p_p^r - c_m)[a_0 + b_0 y(2\bar{y} - y) + a_1 + b_1(y + v)(2\bar{y} + 2\bar{v} - y - v)] - k_M y, \tag{8.66}$$

and

$$\begin{aligned} J_R(v; y) = {} & (p_p^r - p_p^w)[a_0 + b_0 y(2\bar{y} - y)] \\ & + (p_p^r - p_p^w + p_e^r - c_{se})[a_1 + b_1(y + v)(2\bar{y} + 2\bar{v} - y - v)] - k_R v, \end{aligned} \tag{8.67}$$

where k_M and k_R are the per unit level of service provision costs for the manufacturer and the retailer.

Model 8.7-(1) (Nash Game)
In this case, the manufacturer and the retailer determine their levels of after-sales service simultaneously. The optimal values for their decision variables are given by

$$y^* = \bar{y} + \frac{k_R}{2b_0(p_p^r - p_p^w + p_e^r - c_{se})} - \frac{k_M}{2b_0(p_p^r - c_m)} \tag{8.68}$$

and

$$v^* = \bar{v} - \frac{k_R(b_0 + b_1)}{2b_0 b_1(p_p^r - p_p^w + p_e^r - c_{se})} + \frac{k_M}{2b_0(p_p^r - c_m)}. \tag{8.69}$$

Model 8.7-(2) (Stackelberg game)
The manufacturer's decision on base level of service is made first (the manufacturer is the dominant player) followed by the retailer making the optional service-level decision. The optimal values for the decision variables of the manufacturer and retailer are given by

$$y^{**} = \bar{y} - \frac{k_M}{2b_0(p_p^r - c_m)}, \tag{8.70}$$

and

$$v^{**} = \bar{v} - \frac{k_R}{2b_1\left(p_p^r - p_p^w + p_e^r - c_{se}\right)} + \frac{k_M}{2b_0(p_p^r - c_m)}. \tag{8.71}$$

Model 8.8 (Jiang and Zhang)
In Jiang and Zhang (2011), a manufacturer sells a product through a retailer to a group of customers. The manufacturer may or may not offer a BW with the sale of the product and the retailer may or may not offer an additional service plan (EW) that is an optional purchase for the customers.

Assumptions: The manufacturer and retailer are both risk neutral. The customers are assumed to be heterogeneous in their attitude to risk. The risk-aversion parameter γ of a customer is a random variable that is uniformly distributed on the interval $[0, \gamma_m]$. As in Desai and Padmanabhan (2004), customers make their optimal decisions by maximising the mean–variance approximation to their certainty equivalent for a random pay-off [see (4.3)].

Key elements and decision variables: The interaction between the manufacturer and the retailer is modelled as a three-stage Stackelberg game. In the first two stages of the game, the manufacturer is the dominant player (leader) and the retailer is the follower. In Stage 1, the manufacturer chooses w_b the length of the BW, and then, the retailer chooses w_e the length of the EW. In Stage 2, the manufacturer sets the wholesale price p_p^w for the product, and then, the retailer sets the product's retail price p_p^r and the retail price p_e^r for the EW. Finally, the customers decide whether to purchase the product and the optional EW.

Both types of warranty are effective from the date the product is purchased, with the service plan providing longer coverage. Each warranty length is normalised so that $0 \le w_b < w_e \le 1$. If the product fails before time w_b, and a customer has purchased the EW, then the failure is rectified under the manufacturer's BW. Thus, the EW is only used to provide coverage during the residual period between time w_b and w_e.

The monetary benefit to a customer from using the product is m if it works and zero if it fails. The probabilities of these two events occurring are π and $1 - \pi$, respectively. Should the product fail under the BW, a customer receives a refund of $w_b m$ from the manufacturer. Under the EW, a customer will receive a refund of $w_e m$ if the product fails.

The expected servicing cost to the manufacturer to provide a BW is $c_{sb}(w_b; \pi) = (1 - \pi)c_{re}^m(w_b^2)$. The expected servicing cost to the retailer to provide an EW is $c_{se}(w_b, w_e; \pi) = (1 - \pi)c_{re}^r(w_e^2 - w_b^2)$ if the manufacturer offers a BW and $c_{se}(w_e; \pi) = (1 - \pi)c_{re}^r(w_e^2)$ if no BW is offered. (Note: c_{re}^m and c_{re}^r represent the manufacturer's and retailer's cost efficiencies in providing warranty service. If $c_{re}^m < c_{re}^r$, the manufacturer is more efficient in servicing a warranty and vice versa.)

Objective functions and optimal strategies: Stage 1 of the game produces four possible outcomes or subgames to consider in Stage 2. These are denoted by (*N*, *N*) if neither the manufacturer nor the retailer decides to offer a warranty ($w_b = w_e = 0$); (*Y*, *N*) if the manufacturer offers a BW, and the retailer does not offer an EW ($w_b \in (0, 1)$ and $w_e = w_b$); (*N*, *Y*) if the manufacturer does not offer a BW and the retailer offers an EW ($w_b = 0$ and $w_e \in (0, 1)$); and (*Y*, *Y*) if both the manufacturer and retailer decide to offer a warranty ($w_b \in (0, 1)$ and $w_e \in (w_b, 1)$).

In the (*Y*, *Y*) subgame, customer demand has the following structure: Customers with risk-aversion parameter $\gamma \in [0, \gamma_1]$ will purchase the product (with the BW) but will not purchase the EW; customers with $\gamma \in (\gamma_1, \gamma_2]$ will purchase both the product and the EW; and customers with $\gamma \in (\gamma_2, \gamma_m]$ are so risk averse that they do not purchase anything. Different demand structures apply for the other three subgames.

Expressions for the objective functions (expected profits) for the manufacturer and retailer in each of the four subgames are shown in Table 8.3.

The optimal pricing decisions for the manufacturer and retailer in each case are shown in Table 8.4.

Model 8.9 (Heese)
Heese (2012) considers the interactions that occur when two manufacturers (labelled 1 and 2) sell two competing products (also labelled 1 and 2) through the same retailer. The products are sold with different BWs and, in each case, the retailer is the sole EW provider.

Assumptions: A two-dimensional spatial model is used to capture the customers' heterogeneity with respect to their preferences for each product and their

Table 8.3 Objective functions in the four pricing subgames

(*N*, *N*)	(*N*, *Y*)
$J_M\left(p_p^w;p_p^r\right)=\frac{\gamma_3}{\gamma_m}p_p^w$, $J_R\left(p_p^r;p_p^w\right)=\frac{\gamma_3}{\gamma_m}\left(p_p^r-p_p^w\right)$, where $\gamma_3=\frac{m\pi-p_p^r}{m^2\pi(1-\pi)}$.	$J_M\left(p_p^w;w_e,p_p^r,p_e^r\right)=\frac{\gamma_2}{\gamma_m}p_p^w$, $J_R\left(w_e,p_p^r,p_e^r;p_p^w\right)=\frac{\gamma_2}{\gamma_m}\left(p_p^r-p_p^w\right)+\frac{\gamma_2-\gamma_3}{\gamma_m}\left[p_e^r-c_{se}(w_e;\pi)\right]$, where $\gamma_2=\frac{m[1-(1-\pi)(1-w_e)]-p_p^r-p_e^r}{m^2(1-w_e)^2\pi(1-\pi)},\gamma_3=\frac{p_e^r-m(1-\pi)w_e}{m^2\pi(1-\pi)w_e(2-w_e)}$
(*Y*, *N*)	(*Y*, *Y*)
$J_M\left(w_b,p_p^w;p_p^r\right)=\frac{\gamma_1}{\gamma_m}\left[p_p^w-c_{sb}(w_b;\pi)\right]$, $J_R\left(p_p^r;w_b,p_p^w\right)=\frac{\gamma_1}{\gamma_m}\left(p_p^r-p_p^w\right)$, where $\gamma_1=\frac{m[1-(1-\pi)(1-w_b)]-p_p^r}{m^2(1-w_b)^2\pi(1-\pi)}$.	$J_M\left(w_b,p_p^w;w_e,p_p^r,p_e^r\right)=\frac{\gamma_2}{\gamma_m}\left[p_p^w-c_{sb}(w_b;\pi)\right]$, $J_R\left(w_e,p_p^r,p_e^r;w_b,p_p^w\right)=\frac{\gamma_2}{\gamma_m}\left(p_p^r-p_p^w\right)+\frac{\gamma_2-\gamma_1}{\gamma_m}\left[p_e^r-c_{se}(w_b,w_e;\pi)\right]$, where $\gamma_2=\frac{m[1-(1-\pi)(1-w_e)]-p_p^r-p_e^r}{m^2(1-w_e)^2\pi(1-\pi)},\gamma_1=\frac{p_e^r-m\pi(w_e-w_b)}{m^2\pi(1-\pi)(w_e-w_b)(2-w_e-w_b)}$

Table 8.4 Optimal prices in the four pricing subgames

	(*N*, *N*)	(*N*, *Y*)	(*Y*, *N*)	(*Y*, *Y*)
p_p^{w*}	$\frac{m\pi}{2}$	$\frac{m[1-(1-\pi)(1-w_e)]-c_{se}(w_e;\pi)}{2}$	$\frac{m[1-(1-\pi)(1-w_b)]+c_{sb}(w_b;\pi)}{2}$	$\frac{m[1-(1-\pi)(1-w_e)]-c_{se}(w_b,w_e;\pi)}{2}+\frac{c_{sb}(w_b;\pi)}{2}$
p_p^{r*}	$\frac{3m\pi}{4}$	$\frac{3m\pi-[c_{se}(w_e;\pi)-m(1-\pi)w_e]}{4}$	$\frac{3m[1-(1-\pi)(1-w_b)]+c_{sb}(w_b;\pi)}{4}$	$\frac{m[1-(1-\pi)(1-w_b)]}{2}+\frac{m[1-(1-\pi)(1-w_e)]-c_{se}(w_b,w_e;\pi)}{4}+\frac{c_{sb}(w_b;\pi)}{4}$
p_e^{r*}	N/A	$\frac{c_{se}(w_e;\pi)+m(1-\pi)w_e}{2}$	N/A	$\frac{c_{se}(w_b,w_e;\pi)+(w_e-w_b)m(1-\pi)}{2}$

willingness to pay for warranty (BW and EW) coverage. These two dimensions of heterogeneity are assumed to be independent. $d \in [0, 1]$ denotes a customer's preference for product 1 ($1 - d$ is their preference for product 2), and $r \in [0, 1]$ denotes their warranty valuation. For a particular customer, the values of d and r come from uniform distributions.

Customer choice is represented by a point (d, r) lying in a 2D plane and is also influenced by the sales effort of the retailer. For a given r, there is a $d(r)$ such that if a customer's preference $d < d(r)$, then the customer will buy product 1; if $d > d(r)$, then the customer buys product 2; and if $d = d(r)$, then the customer is indifferent between the two products. $d(r)$ is linear in r (similar to the demand function in Desai and Padmanabhan 2004).

Figure 8.5a illustrates the product demand model when product 1 is less liked by customers that do not value warranties but which come with a better warranty than product 2. Customer product taste types are distributed along the horizontal axis, and warranty valuation taste types are along the vertical axis. A customer with (d, r) in region R-1 buys product 1, while a customer with (d, r) in region R-2 buys product 2. The total customer demands for (proportions who buy) the two products are given by

$$D_{p1} = [d(0) + d(1)]/2 \quad \text{and} \quad D_{p2} = 1 - D_{p1}. \tag{8.72}$$

Customers may make their product and EW purchase decisions sequentially. In this case, after a customer has bought one of the two products, the customer then decides whether or not to buy the EW. Customer behaviour is illustrated in Fig. 8.5b. A customer with (d, r) in region R-11 buys product 1 but does not buy the EW; a customer with (d, r) in region R-12 buys product 1 and also the EW; a customer with (d, r) in region R-21 buys product 2 but does not buy the EW; and a customer with (d, r) in region R-22 buys product 2 and the EW. The total customer demand for the EW is given by

$$D_e = (1 - r_1)[d(r_1) + (d(1) - d(r_1))/2] + (1 - r_2)[1 - d(1) + (d(1) - d(r_2))/2] \tag{8.73}$$

Alternatively, customers may decide about product purchase and EW purchase at the same time (simultaneous choice). Figure 8.5c illustrates customer behaviour in this case. Customers with (d, r) in region R-11 buy product 1 but do not buy the EW; customers with (d, r) in region R-12 buy product 1 and also the EW; customers with (d, r) in region R-21 buy product 2 but do not buy the EW; customers with (d, r) in region R-22 buy product 2 and the EW; customers with (d, r) in region R-23 have switched from buying product 1 to product 2 and now also buy the EW; customers with (d, r) in region R-24 have switched from buying product 1 to product 2 and still buy the EW. These changes from Fig. 8.5b show the effect

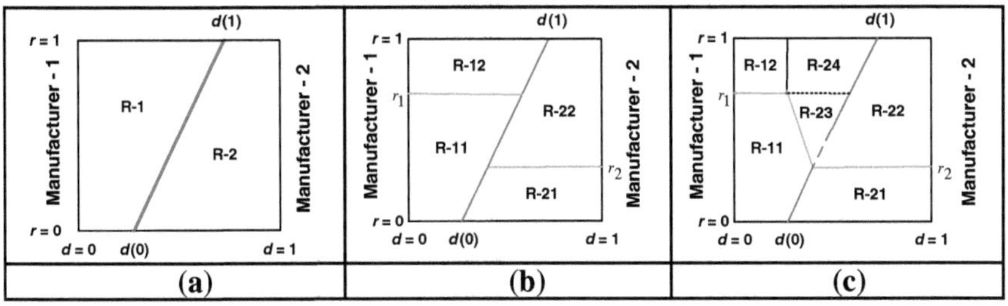

Fig. 8.5 Illustration of product demand and EW demand

that EWs have on product purchase (BW) decisions. The total customer demand for the two products and the EW are now given by

$$\tilde{D}_{p1} = d(0) + r_1[d(r_2) - d(0)]/2, \tilde{D}_{p2} = 1 - \tilde{D}_{p1}, \tag{8.74}$$

and

$$\tilde{D}_e = (1 - r_1) + (r_1 - r_2)[(1 - d(r_2)) + (d(r_2) - d(0))/2]. \tag{8.75}$$

Key elements and decision variables: The retail prices of the two products p^r_{pi}, $i = 1,2$, are assumed to be exogenous. The sets of decision variables for the two manufacturers are $y_i \equiv \{w_i, m_i\}$, $i = 1,2$, where w_i is the length of the BW offered with product *i*, and m_i is the commission paid to the retailer for selling the product (the manufacturer's profit margin on the sale). The per unit manufacturing costs for the two products (including the BW servicing costs) are given by the quadratic functions $c_{mi}(w_i) = c_i + k_i w_i^2$, $i = 1,2$.

The price of the EW offered by the retailer is p_e and the EW coverage is $w_e \geq w_i$, and this begins when the product is purchased rather than when the manufacturer's BW expires. These two variables are assumed to be exogenous (their values are set by an external insurer). The retailer's profit margin per unit EW sold is m_e. The retailer can also exert additional sales effort to influence sales of the two products. If $\varepsilon = e_1 - e_2$ denotes the difference between the sales efforts used to promote the two products, then the cost of the retailer's (distorting) sales effort is given by the quadratic function $c_s(\varepsilon) = k_s \varepsilon^2$. The retailer's single decision variable is $v \equiv \{\varepsilon\}$.

Objective functions and optimal strategies: The two manufacturers simultaneously select the lengths of their BWs (w_i) and then their sales commissions to the retailer (m_i). The retailer then chooses the difference in sales effort (ε) to expend on the two products. Finally, the customers decide which product to purchase and whether to purchase the EW. This sequence of events defines the multistage game that takes place between the various parties in the market.

For the scenario where customers make their product and EW purchase decisions sequentially, the expected profit of manufacturer $i(i = 1, 2)$ is given by

$$J_{Mi}(y; v) = \left[p^{r}_{pi} - c_{mi}(w_i) - m_i\right] D_{pi} \tag{8.76}$$

and the retailer's expected profit is given by

$$J_R(v; y) = m_1 D_{p1} + m_2 D_{p2} + m_e D_e - c_s(\varepsilon). \tag{8.77}$$

When customers make the two purchase decisions simultaneously, the expected profits in (8.76) and (8.77) are modified by replacing the customer demands D_{pi}, $i = 1, 2$, and D_e by $\tilde{D}_{pi}$, $i = 1, 2$, and $\tilde{D}_e$, respectively.

Heese (2012) uses these different objective functions to determine and analyse the optimal strategies for the two manufacturers and the retailer in the two scenarios. The analysis shows that the method the retailer uses to sell EWs substantially affects how customers make their product and EW purchase decisions.

8.3.2 *Two EW Providers and Heterogeneous Customers (Scenario D)*

We now discuss competition between two manufacturers who sell the same product directly to a group of heterogeneous customers. Each manufacturer may choose to offer different purchasing options to the customers, some of which involve EWs.

Model 8.10 (Kameshwaran et al.)
In Kameshwaran et al. (2009), each manufacturer has three possible options to offer customers who purchase their product:

- Purchase only the product (and have the servicing done during its lifetime by a third party)
- Purchase the product and an EW (that provides lifetime servicing) as two separate items
- Purchase the product and the lifetime EW as a bundle.

These options are denoted P, $P + S$ and PS, respectively. Note that, for option $P + S$, a customer may decide to purchase the product but not the EW.

Key elements and decision variables: The manufacturing cost per unit of product is c_m and per unit of EW servicing cost is c_{se}. The decision variable for each customer are

$$x = \begin{cases} 1 & \text{if option } P \text{ is selected,} \\ 2 & \text{if option } P + S \text{ is selected,} \\ 3 & \text{if option } PS \text{ is selected,} \end{cases}$$

and the decision variables for the manufacturers are the prices $y \equiv \{p_p^r\}$ for option P, $y \equiv \{p_p^r, p_e^r\}$ for option $P + S$ and $y \equiv \{p_{pe}^r\}$ for option PS.

The customers are homogenous in terms of the maximum amount $\bar{p}_p$ that they would pay for the product but are heterogeneous in terms of their willingness to pay (WTP) β for the lifetime EW. β is assumed to be uniformly distributed on the interval $[\underline{c}, \bar{c}]$.

Objective functions: All parties (manufacturers and customers) are risk neutral and so want to choose values for their decision variables that maximise their expected profits. The unit profits to a manufacturer and to a customer with WTP β under each purchase option are as follows:

Option	Manufacturer	Customer
P	$p_{pi}^r - c_m$	$\bar{p}_p - p_p^r$
$P + S$	$p_p^r + p_e^r - c_m - c_{se}$	$\bar{p}_p + \beta - p_p^r - p_e^r$
PS	$p_{pe}^r - c_m - c_{se}$	$\bar{p}_p + \beta - p_{pe}^r$

Single Manufacturer—monopoly market

In this scenario, there is only a single manufacturer offering the three purchase options to the customers who are heterogeneous with respect to their WTP β. In this case, we have a two-stage Stackelberg game with the manufacturer acting as the leader (dominant player) and the customers acting as the followers.

In the case of option P, the manufacturer's profit is $J_M(p_p^r; 1) = p_p^r - c_m$. The manufacturer can set any price $p_p^r \leq \bar{p}_p$ to capture all the customers and so the optimal product price and maximum profit are given by

$$p_p^{r*} = \bar{p}_p \text{ and } J_M(p_p^{r*}; 1) = \bar{p}_p - c_m, \tag{8.78}$$

respectively.

In the case of option $P + S$, the manufacturer sells the product and the EW as two separate items. The optimal product price is again $p_p^{r*} = \bar{p}_p$. For a given EW price p_e^r, the proportion of customers who will purchase the EW is $(\bar{c} - p_e^r)/(\bar{c} - \underline{c})$. The manufacturer wishes to find the value of p_e^r that maximises the expected profit $J_M(p_p^{r*}, p_e^r; 2) = (\bar{p}_p - c_m) + (p_e^r - c_{se})(\bar{c} - p_e^r)(\bar{c} - \underline{c})$. The optimal EW price and maximum expected profit are given by

$$p_e^{r*} = (\bar{c} + c_{se})/2 \quad \text{and} \quad J_M\left(p_p^{r*}, p_e^{r*}; 2\right) = \left(\bar{p}_p - c_m\right) + \frac{1}{(\bar{c} - \underline{c})}\left(\frac{\bar{c} - c_{se}}{2}\right)^2, \tag{8.79}$$

respectively.

Table 8.5 Stage 1 Nash offerings game

		Manufacturer 2		
		P	$P+S$	PS
Manufacturer 1	P	0,0	$0, \frac{1}{\bar{c}-\underline{c}}\left(\frac{\bar{c}-c_{se}}{2}\right)^2$	$\frac{1}{\bar{c}-\underline{c}}\left(\frac{\bar{c}-2\underline{c}+c_{se}}{3}\right)^2, \frac{1}{\bar{c}-\underline{c}}\left(\frac{2\bar{c}-\bar{c}-c_{se}}{3}\right)^2$
	$P+S$	$\frac{1}{\bar{c}-\underline{c}}\left(\frac{\bar{c}-c_{se}}{2}\right)^2, 0$	0, 0	$\frac{1}{\bar{c}-\underline{c}}\left(\frac{c_{se}-\underline{c}}{2}\right)^2, 0$
	PS	$\frac{1}{\bar{c}-\underline{c}}\left(\frac{2\bar{c}-\underline{c}-c_{se}}{3}\right)^2, \frac{1}{\bar{c}-\underline{c}}\left(\frac{\bar{c}-2\underline{c}+c_{se}}{3}\right)^2$	$0, \frac{1}{\bar{c}-\underline{c}}\left(\frac{c_{se}-\underline{c}}{2}\right)^2$	(0, 0)

In the case of option *PS*, a customer can only purchase the bundle (product + EW). For a given price p^r_{pe}, the proportion of customers who will purchase the bundle is $\left(\bar{c} - p^r_{pe} + \bar{p}_p\right)/(\bar{c} - \underline{c})$. The manufacturer wishes to find the value of p^r_{pe} that maximises the expected profit $J_M(p^r_{pe};3) = \left(p^r_{pe} - c_m - c_{se}\right)\left(\bar{c} - p^r_{pe} + \bar{p}_p\right)/(\bar{c} - \underline{c})$. The optimal bundle price and maximum expected profit are given by

$$p^{r*}_{pe} = \left(\bar{c} + \bar{p}_p + c_m + c_{se}\right)/2 \quad \text{and} \quad J_M(p^{r*}_{pe};3) = \frac{1}{(\bar{c} - \underline{c})}\left(\frac{\bar{p}_p - c_m + \bar{c} - c_{se}}{2}\right)^2, \tag{8.80}$$

respectively.

Two manufacturers—duopoly market

The two manufacturers (labelled 1 and 2) produce the same product and have identical manufacturing costs and EW servicing costs. The EW provided by a given manufacturer can only be utilised for the product that manufacturer sells.

The strategic interaction between the two manufacturers and the customers is modelled as a three-stage game. In Stage 1, each manufacturer decides which option (*P*, *P* + *S* or *PS*) to offer, and these decisions are made simultaneously. Neither manufacturer knows what their competitor has done until both decisions have been made so this is a Nash game. In Stage 2, each manufacturer selects the price(s) for the option they have chosen in Stage 1, and these pricing decisions are again made simultaneously. Finally, in Stage 3, the customers choose which option to purchase and from which manufacturer to make this purchase.

There are nine possible outcomes for Stage 1, and so there are nine Nash subgames involving prices in Stage 2. Due to the symmetry of the manufacturers in terms of manufacturing costs and servicing costs and options offered, only six distinct Nash pricing games need to be analysed. The NE outcomes (optimal expected profits for the two manufacturers) in these pricing games are then used to solve the Stage 1 Nash "offerings" game. This game, in *normal form*, is shown in Table 8.5. Each row (column) in the table represents a possible decision for manufacturer 1 (2), and the cells contain the objective function values (expected profits) for the two players. The first entry in each cell is the profit for manufacturer 1, and the second entry is for manufacturer 2.

The best responses for each manufacturer to what the other manufacturer chooses are easily identified in the table, and these produce two possible NE strategies. Manufacturer 1 should offer only option *P* and manufacturer 2 should only offer option *PS* (or vice versa) to maximise their individual objective functions.

8.4 Dynamic GT Models for EW Decision-Making

We now describe GT models which allow for the possibility of multiple product failures during the EW period.

Model 8.11 (Stackelberg game)
Assumptions: A monopolist manufacturer sells a product directly to customers for a price p_p^r and included with each sale is a BW of length W. The manufacturing cost per unit of product is c_m. According to the terms of the BW, the manufacturer will rectify each failure that occurs during the BW period at no cost to the customer. The manufacturer is the only maintenance service provider for the product and sells EWs directly to the customers. The customer population is homogeneous in terms of risk attitude, usage intensity, etc. The manufacturer and customers are all risk neutral and have complete information about product reliability and all relevant costs.

Key elements and decision variables: The product has a useful life $L > W$, and a customer always keeps it for this length of time. When it is operating, the product provides the customer with revenue of R per unit time. The manufacturer offers an EW of length T $(0<T\leq L-W)$ to the customer at the time of the product sale. The price of this EW is $p_e^r(T)$, and each failure during the EW period p_e will be rectified by the manufacturer (EWP) at no cost to the customer. During the post EW period $(W+T, L]$, the customer will have to pay the manufacturer to rectify any product failures. The price the manufacturer charges the customer for each repair is p_r whereas the actual cost to the manufacturer is c_r.

The manufacturer always performs a minimal repair at each failure (during the BW period and beyond). All repair times are very small compared with the mean time between failures and so can be ignored. Let $N(t)$ denote the number of failures that occur in the time interval $[0,t)$. Under the minimal repair assumption, $\{N(t); t\geq 0\}$ is a non-homogeneous Poisson process (NHPP) with mean function $\Lambda(t)$.

The set of decision variables for the manufacturer (EWP) is given by $y\equiv \left\{p_p^r, T, p_e^r(T)\right\}$ and the customer's decision variable is

$$x=\begin{cases}0 & \text{if the product is not purchased,}\\ 1 & \text{if the product is purchased but not the EW,}\\ 2 & \text{if both the product and the EW are purchased.}\end{cases}$$

The EWP is the leader and the customer is the follower in the Stackelberg game between the two parties.

Objective functions: For a given value of y chosen by the EWP, the customer's expected profit is given by

$$J_C(x;y) = \begin{cases} 0, & \text{if } x = 0, \\ RL - p_p^r - p_r[\Lambda(L) - \Lambda(W)], & \text{if } x = 1, \\ L - p_p^r - p_e^r(T) - p_r[\Lambda(L) - \Lambda(W+T)], & \text{if } x = 2. \end{cases} \quad (8.81)$$

For a given value of x chosen by the customer, the EWP's expected profit is given by

$$J_M(p_p^r, T, p_e^r(T); x) = \begin{cases} 0, & \text{if } x = 0, \\ p_p^r - c_m + (p_r - c_r)\Lambda(L) - p_r\Lambda(W), & \text{if } x = 1, \\ p_p^r - c_m + p_e^r(T) + (p_r - c_r)\Lambda(L) - p_r\Lambda(W+T), & \text{if } x = 2. \end{cases} \quad (8.82)$$

Customer's optimal strategy: This is determined in exactly the same way as in Model 8.1. The customer will buy the product only if

$$p_p^r \leq RL - p_r[\Lambda(L) - \Lambda(W)]. \quad (8.83)$$

The customer will be indifferent between buying the EW or not when

$$p_e^r(T) = p_r[\Lambda(W+T) - \Lambda(W)]. \quad (8.84)$$

Figure 8.6 shows this indifference curve in the $p_e^r(T)$ versus T diagram and also the customer's optimal decisions.

EWP's optimal strategy: This again is determined in exactly the same way as in Model 8.1. Thc optimal product price is given by

$$p_p^{r*} = RL - p_r[\Lambda(L) - \Lambda(W)]. \quad (8.85)$$

The EWP can choose any EW length $T^* \in [0, L - W]$, and the optimal EW price is then given by

$$p_e^{r*}(T^*) = p_r[\Lambda(W+T^*) - \Lambda(W)]. \quad (8.86)$$

This strategy produces a maximum expected profit to the EWP given by

$$J_M(p_p^{r*}, T^*, p_e^{r*}(T^*); 1) = J_M(p_p^{r*}, T^*, p_e^{r*}(T^*); 2) = RL - c_m - c_r\Lambda(L). \quad (8.87)$$

Thus, the customer will always purchase the product and will then be indifferent between purchasing and not purchasing the EW. The expected profit to the customer is zero under this strategy, so the EWP is able to extract all the consumer surplus.

Model 8.12 (Stackelberg game)
The effect of the presence of risk attitude on customer decision-making is now investigated.

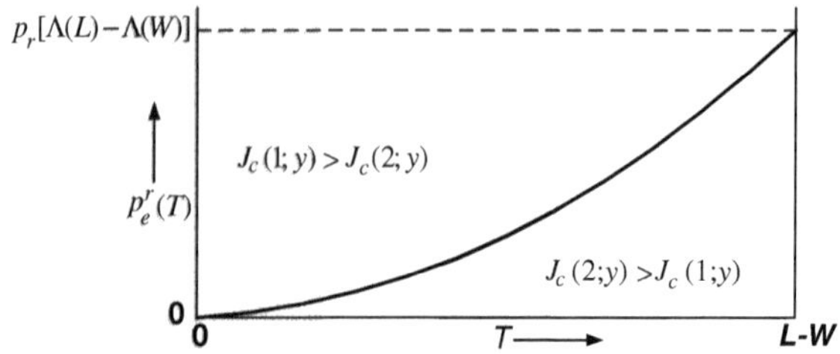

Fig. 8.6 Customer's optimal decisions in $p_e^r(T)$ versus T diagram

Assumptions: As in Model 8.2, the customers are now risk averse with utility function given by (8.11), while the manufacturer (EWP) remains risk neutral. The other assumptions are the same as in Model 8.11.

Key elements and decision variables: These are the same as in Model 8.11.

Objective functions: For a given value of y chosen by the EWP, the profit the customer will earn over the product's useful life is given by

$$Y_C(x;y) = \begin{cases} 0, & \text{if } x = 0, \\ RL - p_p^r - p_r[N(L) - N(W)], & \text{if } x = 1, \\ L - p_p^r - p_e^r(T) - p_r[N(L) - N(W+T)], & \text{if } x = 2. \end{cases} \tag{8.88}$$

The customer's expected utility function $J_C(x;y)$ is derived using (8.11), conditioning on the number of failures that will occur in the intervals $[W, L)$ (if $x = 1$) and $[W + T, L)$ (if $x = 2$) and then removing the conditioning. After some simple manipulation, we find that

$$J_C(x;y) = \begin{cases} 0, & \text{if } x = 0, \\ \frac{1}{\gamma}\left[1 - e^{-\gamma(RL - p_p - p_r'[\Lambda(L) - \Lambda(W)])}\right], & \text{if } x = 1, \\ \frac{1}{\gamma}\left[1 - e^{-\gamma(RL - p_p - p_e(T) - p_r'[\Lambda(L) - \Lambda(W+T)])}\right], & \text{if } x = 2, \end{cases} \tag{8.89}$$

with $p_r' = [e^{\gamma p_r} - 1]/\gamma$. Note that p_r' is increasing in γ and is always $> p_r$ for all $\gamma > 0$.

For a given value of x chosen by the customer, the EWP's expected profit is given by (8.82).

Customer's optimal strategy: This is again determined using the same method as in Model 8.1. The customer will buy the product only if

$$p_p^r \le RL - p_r'[\Lambda(L) - \Lambda(W)] \tag{8.90}$$

The customer will be indifferent between buying the EW or not when

$$p_e^r(T) = p_r'[\Lambda(W + T) - \Lambda(W)]. \tag{8.91}$$

The indifference curve in the $p_e^r(T)$ versus T diagram is shown in Fig. 8.7. It is a steeper convex curve than in the risk-neutral case (Model 8.11) and at $T = L - W$ takes the value $p_r'[\Lambda(L) - \Lambda(W)]$ which is greater than $p_r[\Lambda(L) - \Lambda(W)]$, the risk-neutral equivalent.

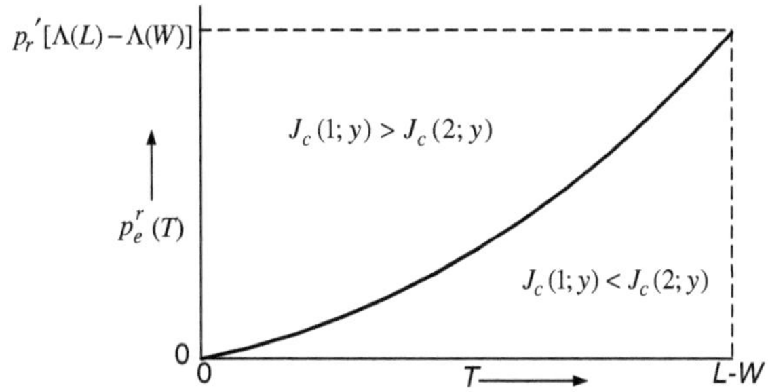

Fig. 8.7 Customer's optimal decisions (risk-averse case)

EWP's optimal strategy: The results are similar to the risk-neutral case (again using the method from Model 8.1) with p_r' replacing p_r. The optimal product price is given by

$$p_p^{r*} = RL - p_r'[\Lambda(L) - \Lambda(W)], \tag{8.92}$$

and this is smaller than the risk-neutral value. The EWP can choose any EW length $T^* \in [0, L - W]$, and the optimal EW price is then given by

$$p_e^{r*}(T^*) = p_r'[\Lambda(W + T^*) - \Lambda(W)], \tag{8.93}$$

which is greater than the risk-neutral value. This strategy produces a maximum expected profit to the EWP given by (8.87). The customer always purchases the product and is then indifferent between purchasing and not purchasing the EW. The customer's expected utility is zero under the EWP's strategy, so once again the EWP is able to extract all the consumer surplus.

Model 8.13 (Jack and Murthy)
Jack and Murthy (2007) discuss EW decision-making when customers have more flexibility in their EW choices. Optimal pricing strategies for the manufacturer (EWP) and optimal maintenance strategies for the customer after the expiry of the BW are derived.

Assumptions: Customers purchase a product directly from a monopolist manufacturer. Included with the sale is a BW of length W and the product has a maximum useful life $L > W$. Under the terms of the BW, the manufacturer will repair each failure of the product that occurs up to age W at no cost to the customer. The customer needs to decide how long to keep the product and how to maintain it from the time the BW expires until it is replaced. For the post-BW period, the customer can either pay the manufacturer to repair the product each time it fails or purchase an EW from the manufacturer. The customer has the flexibility to choose when the EW begins and the length of the cover. The EW terms are identical to those for the BW, so all product failures are again repaired by the manufacturer free of charge to the customer.

Every failure of the product from when it is first purchased until it is replaced by a new one is rectified immediately by the manufacturer performing a minimal repair, and all the repair times are small compared to the times between product failures and so can be ignored. If $N(t)$ denotes the number of product failures that occur in the time interval $[0, t)$ then, under the minimal repair assumption, $\{N(t);\ t \geq 0\}$ is a non-homogeneous Poisson process (NHPP) with mean function $\Lambda(t)$.

The customers are homogeneous in terms of risk attitude, usage intensity, etc. The customers and the manufacturer are risk neutral and have complete information about the reliability of the product and all relevant production and servicing costs.

Key elements and decision variables: The manufacturer generates profits from product sales, EW sales and repairs when a product fails during the post-BW period, and the failure is not covered by an EW. The average cost of each repair to the manufacturer is c_r. The manufacturing cost per unit of product is c_m, and the price that the manufacturer sells the product to the customer is given by

$$p_p^r = c_m + c_r\Lambda(W) + c_0. \tag{8.94}$$

This selling price is exogenous and so the manufacturer's profit per unit sale c_0 is a fixed quantity.

The customer decides to purchase an EW that begins when the product is of age $\tau \geq W$ and ends at age T $(\tau \leq T \leq L)$ when the product is replaced by a new one. If the product fails between age W and age τ, the customer pays the manufacturer to repair the failure, and the price of the repair is p_r. Thus, the average profit that the manufacturer makes on each repair before the EW begins is $c_1 = p_r - c_r$. The price that the manufacturer charges for the EW is given by

$$p_e^r(\tau, T) = c_r[\Lambda(T) - \Lambda(\tau)] + c_2(T - \tau), \tag{8.95}$$

where c_2 is the manufacturer's profit earned per unit time on the EW sale.

The sets of decision variables for the manufacturer and customer are $y \equiv \{c_1, c_2\}$ and $x \equiv \{\tau, T\}$, respectively.

Objective functions: The manufacturer needs to select y optimally in order to maximise the asymptotic expected profit per unit time given by

$$J_M(y; x) = \frac{c_0 + c_1[\Lambda(\tau) - \Lambda(W)] + c_2(T - \tau)}{T}. \tag{8.96}$$

The customer selects the optimal x that minimises the asymptotic expected cost per unit time given by

$$J_C(x;y) = \frac{p_p^r + p_r[\Lambda(\tau) - \Lambda(W)] + p_e^r(\tau,T)}{T} = \frac{c_m + c_r\Lambda(T) + c_0 + c_1[\Lambda(\tau) - \Lambda(W)] + c_2(T-\tau)}{T}. \tag{8.97}$$

Optimal Decisions: The manufacturer (EWP) is the leader, and the customer is the follower in the Stackelberg game between the two parties. For a given y, the manufacturer first determines the customer's best response function $x^*(y)$ for EW purchase and replacement of the product by minimising $J_C(x;y)$. The manufacturer's optimal strategy y^* is found by maximising $J_M(y;x^*(y))$, and the customer's optimal strategy is then given by $x^*(y^*)$.

A complete characterisation of the customer's best response function $x^*(y)$ is obtained when the time to first failure of the product has a Weibull distribution with scale parameter $\theta > 0$ and shape parameter $\beta > 1$. This implies that the number of product failures that occur in the time interval $[0,t)$ is an NHPP with mean function $\Lambda(t) = (t/\theta)^\beta$ and intensity function $\lambda(t) = (\beta/\theta)(t/\theta)^{\beta-1}$. Even in this special case, $x^*(y)$ is such a complicated function of y that it is impossible to derive any analytical results for the manufacturer's optimal strategy y^*.

Jack and Murthy (2007) give a numerical example to illustrate the optimal strategies for the manufacturer and the customer as the customer's budget per unit time for owning and maintaining the product changes. The effect of having risk averse instead of risk-neutral customers on the optimal strategies of both parties is also investigated.

Model 8.14 (Lam and Lam)
In Lam and Lam (2001), customers again purchase a product directly from a monopolist manufacturer and included in the sale of the product is a BW of length W. When the BW expires, the customers have the option to purchase an EW of length W_e from the manufacturer. This purchasing option continues to be available to the customers when each EW expires. Under the terms of the BW or any EW, the manufacturer agrees to repair each product failure at no cost to the customers. Instead of choosing to purchase one of more EWs, the customers may decide to pay the manufacturer to rectify any failure after the BW expires.

Assumptions: As in Model 8.13, the customer population is homogeneous with respect to risk attitude, usage intensity, etc. The customers and the manufacturer are risk neutral, and both parties have complete information about the reliability of the product and all relevant production and servicing costs.

Each customer has two possible options:

1. Adopt a k-renewal policy—purchase the EW from the manufacturer at times $W, W+W_e, \ldots, W+(k-1)W_e$ and then replace the product with a new and identical product at time $W+kW_e$ $(k=0,1,2,\ldots)$.
2. Adopt a k-repair policy—do not purchase any EWs but instead pay the manufacturer to repair the product k times after the expiry of the BW and then

replace the product with a new and identical product at the time of the $(k+1)$th failure $(k = 0, 1, 2, \ldots)$.

Every time a product fails from when it is first purchased until it is replaced, the failure is rectified immediately by the manufacturer performing a "perfect" repair (this restores the item to "as good as new" condition), and all the repair times are small compared to the times between product failures and so can be ignored. If $N(t)$ denotes the number of product failures that occur in the time interval $[0, t)$ then, under the perfect repair assumption, $\{N(t); t \geq 0\}$ is a renewal process with renewal function $M(t) = E[N(t)]$.

Key elements and decision variables: As in Model 8.13, the manufacturing cost per unit of product is c_m, the price that the manufacturer sells the product to the customer is $p_p^r \geq c_m$ and the average cost of each repair to the manufacturer is c_r. The variables c_m, p_p^r and c_r are assumed to be exogenous. The price that the manufacturer charges for an EW is p_e^r. If the product fails after the BW expires and the customer has not purchased an EW then the customer pays the manufacturer the amount p_r to repair the failure.

The set of decision variables for the manufacturer is $y \equiv \{p_e^r, p_r\}$. These two prices (for an EW and a repair) are assumed to satisfy the inequalities

$$p_e^r \leq \alpha p_p^r \text{ and } c_r \leq p_r \leq \beta p_p^r, \ (\alpha, \beta < 1). \tag{8.98}$$

The customer has to decide which maintenance policy (1 or 2) to use after the BW expires together with the best value of k. Thus, the set of decision variables for the customer is $x \equiv \{(j, k), j = 1, 2\}$.

Objective functions: If the customer adopts a *k-renewal policy*, then the customer's asymptotic expected cost per unit time using this policy is given by

$$J_C((1, k); y) = \frac{p_p^r + kp_e^r}{W + kW_e}. \tag{8.99}$$

Under this option from the customer, the manufacturer's asymptotic expected cost per unit time is given by

$$J_M(y; (1, k)) = \frac{c_m + c_r M(W + kW_e) - p_p^r - kp_e^r}{W + kW_e}. \tag{8.100}$$

If a *k-repair policy* is adopted, the customer's asymptotic expected cost per unit time using this policy is given by

$$J_C((2, k); y) = \frac{p_p^r + kp_r}{\mu[M(W) + k + 1]}, \tag{8.101}$$

and the manufacturer's asymptotic expected cost per unit time is given by

$$J_M(y;(2,k)) = \frac{c_m + c_r(M(W)+k) - p_p^r - kp_r}{\mu[M(W)+k+1]}, \tag{8.102}$$

where μ is the mean time between product failures.

Optimal Decisions: The manufacturer is the leader, and the customer is the follower in the Stackelberg game between the two parties. For a given prices for an EW and a repair (y), the manufacturer determines the customer's best response function with regards to maintenance policy adoption ($x^*(y)$) by minimising $J_C(x;y)$. The manufacturer's optimal strategy y^* is found by maximising $J_M(y;x^*(y))$, and the customer's optimal strategy is then given by $x^*(y^*)$.

For a given y, $J_C((1,k);y)$ and $J_C((2,k);y)$ are both minimised at either at $k=0$ or at $k=\infty$. For the *k-renewal policy*, $J_C((1,0);y) = p_p^r \big/ W$ and $J_C((1,\infty);y) = p_e^r/W_e$, so the optimal value of k for the customer is given by

$$k^*(y) = \begin{cases} 0 & \text{if } p_e^r \geq \alpha_1 p_p^r, \\ \infty & \text{if } p_e^r \leq \alpha_1 p_p^r, \end{cases} \quad \text{where } \alpha_1 = W_e/W. \tag{8.103}$$

For the *k-repair policy*, $J_C((2,0);y) = p_p^r \big/ \mu[M(W)+1]$ and $J_C((2,\infty);y) = p_r/\mu$, so the optimal value of k for the customer is given by

$$k^*(y) = \begin{cases} 0 & \text{if } p_r \geq \beta_1 p_p^r, \\ \infty & \text{if } p_r \leq \beta_1 p_p^r, \end{cases} \quad \text{where } \beta_1 = 1/[M(W)+1]. \tag{8.104}$$

Also, since $\mu[M(W)+1] \geq W$, it follows that $J_C((2,0);y) \leq J_C((1,0);y)$ and so a 0-*repair policy* is always preferable to a 0-*renewal policy*.

Thus, the optimal choice for the customer is either a 0-*repair policy*, an ∞-*repair policy* or an ∞-*renewal policy*. The customer is indifferent between an ∞-*repair policy* and a 0-*repair policy* if $p_r = \beta_1 p_p^r$. Indifference between an ∞-*repair policy* and an ∞-*renewal policy* occurs if $p_r/\mu = p_e^r/W_e \Rightarrow p_e^r = \delta_1 p_r$ where $\delta_1 = W_e/\mu$. Indifference between a 0-*repair policy* and an ∞-*renewal policy* occurs if $p_p^r/\mu[M(W)+1] = p_e^r/W_e \Rightarrow p_e^r = \beta_1\delta_1 p_p^r$. The regions where each maintenance policy is optimal for the customer are shown in Fig. 8.8. At the point I, the customer is indifferent between the three policies. The asymptotic expected cost per unit time for the customer at this point is $p_p^r \big/ \mu[M(W)+1]$.

The manufacturer's asymptotic expected costs per unit time in the three optimal policy regions (∞-*renewal*, 0-*repair* and ∞-*repair*) for the customer are

$$J_M(y;(1,\infty)) = \frac{c_r}{\mu} - \frac{p_e^r}{W_e}, \tag{8.105}$$

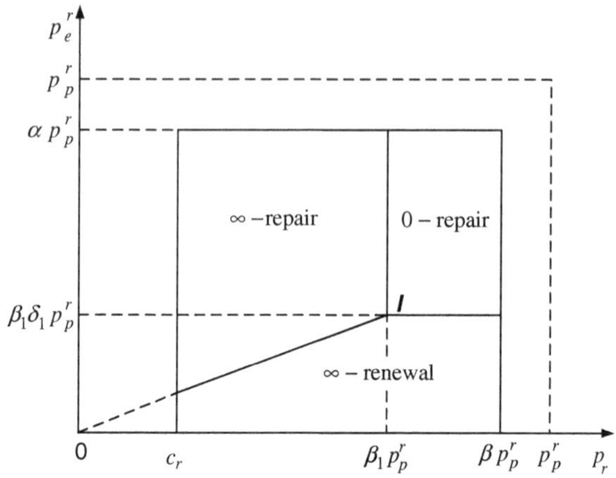

Fig. 8.8 Customer's optimal decisions

$$J_M(y;(2,0)) = \frac{c_m + c_r M(W) - p_p^r}{\mu[M(W)+1]}, \tag{8.106}$$

and

$$J_M(y;(2,\infty)) = \frac{c_r - p_r}{\mu}, \tag{8.107}$$

respectively.

The objective function in (8.106) does not depend on $y = \{p_e^r, p_r\}$, and so its value $d_1 = c_m + c_r M(W) - p_p^r / \{\mu[M(W)+1]\}$ is a constant and cannot be influenced by the manufacturer. The objective functions in (8.105) and (8.107) are minimised when the manufacturer chooses $p_e^r = \beta_1 \delta_1 p_p^r$ and $p_r = \beta_1 p_p^r$, respectively, and their common minimum value is $d_2 = \{c_r/\mu\} - \left\{\beta_1 \delta_1 p_p^r / W_e\right\}$. The manufacturer's optimal pricing strategy $y^* = \{p_e^{r*}, p_r^*\}$ (and hence the customer's optimal maintenance policy) is found by comparing d_1 and d_2.

Model 8.15 (Hartman and Laksana)
Hartman and Laksana (2009) consider EWs for a product which differ in their design according to when and how many times they can be purchased. A Stackelberg game formulation is used to determine the optimal strategies for customers and EW providers.

Assumptions: A product is sold to customers and included with the sale is a BW of length W periods, where a period may be a year, a month or something more frequent. The purchase price of the product is p_p^r, and the product has a maximum useful life of N periods. A customer can purchase an EW of length W_1 periods from a third-party EW provider at price p_e^r and the EW may be renewed.

Each time the product fails, a minimal repair is performed to make the product operational again and repair times are negligible. Under these assumptions, the sequence of product failures follows an NHPP. If the product is not covered by a warranty (either the BW or an EW), then a customer must pay the amount c_r to have the failure rectified. The parameters p_p^r (chosen by the manufacturer) and c_r are exogenous.

Four different types of EW may be offered by the EW provider:

1. An *unrestricted EW*—the customer can purchase an EW at any time after the BW expires.
2. A *restricted, non-deferrable EW*—the customer can only purchase an EW within F periods from the expiry of the BW.
3. A *restricted, non-renewable EW*—the customer cannot purchase an EW after the product reaches age N_1 periods.
4. A *restricted, non-renewable and non-deferrable EW*—the customer can only purchase an EW when the BW expires, and it cannot be renewed.

Key elements and decision variables: The customer uses the product over a time horizon of T periods. The set of decision variables for the customer is $x = \{n_1, n_2, n_3\}$, where n_1 is the first period in which an EW is purchased, n_2 is the last period in which an EW is purchased and n_3 is the age at which the product is replaced by a new one. The decision variable for the EW provider is $y = \{p_e^r\}$.

Objective functions and optimal decisions: The third-party EW provider is the leader, and the customer is the follower in the Stackelberg game between the two parties. For a given value of y chosen by the EW provider, the customer chooses $x^*(y)$, the value of x which minimises total expected discounted costs $J_C(x;y)$ over the time horizon T. The EW provider's optimal strategy y^* is found by maximising expected annual profits $J_{TP}(y;x^*(y))$, and the customer's optimal strategy is then given by $x^*(y^*)$.

Hartman and Laksana (2009) use an unusual method to model the effect of risk attitude on customer decision-making. Instead of making use of utility functions, an adjustment is made instead to the expected number of failures that a customer believes the product will experience in a given time period. If the product is of age n (periods), the expected number of failures that will occur between age n and age $n+1$ is given by

$$\tilde{M}(n) = \sum_{j=0}^{\infty} j^{\gamma} P_n(j), \tag{8.108}$$

where $P_n(j)$ is the probability of j failures occurring and the parameter γ captures a customer's risk attitude. $\gamma < 1$ implies that the customer is risk loving, $\gamma = 1$ implies risk-neutral behaviour and $\gamma > 1$ implies risk aversion.

Consider first the case where an *unrestricted EW* offered by the EW provider. The product is defined to be in state (n, w) at the end of a period if it is of age n (periods), and the remaining amount of warranty coverage is w (periods). Now assume that t periods have elapsed in the time horizon which has length T periods. The "cost-to-go" function $v_t(n, w)$ is defined to be the customer's minimum expected discounted cost when the product is in state (n, w) and optimal decisions are made from time t to the end of the horizon.

If α is the discount factor per period, the values of $v_t(n, w)$ are computed as follows:

$$v_t(n, w) = \alpha v_{t+1}(n+1, w-1), \quad n \geq 0,\ 1 \leq w \leq \max(W, W_1) \tag{8.109}$$

$$v_t(n, 0) = \min \left\{ \begin{array}{l} \mathrm{EW}: p_e^r + \alpha v_{t+1}(n+1, W_1 - 1) \\ K: c_r M(n) + \alpha v_{t+1}(n+1, 0) \\ R: p_p^r + \alpha v_{t+1}(1, W-1) \end{array} \right\}, \quad n < N \tag{8.110}$$

(In the above equation—EW: purchase extended warranty; *K*: keep the product without purchase of EW; *R*: replace the product.)

$$v_t(N, w) = p_p^r + \alpha v_{t+1}(1, W-1), \quad 0 \leq w \leq \max\ (W, W_1) \tag{8.111}$$

$$v_T(n, w) = 0, \quad \forall\ n, w \tag{8.112}$$

If there is an active warranty, Eq. (8.109) defines the transition from state (n, w) to $(n+1, w-1)$. The customer has no decision to make, incurs no immediate cost and the product ages one period. Equation (8.110) applies when the product has not reached its maximum age and a warranty has just expired. In this case, the customer can either keep the product and purchase an EW (EW), keep the product without purchasing an EW (*K*) or replace the product (*R*). Under the first option, the cost of the EW is paid and no further costs are incurred over the next W_1 periods. If the second option is chosen, the expected repair costs for the next period are paid, and the same set of decisions have to be made at the end of the period. Finally, if the third option is chosen, the purchase cost is paid and a BW takes effect, providing cover for the next *W* periods. The minimum of the three cost expressions defines the best decision for the customer. Equation (8.111) applies when the product has reached its maximum age of *N* periods and so must be replaced by the customer. Equation (8.112) is the terminal condition, stating that there is no cost incurred at the end of the time horizon.

Equations (8.109)–(8.112) are termed the "*optimality*" equations. In an example, Hartman and Laksana (2009) compute the functions $v_t(n, w)$ numerically to obtain the customer's optimal strategy $x^*(y) = \{n_1^*(y), n_2^*(y), n_3^*(y)\}$ for different EW prices $y = \{p_e^r\}$, a fixed product price p_p^r and different values of the repair cost c_r and the risk-attitude parameter γ.

The equivalent optimality equations for the customer for the other three types of EW are also given. These equations are similar to those for the unrestricted case with the only modification needed being made to the second equation to deal with the EW restrictions. In an example, the customer's revised optimal strategy is computed numerically for the *restricted, non-renewable and non-deferrable EW*.

The EW provider's optimal strategy $y^* = \{p_e^{r*}\}$ is examined in an example, where each type of EW is offered to the customer. For each EW offered, a different

EW price $y = \{p_e^r\}$ results in a different optimal strategy $x^*(y)$ for the customer. The optimal EW price for the provider (which maximises the expected annual profits $J_{TP}(y; x^*(y))$ is found by evaluating the profit function numerically over a range of values of y.

Finally, the above optimal EW pricing and customer behaviour analysis is extended to the case, where the customer population is heterogeneous in risk attitude. The population is assumed to be divided into a number of distinct groups with different risk attitudes in each group. Using integer programming models in a large number of numerical examples, it is shown that the EW provider can increase profits substantially by offering menus of different types of EWs to the customers.

8.5 GT Models for MSC Decision-Making

There are only a small number of papers that use a game-theoretic approach to MSC decision-making.

Model 8.16 (Murthy and Asgharizadeh)
Murthy and Murthy and Ashgarizadeh (1998) study a simple MSC involving a single service agent and a single customer (equipment owner) who requires the maintenance service.

Assumptions: The purchase price of the equipment is p_p^r. It generates a revenue of R per unit time for the customer when it is operating and has a useful life L. The customer might decide not to purchase the equipment if the purchase cost and maintenance costs exceed the revenue generated (Option 0). If the equipment is purchased, the agent offers two options to the customer for carrying out CM during the period $[0, L)$:

Option 1 (service contract): For a fixed price p_s^r, the agent will repair all failures over the equipment's useful life. If the time taken to carry out a repair T is greater than τ, then the agent will incur a penalty and will have to pay the amount $\alpha(T - \tau)$ to the customer.

Option 2 (no service contract): The customer will pay the agent p_r to repair each failure. Under this option, there will be no penalty incurred for long repair times.

The time to first failure of the equipment is exponentially distributed with mean $1/\lambda$, and all failures are minimally repaired by the agent. The time the agent takes to carry out a repair is exponentially distributed with mean $1/\mu$. The actual cost of each repair to the agent is c_r. This implies that no PM action is needed and that only CM action is carried out on failure.

Let $N(L)$ denote the number of equipment failures that occur over the period $[0, L)$ and T_i $(1 \le i \le N(L))$ denote the time the agent takes to complete the ith repair. We assume that $1/\mu \ll 1/\lambda$, so the total revenue generated by the equipment over its useful life can be approximated by RL. Note also that, under this assumption, $N(L)$ has a Poisson distribution with mean λL.

The agent and the customer both have complete information regarding the model parameters. The agent is risk neutral but the customer is risk averse with utility function given by (8.11). The agent is the leader, and the customer is the follower in the Stackelberg game that takes place between the two parties.

Key elements and decision variables: The set of decision variables for the agent is given by $y \equiv \{p_r, p_s^r\}$, and the customer's decision variable is

$$x = \begin{cases} 0 & \text{if the equipment is not purchased (Option 0),} \\ 1 & \text{if the equipment is purchased and Option 1 is chosen for CM} \\ 2 & \text{if the equipment is purchased and Option 2 is chosen for CM.} \end{cases}$$

Objective functions: For a given value of y chosen by the agent, the profit the customer will earn over the equipment's useful life is given by

$$Y_C(x;y) = \begin{cases} 0, & \text{if } x = 0, \\ RL + \alpha\left[\sum_{i=0}^{N(L)} \max\{0, T_i - \tau\}\right] - p_p^r - p_s^r, & \text{if } x = 1, \\ RL - p_p^r - p_r N(L), & \text{if } x = 2. \end{cases} \tag{8.113}$$

The customer's expected utility function $J_C(x;y)$ is derived using (8.11), conditioning on $N(L)$ and then removing the conditioning. After some manipulation, we find that

$$J_C(x;y) = \begin{cases} 0, & \text{if } x = 0, \\ \frac{1}{\gamma}\left[1 - e^{-\gamma(RL - p_p^r - p_s^r) + \lambda L e^{-\mu\tau}(\mu/(\gamma\alpha+\mu)-1)}\right], & \text{if } x = 1, \\ \frac{1}{\gamma}\left[1 - e^{-\gamma(RL - p_p^r) - \lambda L(1 - e^{\gamma p_r})}\right], & \text{if } x = 2. \end{cases} \tag{8.114}$$

For a given value of x chosen by the customer, the profit the agent will earn by providing the maintenance service for the period $[0, L)$ is given by

$$Y_A(y;x) = \begin{cases} 0, & \text{if } x = 0, \\ p_s^r - c_r N(L) - \alpha\left[\sum_{i=0}^{N(L)} \max\{0, T_i - \tau\}\right], & \text{if } x = 1, \\ (p_r - c_r)N(L), & \text{if } x = 2. \end{cases} \tag{8.115}$$

The service agent's expected profit is obtained by conditioning on $N(L)$ and then removing the conditioning. This gives

$$J_A(y;x) = \begin{cases} 0, & \text{if } x = 0, \\ p_s^r - \lambda L\left[c_r + \frac{\alpha}{\mu} e^{-\mu\tau}\right], & \text{if } x = 1, \\ (p_r - c_r)\lambda L, & \text{if } x = 2. \end{cases} \tag{8.116}$$

Customer's optimal strategy: For a given y chosen by the agent, a comparison of the three expected utilities given in (8.114) indicates which option is optimal for the customer. In the $p_r - p_s^r$ plane, the customer's optimal strategy $x^*(y)$ is characterised by the three regions Ω_0, Ω_1 and Ω_2 shown in Fig. 8.9. Note that $x^*(y) = i$ in region Ω_i.

The line separating Ω_0 and Ω_1 (where the customer is indifferent between Option 0 and Option 1) has equation

$$p_s^r = \bar{p}_s^r = RL - p_p^r + \frac{\alpha\lambda L}{\gamma\alpha + \mu} e^{-\mu\tau}. \tag{8.117}$$

The line separating Ω_0 and Ω_2 (where the customer is indifferent between Option 0 and Option 2) has equation

$$p_r = \bar{p}_r = \frac{1}{\gamma} \ln\left[1 + \frac{\gamma}{\lambda L}\left(RL - p_p^r\right)\right]. \tag{8.118}$$

The curve Γ separating Ω_1 and Ω_2 (where the customer is indifferent between Option 1 and Option 2) has equation

$$p_s^r = \frac{\lambda L}{\gamma}\left[e^{\gamma p_r} - 1 + \left(\frac{\gamma\alpha}{\gamma\alpha + \mu}\right) e^{-\mu\tau}\right]. \tag{8.119}$$

Note that, the region Ω_0 is defined by $p_r > \bar{p}_r$ and $p_s^r > \bar{p}_s^r$.

Agent's optimal strategy: In region Ω_1, the agent's expected profit is maximised when $p_s^r = \bar{p}_s^r$ and $p_r > \bar{p}_r$. These optimal values correspond to all points lying on the horizontal line separating Ω_1 and Ω_0. Similarly, in region Ω_2, the agent's earns maximum expected profit when $p_r = \bar{p}_r$ and $p_s^r > \bar{p}_s^r$. These optimal values correspond to all points lying on the vertical line separating Ω_2 and Ω_0. Thus, the agent's optimal strategy is to either (1) set $p_r^* = \bar{p}_r$ and $p_s^{r*} > \bar{p}_s^r$, or (2) set $p_s^{r*} = \bar{p}_s^r$ and $p_r^* > \bar{p}_r$. The choice that produces the larger expected profit for the agent is the optimal choice. In each case, the customer has zero expected utility so there is no consumer surplus. The agent is a monopolist maintenance service provider and so is able to extract the maximum possible amount from the customer.

Model 8.17 (Ashgarizadeh and Murthy)
Ashgarizadeh and Murthy (2000) extend the initial model of Murthy and Ashgarizadeh (1998) by considering multiple customers for the service agent. In this new model, when a customer's equipment fails its repair will not be started immediately if one or more equipment failures from other customers have already occurred. In this case, the number of customers to service M is an extra decision variable for the agent (in addition to the prices for a service contract and each repair). In Option 1 for maintenance service, the agent now incurs a penalty if the *total time taken* (waiting + repair) to restore failed equipment to the operational

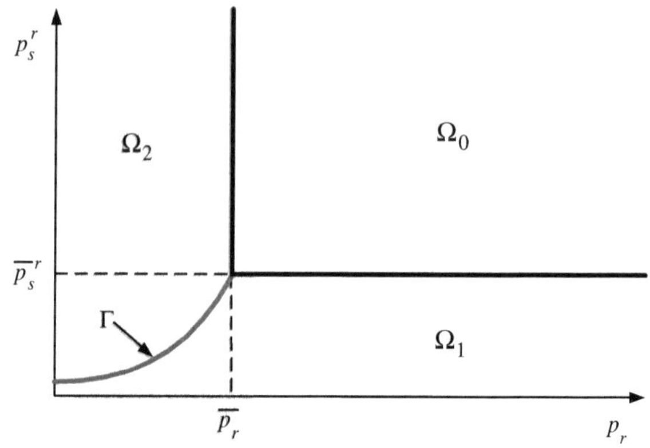

Fig. 8.9 Customer's optimal decisions

state is greater than a specified amount τ. The total time to restore a failed unit is a random variable and its characterisation involves results from an *M*/*M*/1 queue.[1]

The set of decision variables for the agent is $y \equiv \{p_r, p_s^r, M\}$. For each value of M, the customer's optimal strategy $x^*(y)$ is again characterised by the three regions Ω_0, Ω_1 and Ω_2. The equations of the horizontal line separating Ω_0 and Ω_1 and the curve Γ separating Ω_1 and Ω_2 both vary with M, but the equation of the vertical line separating Ω_0 and Ω_2 does not change. As M increases, the horizontal line and the curve both move upwards.

For a fixed M, the agent's optimal strategy is to either (1) set $p_r^*(M) = \bar{p}_r$ and $p_s^{r*}(M) > \bar{p}_s^r(M)$, or (2) set $p_s^{r*}(M) = \bar{p}_s^r(M)$ and $p_r^*(M) > \bar{p}_r$. The one that yields the higher expected profit for the agent is the optimal choice for this value of M. In both cases, the agent is able to extract the maximum amount from the customer, so there is no consumer surplus. The optimal value of M for the agent is determined numerically by comparing expected profits as M varies from 1 to the largest integer $\leq \mu/\lambda$.

Model 8.18 (Murthy and Ashgarizadeh)
One way to reduce customer waiting times for equipment to be repaired is to increase the number of maintenance personnel employed by the service agent (service channels). Failed equipment belonging to more than one customer can then be repaired at the same time, but there are additional (set-up) costs for the agent.

The number of service channels to use S is an extra decision variable for the agent, so $y \equiv \{p_r, p_s^r, M, S\}$. Murthy and Ashgarizadeh (1999) give a complete characterisation of the optimal strategies for the customer and the agent in this extended case. In this case, the characterisation of the time to restore a failed unit involves results from *M*/*M*/*S* queues.

[1] *M*/*M*/*S* is a queue with arrivals occurring according to a Poisson process (or inter-arrival times being exponentially distributed), service time exponentially distributed and S is the number of servers. Further details can be found in most books on queuing theory, see, for example, Gross and Harris (1974).

For each pair of values of M and S, the customer's optimal strategy $x^*(y)$ is once again characterised by the three regions Ω_0, Ω_1 and Ω_2. The equations of the horizontal line separating Ω_0 and Ω_1 and the curve Γ separating Ω_1 and Ω_2 both vary with M and S but the equation of the vertical line separating Ω_0 and Ω_2 does not depend on these values. For a fixed value of M, as S increases, the horizontal line and the curve both move downwards.

A two-stage approach is used to determine the agent's optimal strategy $y^* = \{p_r^*, p_s^{r*}, M^*, S^*\}$. For fixed M and S, the agent's optimal action is the choice between (i) $p_r^*(M,S) = \bar{p}_r$ and $p_s^{r*}(M,S) > \bar{p}_s^r(M,S)$ or (ii) $p_s^{r*}(M,S) = \bar{p}_s^r(M,S)$ and $p_r^*(M,S) > \bar{p}_r$. In the second stage, for a fixed S $(S = 1, 2, \ldots)$, the value $M^*(S)$ which maximises $J_A(p_r^*(M,S), p_s^{r*}(M,S), M, S; x^*)$ is found by using an exhaustive numerical search. Using this value of M, S^* is the value of S which maximises $J_A(p_r^*(M^*(S),S), p_s^{r*}(M^*(S),S), M^*(S), S; x^*)$ and this is also obtained from an exhaustive search. The optimal strategy for the agent is then $y^* = \{p_r^*(M^*(S^*),S^*), p_s^{r*}(M^*(S^*),S^*), M^*(S^*), S^*)$.

Model 8.19 (Murthy and Yeung)
Murthy and Yeung (1995) derive optimal strategies for a customer (equipment owner) and a service agent using a Stackelberg game formulation.

Assumptions: The customer and the service agent are both assumed to be risk neutral. The time to first failure of the equipment has distribution function $F(t)$ and the associated hazard function is $h(t)$ which increases with t. Thus, the likelihood of the equipment failing increases as it ages. The equipment generates a revenue of R per unit time for the customer when it is operating. O denotes the cost per unit time to the customer to use the equipment (whether it is operational or not).

The MSC involves both CM and PM actions carried out by the agent. Under the terms of the contract, the agent is required to replace the equipment by a new one (under PM action) during the interval $[v - \Delta, v + \Delta]$ subsequent to the previous maintenance (CM or PM) action. The cost of this planned replacement is $c_m + c_p$ where c_m is the manufacturing cost of the equipment, and c_p is the additional amount charged by the agent. The parameter $\Delta \geq 0$ measures the agent's service quality with the quality decreasing as Δ increases.

If the equipment fails before the time of the next PM action and its age at failure is less than v_1 $(< v - \Delta)$, then the agent carries out an immediate replacement (under CM action). The agent always has a spare piece of equipment available and charges the customer the amount $c_m + c_p + c_d$ for the failure replacement, and the customer incurs an additional cost of c_f for the equipment failure. If the age at failure is greater than v_1, then the equipment is left in the failed state until the PM action takes place.

It is assumed that the agent actually carries out a PM action when the age of the equipment since the previous maintenance action is given by $v - \Delta + \mathrm{T}$ where T is a random variable which is uniformly distributed over the interval $[0, 2\Delta]$. c_a denotes the agent's administration cost to carry out each maintenance action and c_i is the inventory holding cost per unit time for spare equipment.

Key elements and decision variables: The sets of decision variables for the customer and the agent are given by $x \equiv \{v, v_1\}$ and $y \equiv \{c_p, c_d\}$, respectively. Both parties need to choose these variables optimally in order to maximise their asymptotic expected profit earned per unit time. The agent is assumed to be the leader and the customer is the follower in the Stackelberg game. Given $y = \{c_p, c_d\}$, the customer chooses $x^*(y) = \{v^*(y), v_1^*(y)\}$. The service agent then chooses $y^* = \{c_p^*, c_d^*\}$ and the customer's optimal choice is $x^*(y^*)$.

Objective functions: Every time a maintenance action (CM or PM) occurs this constitutes a renewal point in an ordinary renewal process. The renewal reward theorem (see Appendix B) is then used to obtain the asymptotic expected profit earned per unit time by the customer and the service agent. This is the ratio of the expected profit earned per cycle to the expected cycle length, where a cycle is the time interval between two successive maintenance actions.

The customer's expected cycle profit is given by

$$\begin{aligned}\mathrm{ECP_C}(x;y) = {} & \frac{R}{2\Delta}\int_0^{2\Delta}\left\{\int_0^{v-\Delta+\tau} tf(t)\mathrm{d}t + (v-\Delta+\tau)\bar{F}(v-\Delta+\tau)\right\}\mathrm{d}\tau \\ & - \frac{O}{2\Delta}\int_0^{2\Delta}\left\{\int_0^{v_1} tf(t)\mathrm{d}t + (v-\Delta+\tau)\bar{F}(v_1)\right\}\mathrm{d}\tau \\ & - (c_m + c_p) - \frac{c_f}{2\Delta}\int_0^{2\Delta} F(v-\Delta+\tau)\mathrm{d}\tau - c_d F(v_1)\end{aligned} \tag{8.120}$$

where $f(t)$ and $\bar{F}(t)$ are the density function and the survivor function associated with $F(t)$.

The expected cycle length is given by

$$\mathrm{ECL_C}(x;y) = \frac{1}{2\Delta}\int_0^{2\Delta}\left\{\int_0^{v_1} tf(t)\mathrm{d}t + (v-\Delta+\tau)\bar{F}(v_1)\right\}\mathrm{d}\tau. \tag{8.121}$$

The customer's asymptotic expected profit per unit time is given by

$$J_C(x;y) = \frac{\mathrm{ECP_C}(x;y)}{\mathrm{ECL_C}(x;y)}. \tag{8.122}$$

The agent's expected cycle profit and asymptotic expected profit per unit time are given by

$$ECP_A(y;x) = c_p + c_d F(v_1) - c_a - \frac{c_i}{2\Delta}\int_0^{2\Delta}\left\{\int_0^{v_1} tf(t)\mathrm{d}t + (v - \Delta + \tau)\bar{F}(v_1)\right\}\mathrm{d}\tau \tag{8.123}$$

and

$$J_A(y;x) = \frac{\mathrm{ECP_A}(y;x)}{\mathrm{ECL_A}(y;x)}, \tag{8.124}$$

respectively, where $\mathrm{ECL_A}(y;x) \equiv \mathrm{ECL_C}(x;y)$.

Optimal Decisions: Murthy and Yeung (1995) do not derive the conditions which will ensure the existence of optimal strategies $x^* = \{v^*, v_1^*\}$ for the customer and $y^* = \{c_p^*, c_d^*\}$ for the service agent. They only give qualitative statements about the optimal results, stating that the agent's optimal strategy must result in zero asymptotic expected profit per unit time for the customer. Also, the agent's optimal strategy is not unique with different combinations of c_p^* and c_d^* yielding the same maximum asymptotic expected profit per unit time.

Model 8.20 (Jackson and Pascual)
Jackson and Pascual (2008) formulate an MSC GT model, where the objective of the service agent is to determine the pricing structure of the contract and the number of customers to service, whereas each customer (equipment owner) needs to specify the interval between PM actions, and the time when the equipment should be replaced.

Assumptions: The price a customer pays to purchase the equipment is p_p^r. The equipment generates a revenue of R per unit time for a customer when it is operating and it has a useful life L.

Under the MSC offered by the agent, CM and PM actions will be performed over the equipment's useful life for a fixed price p_s^r. The agent will incur a penalty if the total time taken T (= waiting time + repair time) to restore the failed equipment to its operational state is greater than a specified amount τ. If this happens (under CM action), then the agent will pay the amount $\alpha(T - \tau)$ to the customer.

The time to first failure of new equipment has hazard function $h_0(t) = a + bt, (a, b > 0)$. During each CM action, the failure is rectified by the agent performing a minimal repair. The time taken to complete each repair is exponentially distributed with mean $1/\mu$ and the repair cost to the agent is c_r. The agent also carries out periodic, imperfect PM actions (overhauls) at times $z, 2z, \ldots, (n-1)z$, and then, the equipment is replaced by new equipment at time $L = nz$. The equipment's hazard function for time to first failure after the jth $(j \geq 1)$ overhaul is given by

$$h_j(t) = ph_{j-1}(t - z) + (1 - p)h_{j-1}(t). \tag{8.125}$$

$p(0 < p < 1)$ is a factor that characterises the quality of the overhauls with the extreme cases $p = 0$ and $p = 1$ implying "minimal" overhauls and "perfect" overhauls, respectively. The cost of an overhaul to the agent is c_o.

The agent and the customers are risk neutral, and both parties have complete information regarding the model parameters.

Key elements and decision variables: The set of decision variables for the agent is given by $y \equiv \{p_s^r, M\}$, where M is the number of customers to service. For each customer, the set of decision variables is given by $x \equiv \{z, n\}$.

For customer j $(j = 1, 2, \ldots, M)$, let N_j denote the number of equipment failures that occur over the period $[0, L)$ and let T_{ij} $(i = 1, 2, \ldots, N_j)$ denote the time the agent takes to complete the ith repair. Failed equipment is repaired by the agent on a "*first-failed, first-repaired*" basis. The equipment's useful life $L = nz$ is sufficiently large so that the steady-state distribution for T_{ij} can be used in the analysis. We also assume that $\mu > aM$, so the queue of failed machines will not increase indefinitely with time.

Objective functions: The profit that the jth customer will earn over the equipment's useful life is given by

$$Y_C(x; y) = R\left(nz - \sum_{i=0}^{N_j} T_{ij}\right) + \alpha\left(\sum_{i=0}^{N_j} \max\{0, T_{ij} - \tau\}\right) - p_p^r - p_s^r. \tag{8.126}$$

The agent's profit earned by providing the maintenance service for the M customers is given by

$$Y_A(y; x) = \sum_{j=1}^{M}\left[p_s^r - c_r N_j - c_o(n - 1) - \alpha\sum_{i=0}^{N_j} \max\{0, T_{ij} - \tau\}\right]. \tag{8.127}$$

The expected profits for the jth customer and the agent are both obtained by conditioning on N_j and then removing the conditioning. After some extensive manipulation, the results are

$$\begin{aligned} J_C(x; y) = {} & R\left(nz - \hat{H}\sum_{k=0}^{M-1}\frac{(k+1)P_k}{\mu}\right) \\ & + \alpha\hat{H}\left[\sum_{k=0}^{M-1} P_k \mu^k e^{-\mu\tau}\left(\sum_{l=0}^{k}\frac{\tau^{k-l}(k+1-\mu\tau)}{\mu^{l+1}(k-l)!} + \frac{\tau^{k+1}}{k!}\right)\right] - p_p^r - p_s^r, \end{aligned} \tag{8.128}$$

and

$$J_A(y;x) = M\left[p_s^r - c_r\hat{H} - c_o(n-1)\right] - \alpha\hat{H}\left(\sum_{k=0}^{M-1} P_k\mu^k e^{-\mu\tau}\left(\sum_{l=0}^{k}\frac{\tau^{k-l}(k+1-\mu\tau)}{\mu^{l+1}(k-l)!} + \frac{\tau^{k+1}}{k!}\right)\right) \quad (8.129)$$

respectively. $\hat{H}$ is the expected number of failures over an equipment's useful life $L = nz$ and is given by

$$\hat{H} = anz + bz^2\left(\frac{n^2(1-p)+np}{2}\right). \quad (8.130)$$

P_k is the probability that k out of the M pieces of equipment have failed ($k = 0, 1, \ldots, M-1$) and is given by

$$P_k = \frac{(M-k)\left(\bar{h}/\mu\right)^k\{M!/(M-k)!\}}{\sum_{k=0}^{M-1}\left[(M-k)\left(\bar{h}/\mu\right)^k\{M!/(M-k)!\}\right]} \quad (8.131)$$

where $\bar{h} = \hat{H}/nz$.

Optimal Decisions: The case of a single customer ($M = 1$) is considered first. The agent has the single decision variable $y = \{p_s^r\}$ and the two parties/players (customer and agent) play a cooperative game. The *Nash bargaining solution* is obtained by equating the players' expected profits. The optimal price p_s^{r*} for the MSC satisfies the condition

$$R\left(nz - \frac{\hat{H}}{\mu}\right) + \alpha\hat{H}\frac{e^{-\mu\tau}}{\mu} - p_p^r - p_s^{r*} = p_s^{r*} - c_r\hat{H} - c_o(n-1) - \alpha\hat{H}\frac{e^{-\mu\tau}}{\mu} \quad (8.132)$$

which implies that

$$p_s^{r*} = \frac{R}{2}\left(nz - \frac{\hat{H}}{\mu}\right) + \alpha\hat{H}\frac{e^{-\mu\tau}}{\mu} + \frac{c_r}{2}\hat{H} + \frac{c_o}{2}(n-1) - \frac{p_p^r}{2}. \quad (8.133)$$

Substituting this optimal MSC price into the expression for the customer's expected profit gives

$$J_C\left(n,z;p_s^{r*}\right) = \frac{R}{2}\left(nz - \frac{\hat{H}}{\mu}\right) - \frac{c_r}{2}\hat{H} - \frac{c_o}{2}(n-1) - \frac{p_p^r}{2}. \quad (8.134)$$

The optimal values of the customer's decision variables n^* and z^* are found by maximising the modified objective function (asymptotic expected profit earned per unit time)

$$\frac{J_C(n,z;p_s^{r*})}{nz} = \frac{R}{2}\left(1-\frac{\bar{h}}{\mu}\right) - \frac{c_r}{2}\bar{h} - \frac{c_o}{2}\left(\frac{1}{z}-\frac{1}{nz}\right) - \frac{p_p^r}{2nz}. \tag{8.135}$$

This two-variable optimisation needs to be done numerically using an exhaustive search.

In the case of $M > 1$ customers, the Nash bargaining solution for the optimal MSC price p_s^{r*} satisfies the condition

$$J_C(n,z;p_s^{r*},M) = \frac{J_A(p_s^{r*},M;n,z)}{M}. \tag{8.136}$$

Using (8.128) and (8.129) in (8.136) and simplifying gives

$$\begin{aligned} p_s^{r*} &= \frac{R}{2}\left(nz - \hat{H}\sum_{k=0}^{M-1}\frac{(k+1)P_k}{\mu}\right) \\ &\quad + \alpha\hat{H}\left[\sum_{k=0}^{M-1}P_k\mu^k e^{-\mu\tau}\left(\sum_{l=0}^{k}\frac{\tau^{k-l}(k+1-\mu\tau)}{\mu^{l+1}(k-l)!}+\frac{\tau^{k+1}}{k!}\right)\right] + \frac{c_r}{2}\hat{H} + \frac{c_o}{2}(n-1) - \frac{p_p^r}{2}. \end{aligned} \tag{8.137}$$

Substituting this optimal MSC price into the expression for the agent's expected profit gives

$$J_A(p_s^{r*},M;n,z) = M\left[\frac{R}{2}\left(nz - \hat{H}\sum_{k=0}^{M-1}\frac{(k+1)P_k}{\mu}\right) - \frac{c_r}{2}\hat{H} - \frac{c_o}{2}(n-1) - \frac{p_p^r}{2}\right]. \tag{8.138}$$

The optimal values of the remaining decision variables M^*, n^* and z^* are found by maximising the modified objective function (asymptotic expected profit earned per unit time)

$$\frac{J_A(p_s^{r*},M;n,z)}{nz} = M\left[\frac{R}{2}\left(1 - \bar{h}\sum_{k=0}^{M-1}\frac{(k+1)P_k}{\mu}\right) - \frac{c_r}{2}\bar{h} - \frac{c_o}{2}\left(\frac{1}{z}-\frac{1}{nz}\right) - \frac{p_p^r}{2nz}\right]. \tag{8.139}$$

This three-variable optimisation must be done numerically using an exhaustive search.

References

Ashgarizadeh E, Murthy DNP (2000) Service contracts—a stochastic model. Math Comput Model 31:11–20

Desai PS, Padmanabhan V (2004) Durable good, extended warranty and channel coordination. Rev Mark Sci 2(2):1–23

Gross D, Harris CM (1974) Fundamentals of queuing theory. Wiley, New York

Hartman JC, Laksana K (2009) Designing and pricing menus of extended warranty contracts. Naval Res Logistics 56:199–214

Heese SH (2012) Retail strategies for extended warranty sales and the impact of manufacturer base warranties. Decis Sci 43:341–367

Jack N, Murthy DNP (2007) A flexible extended warranty and related optimal strategies. J Oper Res Soc 58:1612–1620

Jackson C, Pascual R (2008) Optimal service contract negotiation with aging equipment. Eur J Oper Res 189:387–398

Jiang B, Zhang X (2011) How does a retailer's service plan affect a manufacturer warranty? Manage Sci 57:727–740

Kameshwaran S, Viswanadham N, Desai V (2009) Bundling and pricing of product with after-sale services with product. Int J Oper Res 6:92–109

Kurata H, Nam SH (2010) After-sales service competition in a supply chain: optimisation of customer satisfaction level or profit or both? Int J Prod Econ 127:136–146

Lam Y, Lam PKW (2001) An extended warranty policy with options open to consumers. Eur J Oper Res 131:514–529

Li K, Mallik S, Chhajed D (2012) Design of extended warranties in supply chains under additive demand. Prod Oper Manage 21:730–746

Murthy DNP, Ashgarizadeh E (1998) A stochastic model for service contract. Int J Reliab Qual Saf Eng 5:29–45

Murthy DNP, Ashgarizadeh E (1999) Optimal decision making in a maintenance service operation. Eur J Oper Res 116:259–273

Murthy DNP, Yeung V (1995) Modelling and analysis of maintenance service contracts. Math Comput Modell 22:219–225

Part III
Lease Contracts

Chapter 9
Leasing and Maintenance of Leased Assets

9.1 Introduction

As discussed in Chap. 1, individuals, businesses and governments use a variety of engineered objects (products, plants and infrastructure) in their daily operations. The traditional approach has been to acquire an object using one of the following two methods:

1. Sale for cash: Outright purchase.
2. Conditional sales contract on a deferred payment plan.

There is a growing trend towards the use of alternative means to derive the benefits of an object using one of the following two methods:

1. Lease without an option to purchase (may also include a renewal option).
2. Lease and final purchase.

The findings of a survey conducted by the Equipment Leasing Association (ELA) in the USA in 2002 (ELA 2002a) report the following:

- 80 % of businesses acquire equipment through leasing.
- Leasing accounts for roughly 30 % of business capital investment.
- Nearly 50 % of office equipment is leased.
- Leasing companies own more equipment than companies in other US industries.

The ELA Online Focus Group Report (ELA 2002b) states that 60 % of leasing benefits come from maintenance options. This is because some leases come with maintenance as an integral part of the lease so that the physical equipment is bundled with maintenance service and offered as a package under a LC.

The maintenance of leased assets raises some new and interesting issues. The responsibility for the maintenance can be either with the owner or with the user, and they can either do it in-house or outsource it to some third-party external service agents. Maintenance decisions need to take into account the terms in the lease contract.

D. N. P. Murthy and N. Jack, *Extended Warranties, Maintenance Service and Lease Contracts*, Springer Series in Reliability Engineering,
DOI: 10.1007/978-1-4471-6440-1_9,

In this chapter, we focus on leasing and the maintenance of leased items. The outline of the chapter is as follows. Section 9.2 deals with an overview of leasing and discusses issues such as the key elements, reasons and advantages and disadvantages of leasing. There are many different types of LCs, and Sect. 9.3 looks at the classification and structure of these contracts. Leasing has been studied by researchers from many different disciplines, and Sect. 9.4 gives a brief review of the literature. Sections 9.5–9.7 deal with leasing of consumer products, industrial and commercial plants, and infrastructures, respectively. Section 9.8 looks at maintenance of leased items.

9.2 Leasing

A lease is a contractual agreement under which one party (the owner who is also referred to as the *lessor*) leases to another party (also referred to as the *lessee*) an engineered object (product, plant, infrastructure) for use as per the terms of the LC.[1]

According to Fishbein et al. (2000), there are several reasons for leasing and they include the following:

- Rapid technological advances have resulted in improved equipment appearing on the market, making the earlier generation equipment obsolete at an ever-increasing pace.
- The cost of owning equipment has been increasing very rapidly.
- Businesses viewing maintenance as a non-core activity.
- It is often economical to lease equipment, rather than buy, as this involves less initial capital investment, and often there are tax benefits that make it attractive.

According to Baker and Hayes (1981), some of the pioneers in business equipment leasing were IBM and Xerox. Since then, the number of businesses that lease business equipment has grown significantly, and many kinds of equipment are leased. ELA (2005) gives a list of some of the businesses leasing their products under operating leases.

9.2.1 *Key Elements of Leasing*

The four key elements of leasing are shown in Fig. 9.1. A brief discussion of each of these is as follows.

[1] The term *asset* is often used in the lease literature instead of engineered object and can include real estate such as houses, apartments, buildings, etc.

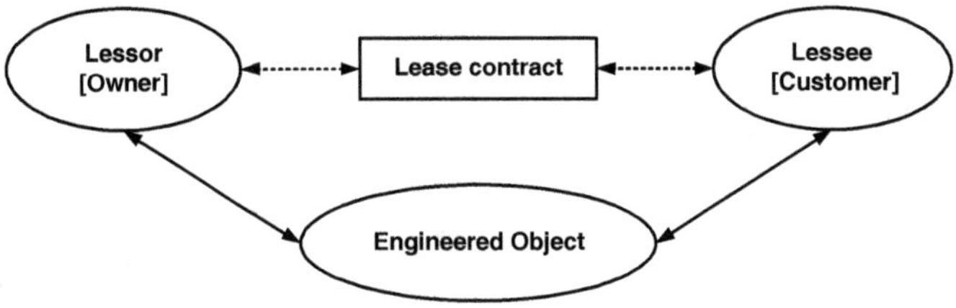

Fig. 9.1 Key elements of leasing

9.2.1.1 Engineered Object

Engineered objects that are leased include a range of

Consumer products Such as cars, white goods, furniture, computers as well as infrastructures.

Commercial and industrial products Used in various industry sectors and the following is a small illustrative sample:

- Mining: Trucks, pumps, fencing, mining machinery, etc.
- Transport: Trucks, ships, aircraft, buses, cars, etc.
- Manufacturing: Various kinds of machinery, forklifts, etc.
- Hospital: Washing machines, machines used for diagnostics, monitoring, etc.
- Restaurants/Hotels: Kitchen appliances, furniture, etc.

Infrastructures In various sectors as illustrated by the following sample:

- Buildings.[2]
- Transport: Rail, road, etc.
- Utilities: Power, water, gas, etc.

9.2.1.2 Lessor

The number of companies in the leasing business has increased dramatically due to the increase in both individuals and businesses acquiring various kinds of engineered objects through lease arrangements. Lessors include the following:

- Financial institutions: Commercial banks, insurance companies and finance companies do most of the leasing. Many of these organisations have subsidiaries that are primarily concerned with equipment leasing and make lease arrangements for almost anything.
- Companies that specialise in leasing. Some are engaged in general leasing (dealing with all kinds of equipment) while others specialise in particular

[2] This also includes rental of apartments and houses by individuals as well as buildings leased for commercial and industrial operations.

equipment (e.g., trucks, computers) and others deal with specific industry sectors (e.g., mining, health, transport).
- Equipment manufacturers leasing the equipment they manufacture.

9.2.1.3 Lessee

The lessee is the party leasing the equipment from the lessor. A lessee can be

- An individual household (for consumer products).
- A business (for products, plants and infrastructures).
- A government agency (for products, plants and infrastructures).

9.2.1.4 Lease Contract

There are several different types of contracts and we will discuss them in the next section.

9.2.2 Reasons for Leasing

According to Schallheim (1994), the reasons for leasing (as promoted by leasing companies and from the lessee perspective) can be broadly grouped into four categories:

1. *Tax savings* Leasing offers tax advantages for businesses with excess tax shields and/or low taxes. Depending on the tax system, leasing may also permit more rapid amortisation than depreciation.
2. *Pure financial cost savings* The advantages include 100 % financing, off-balance sheet financing, lower initial outlay and a cheaper way to acquire an asset.
3. *Transaction and information cost savings* Leasing requires less book keeping, avoids a purchase transaction and offers greater convenience and flexibility, etc.
4. *Risk sharing* Protection against asset obsolescence, hedging against inflation and business risk avoid the sale of the asset when no longer needed, etc.

9.2.3 Advantages and Disadvantages of Leasing

There are certain advantages and disadvantages in leasing for both the lessor and the lessee. We list a few of these.

9.2.3.1 Lessee Perspective

The advantages are as follows:

- The lessee obtains new equipment (based on the latest technologies) and thus avoids the risks associated with equipment obsolescence.
- The lessee is able to spread the payments over the lease period (no need for initial capital).
- Leasing offers greater flexibility as the lessee can choose from a range of lease options.
- Depending on the LC, the lessor provides the maintenance and other supports and pays the taxes, insurance, etc.

The disadvantages are as follows:

- If the lessee's needs change over the lease period, then premature termination of the lease agreement can incur penalties.
- The risks to the lessee should the lessor not provide the level of maintenance needed.
- If the lessee fails to make lease payments as per schedule, the leased equipment can be repossessed.
- If maintenance is not a part of the lease agreement, then the lessee has to provide for this separately.
- The overall cost to the lessee is significantly higher than the purchase price of the equipment because the lessor needs to allow for not only the financing costs but also pay for other costs associated with insurance, taxes, etc.

9.2.3.2 Lessor Perspective

The advantages are as follows:

From the lessor's perspective, leasing implies investment (in financing in the case of a finance lease or in equipment in the case of an operational lease) with the expectation of sufficient revenue and profits.

The disadvantages are as follows:

A critical issue is the risk for the lessor and there are several different types of risks. These are discussed in Chap. 11.

9.3 Leases: Classification and Contracts

There are several types of leases but, unfortunately, there is no standard terminology in English-speaking countries. The terms used in the USA often differ from those used in the UK. There are two ways of classifying leases and they are as follows.

9.3.1 Accounting (Lessee) Perspective

From an accounting perspective, leases are classified as either (1) capital leases or (2) operating leases, and they have an implication for the balance sheet of the business that is leasing. As such, this can also be viewed as the lessee perspective.

Capital leases: These are leases that meet *one or more* of the following criteria:

1. The lease transfers ownership of the property to the lessee by the end of the lease.
2. The lease contains a bargain purchase option.
3. The lease is equal to 75 % or more of the estimated economic life of the leased asset.
4. The present value at the beginning of the lease term of the minimum lease payments is at least 90 % of the fair value of the leased asset to the lessor at the inception of the lease (over and above any related investment tax credit retained by the lessor).

Operating leases: All leases meeting none of the four criteria that define a capital lease.

9.3.2 Lessor Perspective

From the lessor's perspective, there are several types of leases.

9.3.2.1 Finance Leases

In a finance lease, the lessee pays the lessor for the use of equipment over a specified period. At the end of the lease period, the lessee acquires the ownership of the equipment either at no cost or at a previously established price. The type of equipment sold with this type of lease can vary from very expensive industrial and commercial equipment (such as a financial institution leasing aircraft to an airline operator) to less expensive consumer products (banks or retailers leasing domestic appliances, cars, etc. to households).

The main characteristics of a finance lease are as follows:

1. The primary lease period is usually a significant portion of the useful life of the equipment.
2. The cost of the equipment is recovered during the primary lease period through periodic payments.
3. The lease is non-cancellable and the lessee has a legal obligation to continue payments to the end of the term.
4. Service, maintenance, taxes, insurance, etc. are the responsibility of the lessee.
5. Ownership of the equipment reverts to the lessor at the end of the lease term.

9.3.2.2 Operating Leases

There is no exact definition of an operating lease. In an operating lease, the lessee pays the lessor for the use of equipment over a specified period. Both new and used products (consumer, commercial and industrial) are leased with operating leases. At the end of the lease period, the lessor retains ownership of the item and can renew the lease contract (if the lessee is interested), lease the item to some other lessee or sell the equipment as a second-hand item. Additional services, such as operator training (to ensure that the leased item is operated properly—for example, the leasing of specialised industrial equipment) and maintenance (to ensure that the equipment is in a proper operating condition and meets the requirements stated in the LC), taxes and insurance, are provided by the lessor as part of the lease contract.

The main characteristics of an operating lease are as follows:

1. The primary lease period is relatively short in comparison to the useful life of the equipment.
2. The contract can be cancelled during the primary lease period (under conditions defined in the LC).
3. Maintenance is performed by the lessor.
4. From a taxation point of view, the lessor is allowed to claim depreciation and the lessee to claim rental payments as tax deductions.

9.3.2.3 Leveraged Leases

A leveraged lease involves a third-party lender (consisting of one or more creditors) who supplies most of the funds for the lessor to finance the purchase of an item and then lease it out. According to Watts (1971):

> A leveraged lease is so named because the size of the financing involved can fundamentally alter the leverage ratio of the firm when looked at purely as credit.

This is a more complex form of lease often involving the pooling together of various lessors and lenders. It is discussed further in Sect. 9.7.

9.3.2.4 Sale and Leaseback Leases

Under a sale and leaseback lease, the owner sells the equipment to a lessor (usually a finance company) and leases it immediately without ever surrendering the use of the equipment.[3] The maintenance can be carried out either by the lessee or by some third-party independent service agents. This type of lease is used mainly for

---

[3] For more on this type of lease, see Sizer (1987).

infrastructure assets such as rail transport, electricity, sewerage and water pipe networks and buildings. The new owner (lessor) assumes the rights and benefits of ownership, including tax benefits of depreciation, tax credits and any residual value.

9.3.2.5 Other Types of Leases[4]

One important segment of the leasing market is tax-motivated leasing.

- *Venture lease* This is aimed at start-up firms who have limited access to capital markets.
- *Across national borders lease* These leases exploit complex tax laws in two or more countries.
- *Net lease* In a net lease, the lessee is responsible for expenses such as those for maintenance, taxes and insurance.
- *Full-payout lease* In a full-payout lease, the lessor pays expenses such as those for maintenance, taxes and insurance. Under this lease, the lessor recovers the original cost of the asset during the term of the lease.

9.3.3 Terms of Lease Contract

The LC needs to take into account the interests of both the lessor and the lessee. The contract spells out the precise provisions of the agreement. Agreements may differ, but most would include the following items[5]:

- Equipment: Description, model, serial number, date of manufacture, etc.
- Lease term: The start and end dates.
- Renewal options: If applicable.
- The specific nature of the financing agreement.
- Lease payments: Amount to be paid; frequency of payment (monthly, quarterly, etc.) and due date.
- Late charges: If lease payments are not made by due date.
- Security deposit: The lessor can use this amount to repair any damage to the equipment caused by the lessee. Should the lessee breach any terms of the contract the deposit is forfeited (subject to it not violating any law of the land).
- Delivery: The costs of delivery—borne by one party (lessor or lessee) or shared by both.

[4] For a discussion of other types of leases, see Coyle (2000) and ELA (2005).

[5] It can include other items (mostly legal terms) such as, Encumbrances, Lessor Representation, Severability, Assignment, Binding effect, Governing Law, Entire Agreement, Cumulative Rights, Waivers, Indemnification, etc.

- Default: This occurs when the lessee fails to meet the obligations under the contract. The contract defines the options available to the lessor, and this can include repossession of the equipment.
- Possession and surrender of equipment: Obligation of the lessee to return the equipment in good working condition accounting for normal wear and tear.
- Use of equipment: Rules and regulations with which the lessee needs to conform.
- Maintenance: Defines the maintenance to be carried out and the party who is responsible.
- Insurance: Defines the party who is responsible for covering various kinds of risks (fire, theft, collision, damage, etc.).
- Schedule of the value of the equipment for insurance and settlement purposes in case of damage or destruction as a function of age and/or usage.
- Additional terms and conditions: Relating penalties and/or incentives based on equipment performance.

A LC is divided into several sections. The following two are illustrative cases.

Contract 1
The contract is comprised of eight sections and the section headings are as given below[6]:

1. Terms of lease payment
2. Equipment procurement and delivery
3. Use, maintenance and insurance of equipment
4. Expiration or termination of lease, return of equipment
5. Warranties
6. Default and remedies
7. Financial information
8. Miscellaneous.

Contract 2
The contract is comprised of 3 main sections with several subsections as indicated below.[7]

Financial

1. Basis of rental
2. Initial payment in excess of rental
3. Installation charges
4. Taxes and assessment
5. Surcharge above basic rental and above normal operations
6. Maintenance[8]
7. Liability

---

[6] For more details of each section, see Schallheim (1994).

[7] For further details, see MAPI (1965).

[8] It is common for either the lessor or lessee to assume the responsibility of maintenance.

Operating

1. Supplies and accessories
2. Repair, alteration and removal of equipment

Legal

1. Title
2. Period of agreement
3. Purchase options.

9.3.4 Residual Value

The *residual value* (*salvage value*) is the amount a leased item is worth—its market value—at the maturity (end) of the lease. An important issue of concern for the lessor is the *residual value risk*.[9] This depends on usage and maintenance and is discussed in a later section. The management of risk is an important issue for both lessor and lessee and is discussed in Chap. 11. A lessee who is unwilling or unable to assume the risk can transfer the risk to the lessor. However, this will be at the expense of an increased lease payment.

9.4 Study of Leasing

Leasing has been studied by researchers from many different disciplines, and the literature on leasing is extensive.[10] The different topics studied can be categorised into the following groups: (1) legal, (2) accounting and finance, (3) economics, (4) marketing, and (5) management. A brief review of the literature is given below. We discuss each briefly and give some references where interested readers can get further details.

9.4.1 Legal

As mentioned previously, a lease is a contract. For consumer products, the contract can be fairly simple to complex. For industrial and commercial products and infrastructures, the complexity of the LC increases significantly (especially for

[9] There are many other kinds of risks in leasing. These include demand risk, financing risk, operating risk, regulatory risk, systematic risk and technological risk.

[10] There are many books, for example, Coyle (2000), Elgers and Clark (1980), Kaster (1979a, b), MAPI (1965), Schallheim (1994) and Wainman (1991) is a small sample.

leveraged leases). The drafting of the LCs requires a specialised legal background as a poorly drafted contract can lead to lots of legal problems. See Schallheim (1994) for more on tax rules for leasing and Chemmanur et al. (2010) deal with the theory of contractual provisioning.

9.4.2 Accounting and Finance

The main focus of accounting (for the lease of industrial and commercial products and of infrastructures) is the reporting of costs, revenues, depreciation and tax benefits of leasing in the balance sheets of both the lessor and the lessee. There are several books [for example, Baker and Hayes (1981); Elgers and Clark (1980); MAPI (1965); Kaster (1979a, b) and Schallheim (1994)] that deal with issues such as accounting for leases, financing of leases, guidelines to maximising financial and tax advantages.[11] More recent papers include Krishnan and Moyer (1994), Kleiman (2001) and Kong and Long (2001).

9.4.3 Economics

Economic theory deals with the behaviour of markets. A market for goods (products and/or services) is comprised of two parties—buyers and sellers. According to Schallheim (1994):

> The leasing market brings together buyers of lease contracts, the lessors (also known as owners), with sellers of lease contracts, lessees (also known as users). From the economic perspective, the lessor is the purchaser of the lease contract just as the lender is referred to as a purchaser of the debt contract.

The leasing market is influenced by the market for new and used products as shown in Fig. 9.2 for a vehicle lease where there are two types of customers—individuals (households leasing one vehicle) and corporate (businesses leasing a fleet of vehicles).

There is a vast literature dealing with leasing markets and an illustrative sample is the following: Chen and Huang (2005), Gavazza (2005), Gerety (1995), Handa (1991), Johnson and Waldman (2003), Lewellen et al. (1976), Sharpe and Nguyen (1995), Stremersch et al. (2001), and Waldman (1997).

[11] The *Journal of Equipment Lease Financing* is a journal devoted to the financing issues relating to leasing.

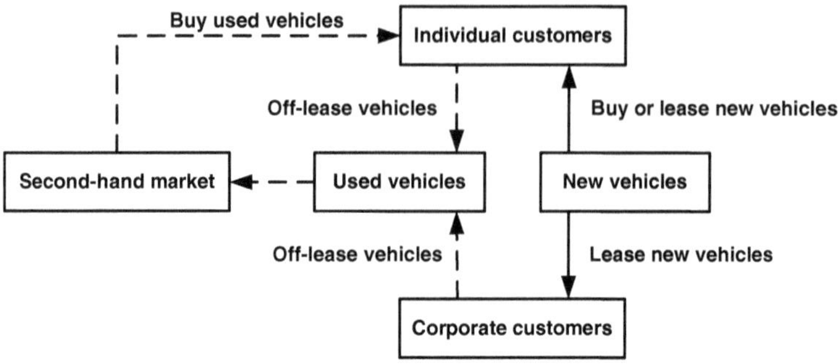

Fig. 9.2 Some of the key elements in the leasing market for vehicles

9.4.4 Marketing

The literature on the marketing of leases has focussed on two topics: (1) prices and (2) channels for the leasing of products. Prices, or lease payments in the economic context, are determined by supply and demand. As a result, a more competitive market (with several lessors) will result in lower prices for the customers (lessees).

A manufacturer who also is a lessor can lease the product either directly to the customer (lessee) or through a retailer. Often, the retailer might also act as a lessor so that there are multiple channels for product leasing. Figure 9.3 is one such scenario where the manufacturer and retailer are both lessors, and the manufacturer's lease involves the retailer in the channel.

There is considerable literature dealing with both pricing and channels for leasing and the following is an illustrative sample: Anderson and Bird (1980), Anderson and Lazer (1978), Aras et al. (2011), Bhaskaran and Gilbert (2005), Desai and Purohit (1998), Huang and Yang (2002), Purohit (1994, 1997), Purohit and Staelin (1994) and Tilson et al. (2006).

9.4.4.1 Remanufacturability

The returned off-lease goods (see Fig. 9.3) are sold in the second-hand markets. They can either sold in the condition they are in or subjected to a remanufacturing process and sold as remanufactured goods. In general, the remanufactured items are not as good as new but better than before. In a sense, this is similar to the returned items being subjected to imperfect maintenance resulting in an improvement in reliability. This topic has become important in the context of sustainability.

As such, both remanufacturing and second-hand markets are important in the context of leasing. There are many papers dealing with these topics including Debo et al. (2005, 2006), Ferguson and Toktay (2006), Ferrer and Swaminathan (2006), Groenevelt and Majumder (2001) and Mitra (2007).

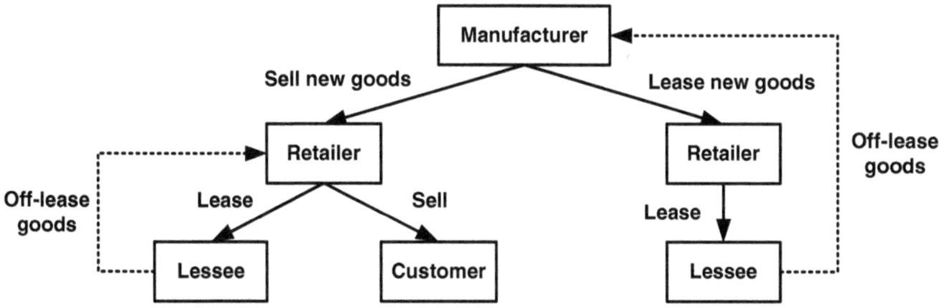

Fig. 9.3 An illustrative channel for the leasing of a product

9.4.5 Management

The management-oriented literature deals with the following leasing-related issues:

- Buy versus lease options through proper cost and benefit analysis. For more details, see Anderson and Martin (1977), Elgers and Clark (1980), Ezzel and Vora (2001), Miller and Upton (1976), Schallheim (1994) and Vargas and Saaty (1981).
- Selection of the optimal lease option when there are several alternatives: For more details, see Mollaghasemi et al. (1995).
- Designing, negotiating and managing lease schemes. For more details, see Deelen et al. (2003), Pfrang and Wittig (2008) Smith and Wakeman (1985).
- Administration of LCs.
- Debt management: For more details, see Eisfeldt and Rampini (2009), Fawthrop and Terry (1975).
- Risks for both lessor and lessee: This topic is discussed in Chap. 11.

9.5 Illustrative Examples of Lease Contracts

LCs vary in their complexity depending on whether they are for products, plants or infrastructures. In this section, we present some illustrative LCs.

9.5.1 Consumer Products

9.5.1.1 Household Appliances

Household appliances (white and brown goods, televisions, computers, etc.) are bought by individuals for their personal use. The LCs (operating or finance) are

Table 9.1 A simple standard lease contract

Lease Contract

_______________________ (hereafter "Lessor") and ___________________________ (hereafter "Lessee") hereby enter into a lease agreement under the following terms:

Lessor shall convey to Lessee full possession and use of the following product:

____

The term of this lease shall be from **MM/DD/YYYY** until **MM/DD/YYYY** at midnight on each date.

The Lessee is obliged to pay Lessor a total of **$..........** for the rights conveyed under this lease.

Upon expiration of this lease, Lessee shall have the option to purchase the product for the price of **$..........** If Lessee exercises this option to buy the property, **......** percent of all monthly payments made by Lessee shall be applied towards the purchase price.

Lessee shall pay to Lessor **$........** upon or before taking possession of the property. Thereafter, Lessee shall pay Lessor the sum of **$........** on or before the **......** day of each month until the expiration of this lease.

If Lessee fails to make a payment on or before its due date, a late fee of **$........** shall be due and payable immediately to Lessor.

If Lessee fails to pay all amounts due within **X** days of their due dates, then Lessor may terminate Lessor's obligations under this lease and take back possession and control of the asset. In the event of termination for non-payment, Lessee shall remain liable for the balance due under this lease.

Lessee shall be responsible for maintaining the property in clean working order at Lessee's expense during the term of this lease.

Upon expiration or termination of this lease, Lessee shall return the property to Lessor in substantially the same condition in which the property was received by Lessee, taking into account normal wear and tear.

In witness to their agreement to the terms of this contract, the parties affix their signatures below:

____

Lessor, signature & date	Lessee, signature & date
Address______________________________	Address____________________________
City, state, ZIP __________________	City, state, ZIP _______________

fairly simple and drafted by the lessor. There are many internet sites which have contract templates that can be downloaded and used by both the lessor and lessee. Table 9.1 is one such contract.

9.5.1.2 Motor Vehicles

The lessors in the motor vehicle leasing industry are manufacturers, retailers and third parties (such as financial institutions). Table 9.2 indicates the 35 sections of the Ford LC and details of each of these can be found in Appendix D.

Table 9.2 Ford lease contract

1. Amount due at lease signing or delivery
2. Monthly payments
3. Other charges
4. Total of payments
5. Amounts due at lease signing or delivery (itemisation)
6. How the amount due at lease signing or delivery will be paid
7. Your monthly payment is determined as shown below
8. Excess wear and use
9. Extra mileage option credit
10. Purchase option at end of lease term
11. Warranty
12. Official fees and taxes
13. Lessor services
14. Late payments
15. Life, disability and other insurance
16. Itemisation of gross capitalised cost
17. Vehicle use and subleasing
18. Vehicle maintenance and operating costs
19. Damage repair
20. Vehicle insurance
21. Termination
22. Return of vehicle
23. Standards for excess wear and use
24. Odometer statement
25. Voluntary early termination and return the vehicle
26. Default
27. Loss or destruction of vehicle
28. Assignment and administration
29. Taxes
30. Titling
31. Life insurance
32. Indemnity
33. Security deposit
34. Consumer reports
35. General
Rights you and we agree to give up
Rights you and we do not give up

TRAC

Motor vehicle leases (in the USA) contain a terminal rental adjustment clause (*TRAC*) which states that, on the termination of the lease, the lessee is required to pay the lessor the difference, if positive, between the expected value, as used to calculate payments for the lease agreement, less the actual wholesale value of the vehicle. If the difference is negative, then the actual value is greater than the expected value, the lessee keeps the gains. TRAC effectively shifts the risks and rewards of the ownership to the lessee.

Table 9.3 Wendt equipment lease contract

Article	1. The parties
Article	2. The rental period
Article	3. Rent
Article	4. Overtime rate basis
Article	5. Terms of payment
Article	6. Loading and freight charges
Article	7. Notice of return or recall
Article	8. Subleasing
Article	9. Relocation equipment
Article	10. Repairs and maintenance
Article	11. Inspection
Article	12. Insurance and indemnification
Article	13. Title
Article	15. Waivers
Article	16. Limited liability
Article	17. Indemnity

9.5.2 Industrial and Commercial Products and Plants

Businesses lease industrial and commercial products for a variety of reasons. In this case, both the lessor and the lessee are businesses. The items leased can be either single or a fleet (for example, cars, trucks).

For many products (for example, complex machinery), proper installation and training of personnel is required to use the product. Also, if the leased product does not perform satisfactorily, then this can have a major impact on the operations of the lessee. In this context, proper maintenance becomes an important issue. Risk and the coverage of risk through insurance are important elements that need to be addressed in the LC. As such, these contracts are more complex and contain several articles.

9.5.2.1 Equipment Lease

Table 9.3 lists the 17 articles of the Wendt equipment LC, and further details of each article can be found in Appendix D.

9.5.2.2 Leveraged Lease

Elgers and Clark (1980) discuss various special cases of leveraged leases, and Fig. 9.4 is an example of one of these. There are several parties involved and the relationships between them (numbered 1–12) are listed below.

1. Investment
2. Debt Certificates

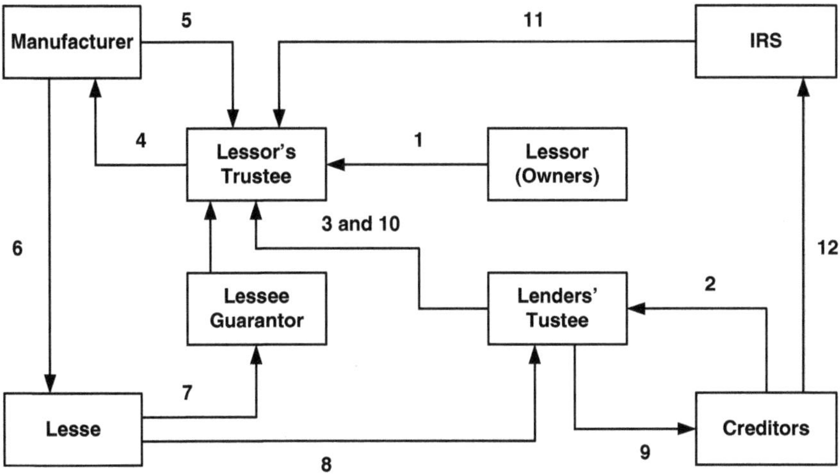

Fig. 9.4 Different parties in leveraged leasing

3. Non-recourse Loan
4. Purchase of Equipment
5. Title of Equipment
6. Transfer of Equipment
7. Guarantee
8. Rental Payments
9. Debt Service
10. Excess Rents over Debt Service
11. Tax benefits (Savings)
12. Interest Taxable Income (ITI).

The lessor's trustee (also called the owner trustee) holds the title to the asset and is responsible for raising the extra capital by selling creditor instruments to third parties. The financial condition of the lessee is an important factor in the determination of the lease payments. The lessee guarantor is to back up lessee's undertaking if the credit rating of the lessee is not strong enough and this is discussed further in Chap. 11.

This type of LC is much more complex and can often consist of tens (or hundreds) of pages.

9.6 Maintenance of Leased Assets

If an asset is owned by the user, then there is an incentive for the user to take proper care of the asset because the residual (or salvage) value belongs to the user. This involves using due care and maintenance resulting in normal wear and tear of

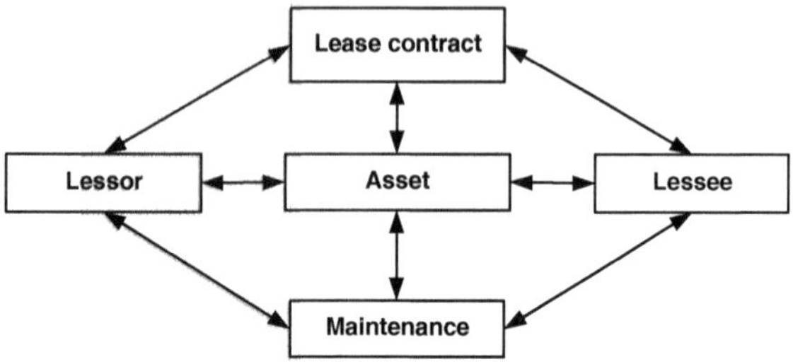

Fig. 9.5 Key elements for maintenance of leased items

Table 9.4 Different scenarios for maintenance

	Responsibility	
	Lessor	Lessee
In-house	A	B
Outsource	C	D

the asset. If the asset is leased, the lessee has an incentive to spend less on maintenance because the residual savings belong to the lessor/owner. Also, if the usage intensity is high, then this can lead to faster degradation. These factors can lead to excessive wear and tear, and this is referred to as the *abuse problem*.

There are several ways to handle a potential asset abuse problem. The lessor can:

1. charge a larger lease payment to cover the losses from asset abuse,
2. provide maintenance through a service lease (If the lease includes some type of maintenance agreement, sometimes it is called *a service lease or a full-service lease*. The service lease may offer maintenance services at a lower transaction cost than in the case of a purchase with a separate maintenance contract.), and
3. cover specific problems through provisions in the lease agreement (e.g., mileage limits in case of cars).

The level of maintenance needed in the context of leased items is an important issue for both the lessor and the lessee. The LC specifies the responsibility for maintenance—it can be either the lessor or the lessee. Figure 9.5 shows the key elements for effective maintenance decision-making

The maintenance can be either done in-house or outsourced. As a result, we have four different scenarios for the maintenance of an asset as indicated in Table 9.4. In the Scenarios C and D, the maintenance is carried out by an external service agent under a MSC with either the lessor or the lessee being responsible for the maintenance. Optimal maintenance of leased assets is discussed in Chap. 10.

9.7 Decision Problems in Leasing

The two main parties in a lease arrangement are the lessors and the lessees. Their goals and objectives are different and as such the decision problems that they need to address are also different.

Lessor's Perspective
The lessor is a business (either the manufacturer or some other entity) and as such has certain business objectives. At the strategic level, these can include issues such as ROI, market share and profits. In order to achieve these objectives, the lessor needs to have proper strategies in place at the strategic level (to deal with issues such as type and number of equipment to lease, upgrade options to compensate for technological obsolescence, etc.) and at the operational level (maintenance servicing, inventory of spares, crew size, etc.).

Lessee's Perspective
It is necessary to differentiate between the two types of lessees—individual households and businesses.

For consumer products, the lessee is an individual household and items are leased to meet specific needs (such as kitchen appliances for cooking and washing; televisions for entertainment; and automobiles for transport.).

A business leases equipment (commercial and/or industrial products) to produce outputs—goods and/or services. The lessee has to choose which equipment to lease when there are several competing brands, the best lease arrangement from the set of lease options available, the terms of the lease, etc. Critical to this decision-making are issues such as equipment availability and cost. Also, the lessee needs to take into account the effect of failures on the operations of the business and their subsequent impact on customer satisfaction.

9.7.1 Framework for Decision-Making

The systems approach provides the framework to evaluate the outcomes of different decisions and for choosing the optimal decision. Figure 9.6 shows the key elements for a LC where the lessor is responsible for maintenance.

9.8 Game-Theoretic Approach to Decision Problems in Leasing

The interaction between lessors and lessees defines the lease market. Both lessor and lessee need to find the solution to a variety of decision problems. For the lessor, the problem could be to decide on the terms of the LC (price, duration,

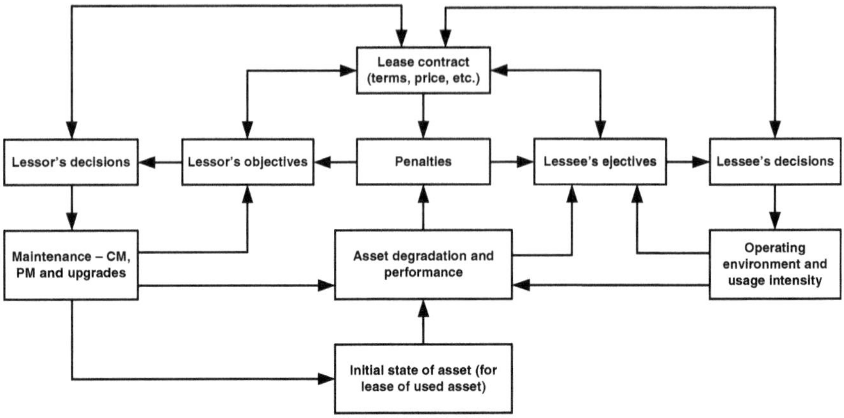

Fig. 9.6 Framework for decision-making in leasing

etc.), and for the lessee, it could be to decide on whether to lease or buy a product; to choose the best among alternate lease options if the decision is to lease, etc. Game theory (GT) provides the most appropriate framework for finding the optimal solutions to these decision problems. The decisions need to take into account the different decision makers (players) in the lease market and the power structure between the players.

9.8.1 Characterisation of the Lease Market

The most general characterisation of a lease market is as shown in Fig. 9.7, and this involves several interacting elements. We discuss each of these briefly using the following terminology.

Parties These are distinct groups (manufacturers, lessors, lessees, retailers and service agents, etc.) or other parties (such insurers, creditors, etc.) in the market. An example is the case where the manufacturer is the sole lessor and leases and services the leased items through manufacturer owned service centres. In this case, there are only two parties—manufacturer and customers.

Players There can be one or more players making up each party. In the above-mentioned example, there is only one manufacturer but there can be several customer groups with different characteristics (for example, individual or corporate customers).

Lessors: Lessors can be divided into the following three distinct groups.

- Manufacturers
- Retailers
- Third parties.

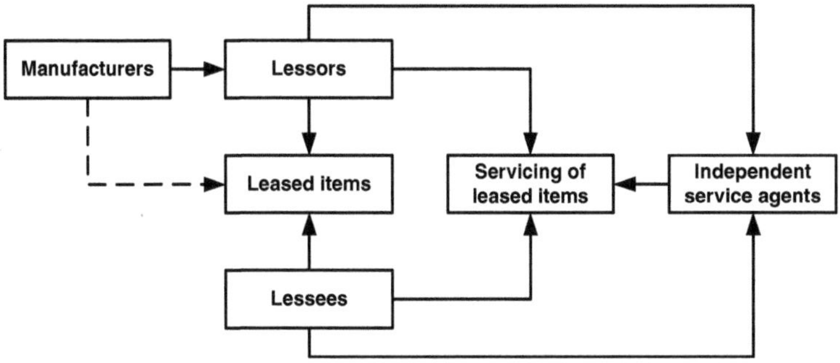

Fig. 9.7 Key elements of the lease market

In some lease markets, one or more of these groups of providers may not be present. Also, there may be one or more players in each group. For example, in a monopolistic market, there is only one lessor, whereas in an oligopolistic lease market there are two or more lessors.

Retailers: There can be one or several retailers in the lease market. The retailers sell the product and also lease it (either as lessors or as marketing agents for other lessors).

Lease Contracts (LCs): The leases can be divided into four distinct groups based on the type of lessor and whether the leases are marketed directly by the lessor or through retailers. The different groupings are as indicated below.

- LC-1: Manufacturer leasing items directly to customers
- LC-2: Manufacturer leasing items through retailer to customers
- LC-3: Retailer leasing items directly to customers
- LC-4: Independent lessors leasing directly to customers.

Within each group, a lessor may offer one or more types of LCs which differ in their terms and price.

Lessees: There can be one lessee or several lessees, and the lessee population can be either homogeneous or heterogeneous. In the latter case, the population can be divided into several groups based on characteristics such as attitude to risk and usage intensity.

Service agents: The service agents (SAs) can be divided into three groups as indicated below.

- Lessor owned
- Lessee owned
- Independently owned.Note that in the first two cases, the maintenance service is done "in-house" by the lessor or the lessee (depending on the LC), whereas in the last case, the maintenance service is outsourced and the responsibility is either with the lessor or the lessee (depending on the LC).

The EW/MSC market scenarios (indicated in Table 6.1) are also equally applicable for the lease market so that the market can be monopolistic, oligopolistic or competitive, and the number of customers (lessees) can be one, few or many.

9.8.2 Illustrative GT Scenarios for Lease Decision-Making

There are many possible scenarios based on different combinations of parties (lessors, retailers, lessees and service agents) in the market, the number of players in each of the groups and the power structure between the parties/players. As discussed in Sect. 4.5, there are two kinds of power structure between any two players—dominance (which we denote by → in our schematic representations) and equal or no dominance (which we denote by ↔ in our schematic representations). In the case where there is a dominance relationship between two players, the follower's decisions depend on the decisions of the leader. In the equal or no dominance case, the players' decisions are assumed to be made simultaneously. A player's response function to the decisions made by another player is indicated by a broken arrow.

It is not possible to discuss all possible scenarios. Instead, we look at a few and some of these will be discussed further in Chap. 10.

Scenario 1: Two Parties (Monopolistic Lease Market)
The two parties are a single lessor (the manufacturer who leases the items directly) and customers. Depending on the number of customers, the lease market is M-11 (single customer), M-12 (few customers) or M-13 (many customers). The lessor is the dominant player (leader) and the lessees are the followers.

The lessor's decision variables are as follows: (1) the number of different LC-1s to offer, (2) the terms (e.g., duration) and conditions (e.g., exclusions) of each LC-1, and (3) the price of each LC-1. For a given set of LC-1s, the customer's (s') decision variables are as follows: (1) whether to lease or purchase, and (2) the best LC-1 to select if there are two or more LC-1s to choose from.

For a given set of LC-1 options (price, duration, etc.), the lessee chooses the best option (discussed further in Chap. 10) and this defines their response function. The manufacturer then makes the optimal decision taking into account the response function. This is shown schematically in Fig. 9.8 and is a two-stage Stackelberg game.

Scenario 2: Three Parties (Oligopolistic Lease Market)
The three parties are the two lessors: (1) the manufacturer (who offers LC-2 through the retailer), (2) the retailer (who offers LC-3), and (3) the lessees.

There are two sets of LCs—LC-2s and LC-3s. For an LC-2, there is the "wholesale" price (the price that the retailer/dealer pays to the manufacturer) for each LC sold and the "retail" price (the price charged by the retailer to customers). The difference between the two is the markup on the LC-2 price—a decision variable for the retailer/dealer. The maintenance servicing is done by the dealer for

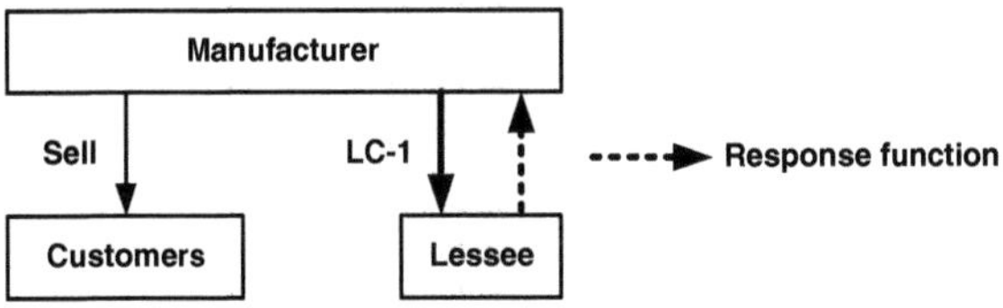

Fig. 9.8 Scenario 1 (single lessor)

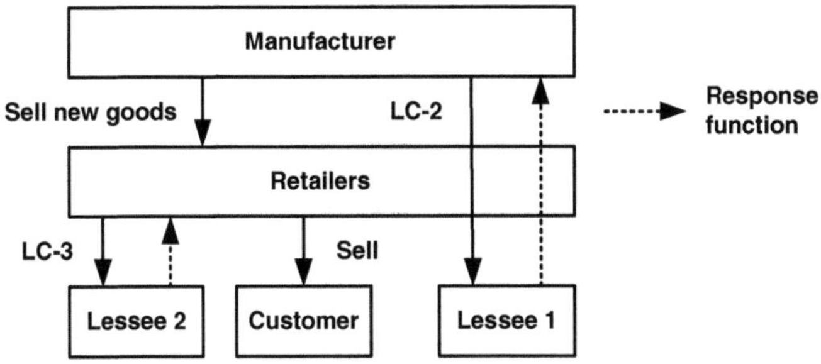

Fig. 9.9 Scenario 2 [two lessors]

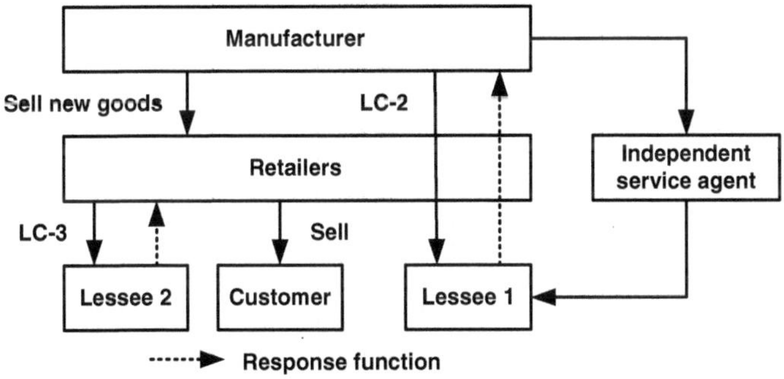

Fig. 9.10 Scenario 3 (two lessors and an independent service agent)

both the LC-2 and LC-3 cases. Also, we have two types of lessees—Lessee 1 who leases the product from the manufacturer and Lessee 2 who leases from the retailer.

As a result, we have a three-stage Stackelberg game with two separate vertical response functions as shown in Fig. 9.9. The retailer's optimal decision is obtained as the solution of the lower level game taking into with the wholesale price for LC-2 charged by the manufacturer and the response function of customers. The manufacturer's optimal decision is obtained as the solution of the higher level game which takes into account the response function of the retailer.

Scenario 3: Four Parties
This is an extension of Scenario 2 with all maintenance servicing for the LC-2 being carried out by independent service agent under a contract (similar to maintenance outsourcing contract), whereas the maintenance servicing for LC-3 is done by the retailer. The contract between the manufacturer and the service agent introduces new decision variables (charges for different kinds of repair), and the game-theoretic characterisation which is shown in Fig. 9.10 is more complex. Note here that the independent service agent is a follower in the game.

References

Anderson PF, Bird MM (1980) Marketing to the industrial lease buyer. Ind Mark Manage 9:111–116
Anderson PF, Lazer W (1978) Industrial lease marketing. J Mark 42:71–79
Anderson PF, Martin JD (1977) Lease versus purchase decisions: a survey of current practice. Finance Manage 6:41–47
Aras N, Gullu R, Yurulmez S (2011) Optimal inventory and pricing policies for remanufacturable leased products. Int. J. Prod Econ 133:262–271
Baker CR, Hayes RS (1981) Lease financing—a practical guide. Wiley, New York
Bhaskaran SR, Gilbert SM (2005) Selling and leasing strategies for durable goods with complementary products. Manage Sci 51:1278–1290
Chemmanur T, Jiao Y, Yan A (2010) A theory of contractual provisions in leasing. J Finan Intermediation 19:116–142
Chen KY, Huang S (2005) Durable goods lease contracts and used-goods market behaviour: an experimental study. In: Rapport A, Zwick R (eds) Experimental business research, vol II. Springer, New York
Coyle B (2000) Leasing. Glenlake, Chicago
Debo LG, Toktay LB, Wassenhove LNV (2005) Market segmentation and product technology selection for remanufacturable products. Manage Sci 51:1193–1205
Debo LG, Toktay LB, Wassenhove LNV (2006) Joint life-cycle dynamics of new and remanufactured products. Prod Oper Manage 15:498–513
Deelen L, Dupleich M, Othieno L, Wakelin O (2003) Leasing for small and micro enterprises—a guide for designing and managing leasing schemes in developing countries. ILO, Geneva, Switzerland
Desai P, Purohit D (1998) Leasing and selling: optimal marketing strategies for a durable goods firm. Manage Sci 44(11):19–34
Eisfeldt A, Rampini A (2009) Leasing, ability to repossess, and debt capacity. Rev Financ Stud 22:1621–1657
ELA (2002a) Equipment Leasing and Financial Foundation 2002 state of the industry report. Price water house coopers. Available on http://www.leasefoundation.org/pdfs/2002SOIRpt.pdf
ELA (2002b) Equipment Leasing Association online focus groups report. Available on http://www.chooseleasing.org/Market/2002FocusGroupsRpt.pdf
ELA (2005) The economic contribution of equipment leasing to the U.S. economy: growth, investment & jobs—update. Equipment Leasing Association, Global Insight, Advisory Services Group. Available on http://www.elaonline.org/press/
Elgers PT, Clark J (1980) The lease/buy decision. The Free Press, New York
Ezzel JR, Vora PP (2001) Leasing versus purchasing: direct evidence on a corporation's motivation for leasing and consequences of leasing. Quart Rev Econ Finan 41:33–47

Fawthrop RA, Terry B (1975) Debt management and the use of leasing in UK corporate financing strategies. J Bus Finan Acc 2:295–314
Ferguson ME, Toktay LB (2006) The effect of competition on recovery strategies. Prod Oper Manage 15:351–368
Ferrer G, Swaminathan JM (2006) Managing new and remanufactured products. Manage Sci 52:15–26
Fishbein BK, McCarry LS, Dillon PS (2000) Leasing: a step toward producer responsibility. Available on http://www.informinc.org
Gavazza A (2005) Leasing and secondary markets: theory and evidence from commercial aircraft. Working paper, Yale University
Gerety VE (1995) Understanding the equipment leasing marketplace. J Equip Lease Finan 13:3–12
Groenevelt H, Majumder P (2001) Competition in remanufacturing. Prod Oper Manage 10:125–141
Handa P (1991) An economic analysis of leasebacks. Rev Quant Financ Account 1:177–189
Huang S, Yang Y (2002) Pricing lease contracts with options in imperfect markets for durable goods. Technical report, Ford Research Laboratory
Johnson JP, Waldman M (2003) Leasing, lemons, and buybacks. Rand J Econ 34:247–265
Kaster LR (ed) (1979a) Sale-leasebacks—economics, tax aspects and lease terms. Practising Law Institute, New York
Kaster LR (ed) (1979b) Sale-leasebacks—economics, tax aspects and lease terms—appendix 1. Practising Law Institute, New York
Kleiman RT (2001) The characteristics of venture lease financing. J Equip Lease Financ 19:1–10
Kong S, Long MS (2001) The fixed payment finance decision to borrow or lease. Rev Financ Econ 10:41–55
Krishnan SV, Moyer CR (1994) Bankruptcy costs and the financial leasing decision. Financ Manage 23:31–42
Lewellen WG, Long MS, McConnell JJ (1976) Asset leasing in competitive capital markets. J. Finan 31:737–798
MAPI (1965) Leasing of industrial equipment. Machinery and Allied Products Institute, Washington, DC
Miller M, Upton C (1976) Leasing, buying, and the cost of capital services. J Finan 31:761–786
Mitra S (2007) Revenue management for remanufactured products. Omega 35:553–562
Mollaghasemi M, Pet-Edwards J, Gupta U (1995) A multiple criteria buy versus lease analysis for government contracts. IEEE Trans Eng Manage 42:278–287
Pfrang DC, Wittig S (2008) Negotiating office lease contracts: from a game-theoretical towards a behavioural view. J Eur Real Estate Res 1:88–105
Purohit D (1994) Rental, sales and buybacks: managing secondary distribution channels. J Mark Res 31:325–338
Purohit D (1997) Dual distribution channels: the competition between rental agencies and dealers. Mark Sci 16:228–245
Purohit D, Staelin R (1994) Rental, sales, and buybacks: managing secondary distribution channels. J Mark Res 2:161–191
Schallheim JS (1994) Lease or buy? principles for sound decision making. Harvard Business School Press, Cambridge
Sharpe SA, Nguyen HH (1995) Capital market imperfections and incentive to lease. J Financ Econ 39:271–294
Sizer J (1987) An Insight into Management Accounting. Penguins Books Ltd., Middlesex
Smith CW, Wakeman LM (1985) Determinants of corporate leasing policy. J. Finan 40:895–908
Stremersch S, Wuyts S, Rambach RT (2001) The purchasing of full-service contracts: an exploratory study within the industrial maintenance market. Ind Mark Manage 30:1–12
Tilson V, Wang Y, Yang W (2006) Channel strategies for durable goods: coexistence of selling and leasing to individual and corporate consumers. Technical memorandum # 813, Department of operations, Case Western Reserve University, Cleveland, OH

Vargas LG, Saaty TL (1981) Financial and intangible factors in fleet lease or buy decision. Ind Mark Manage 10:1–10
Wainman D (1991) Leasing. Waterlow, Sittingbourne
Waldman M (1997) Eliminating the market for second-hand goods: an alternative explanation for leasing. J Law Econ 40:61–92
Watts B (1971) Industry and finance. Macdonald and Evans, London

Chapter 10
Models for Lease and Maintenance Decisions

10.1 Introduction

There are two types of models for making lease decisions. The first type consists of models dealing with lease versus buy decisions and is based on cash flows taking into account the tax issues, discounting, etc. The second type consists of GT models involving the lessor(s) and the lessee(s) and possibly other parties (such as retailers, external service agents). Models dealing with maintenance decisions for leased items are extensions of the models discussed in Chaps. 3 and 7 and take into account the terms of the lease (such as penalties) and lease duration. In the context of the leasing of used items, the models also deal with upgrades. In this chapter, we look at the various models for leasing and maintenance that have been reported in the literature. The outline of the chapter is as follows. Section 10.2 deals with the framework for building models for lease and maintenance decision-making with the focus on GT models. We discuss the key elements, different scenarios and model formulations. Section 10.3 deals with models where customers choose between leasing and buying. Section 10.4 looks at the different GT models and the optimal decisions from both the lessor and the lessee perspectives. Section 10.5 deals with models for maintenance decisions for leased items.

10.2 Framework for Modelling

The lease process and the lease market involve several interacting elements as shown in Figs. 9.6 and 9.7, respectively. The characterisation and modelling of each element can be done in several ways, leading to many different scenarios and several different GT models that are discussed in a later section of the chapter. We confine our attention to the leasing of products.[1]

[1] The leasing of complex plants and infrastructures is more complicated as it involves many other parties depending on the lease. Figure 9.4 shows some the parties in the case of a leveraged lease.

D. N. P. Murthy and N. Jack, *Extended Warranties, Maintenance Service and Lease Contracts*, Springer Series in Reliability Engineering,
DOI: 10.1007/978-1-4471-6440-1_10,

10.2.1 Key Elements and Their Characterisations

Parties: The distinct parties (or groups) are (1) lessors (manufacturers, dealers/ retailers and other independent providers), (2) lessees (customers) and (3) service agents (if the maintenance of the leased item is outsourced). The customers may be either homogeneous or heterogeneous. In the latter case, the customer heterogeneity may be due to differences in usage, risk attitude, income and information. Thus, each party in a lease market may consist of one or more *players* and this leads to different market structures as illustrated through examples in Sect. 9.8.2.

Product: As discussed in Chap. 2, every product is unreliable. The reliability of a product changes with its age, this being the dynamic characterisation of reliability. Many GT models (especially those from the economic and marketing literature) model reliability in a static sense—the product either fails or does not fail over the interval of interest (lease period). Two other important variables are the salvage value and residual life of the product at the end of the lease period.

Demand for lease: This can be treated as either exogenous (an external variable) or endogenous (a function of other model variables—such as price and duration of the lease, renewal and/or termination options). The lease market for new products needs to take into account the second-hand market and buyers of new and used items. Figure 10.1 shows three groups of lessees, depending on whether the manufacturer leases the product directly (LC-1) or through dealers/ retailers (LC-2) and whether the retailer leases directly without the manufacturer being involved (LC-3). There are two groups of buyers—those buying directly from the manufacturer or through the retailer as shown in the figure.

Lease contracts (LCs): These are characterised by variables such as price, duration and other terms (such as renewal and/or termination options, extension options).

There are two kinds of options (Gamba and Rigon 2008):

- *American put option* (cancellation option): The lessee has the right to extinguish the contract before expiration (with a penalty incurred for exercising the option).
- *European call option*: The lessee has the option to buy the product for a predetermined residual value at the final date.

The valuation of LCs is important for the lessee's decision-making process. The valuation of LCs with no options involves single-period formulations, and those with options involve multiperiod formulations. We discuss these further later in the section.

Maintenance (PM and CM): The maintenance can be the responsibility of the lessor or the lessee, depending on the LC. As mentioned earlier, if maintenance is outsourced to a third party, we have an external service agent.

Power Structure: As discussed in Sect. 6.5, the two possible types of power structure that can occur between any two players A and B in the lease market are *dominance* which we indicate by A $\rightarrow$ B and *no dominance* (equal power) indicated by A $\leftrightarrow$ B. In the former case, the dominant player A's decisions are known

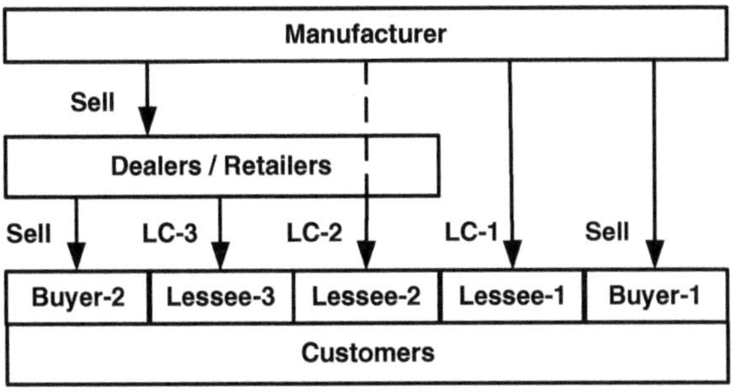

Fig. 10.1 Market for lease and sale of new products

to and influence the decisions made by the dominated player B. A is known as the *leader* and B the *follower* in this type of power structure. In the latter case, the two players are assumed to make their decisions simultaneously or at least are unaware of each others' decisions.

Decision Problems: The decision problem for each player is different. It is characterised by an objective function that may involve expected cost, expected utility, expected revenue, sales, profits, etc. The decision variables can be the choice between two or more alternatives (for lessees), price and duration (for lessors) and actions such as repair versus replace and type of repair (for service agents). The effects of uncertainty and risk need to be taken into account.

Information: There are three types of information—(1) product related (reliability of the leased item), (2) customer related (homogeneous or heterogeneous, attitude to risk, income, etc.) and (3) service related (service delivery, etc.). Other issues include symmetry versus asymmetry in information between players, perfect (complete) or imperfect (incomplete and uncertain) information.

10.2.2 Different Scenarios

The possible scenarios occur as a result of the various combinations of the elements discussed earlier and their characterisation. A multilevel characterisation may be used with two or more levels. An example of the first-level characterisation is indicated in Table 10.1.

For each of these nine scenarios, we can have several higher-level classifications based on the characterisation of customers—buyers and leasers, attitude to risk (averse or neutral), information available, etc.

10.2.3 Model Formulations

Model formulations can be either static or dynamic. Most of the GT models reported in the economics and marketing literature are static and single period.

Table 10.1 First-level characterisation

		Number of periods		
		One	Two	Three
Number of parties	One	A	D	G
	Two	B	E	H
	Three	C	F	I

A few models deal with multiperiod problems with a static formulation in each period. Proper maintenance models need to allow for the possibility of multiple failures occurring over time, and so dynamic formulations are required. The operational research and reliability literature contain models which are dynamic in nature.

10.3 Models for Lease Versus Buy Decisions

The customer is a business that needs to decide between leasing and purchasing an expensive asset. In the one-period case, there is no option of renewing or extending the lease. There are two approaches to building models to assist in the decision-making, and these are as follows:

- Approach 1: Discounted cash flow (DCF) analysis (based solely on financial factors).
- Approach 2: The analytic hierarchy process (AHP).[2]

In this section, we confine our attention to Approach 1 with Approach 2 being discussed in Chap. 11. We first look at the single-period formulations and then discuss briefly the multiperiod formulations.

10.3.1 Single-Period Formulations (Lease with No Option)

We begin by defining the parameters and variables used.

N Duration of Period 1 (years)

t Discrete time $t = 1, 2, \ldots, N$ (years)

k Marginal cost of capital for the lessee (This is the required return on new investments that will leave the market value of the lessee's equity unchanged, and it depends on the financial status of the lessee to borrow capital.)

r Marginal cost of debt to the lessee (this is the pre-tax rate of borrowing)

[2] The AHP is the creation of Thomas Saaty. For more details, see Saaty (1980).

T Tax rate of the lessee
d Discount rate for tax shelters on contractual payments $[(1 - T)r \leq d \leq k]$
s Risk level associated with sale of the asset $[s \geq k]$
I_t Interest payment in time period t for any loan taken
A_t Amortisation of loans for purchase of asset in time period t
D_t Depreciation of asset purchased in time period t
E_t Maintenance and operating cost in time period t
O_t Additional expenses that would not exist (such as administration, insurance) in the case of lease in time period t
L_t Lease payment in time period t
V_0 Book value of the asset before purchase or lease
V_N Book value of the asset at the end of the lease period

Comments: All expenses except A_t are tax deductible and hence provide tax shelter. In the simplest characterisation, $r = k = s$ so that we have only one parameter as opposed to three.

Model 10.1 (Deterministic Model)[3]

Assumptions: There is no uncertainty in the model. All the parameters and variables are deterministic, and the values of these quantities are known to the customer so there is complete information.
Decision: The customer must decide whether to buy or lease the asset.
Objective function: The customer makes the decision by comparing the net present value (NPV) of both options.

If the asset is purchased, the NPV of cash flows (expenses) to the customer is the sum of the following four components:

1. Net *resale* value: $C_P^R = V_0 - \frac{V_N}{(1+k)^N}$
2. *Cash expense*: $C_P^E = \sum_{t=0}^{N} \frac{O_t}{(1+k)^t} + \sum_{t=0}^{N} \frac{I_t}{(1+r)^t} + \sum_{t=0}^{N} \frac{A_t}{(1+r)^t}$
3. *Tax shelter*: $C_P^T = \sum_{t=0}^{N} \frac{O_t T}{(1+k)^t} + \sum_{t=0}^{N} \frac{I_t T}{(1+d)^t} + \sum_{t=0}^{N} \frac{D_t T}{(1+d)^t}$
4. *Maintenance and operation*: $C_P^M = \sum_{t=0}^{N} \frac{E_t(1-T)}{(1+k)^t}$.

Thus, the NPV of cash flows if the asset is purchased is given by

$$C_P = C_P^M + C_P^R + C_P^E - C_P^T \tag{10.1}$$

The NPV of cash flows (expenses) to the customer in the case of leasing is the sum of the following three components:

[3] For more details of the model formulation, see Vargas and Saaty (1981).

1. *Cash expense*: $C_L^E = \sum_{t=0}^{N} \frac{L_t T}{(1+k)^t}$
2. *Tax shelter*: $C_L^T = \sum_{t=0}^{N} \frac{L_t T}{(1+d)^t}$
3. *Maintenance and operation*: $C_L^M = \sum_{t=0}^{N} \frac{E_t(1-T)}{(1+k)^t}$

Thus, the NPV of cash flows to the customer if the asset is leased is given by

$$C_L = C_L^M + C_L^E - C_L^T. \tag{10.2}$$

The optimal decision for the customer is as follows: If $C_P < C_L$, then buy the asset; if $C_P > C_L$, then lease; and if $C_P = C_L$, then the customer is indifferent between the two options.

Model 10.2 (Stochastic Model)

In Model 10.1, all the parameters and variables are deterministic quantities and known (complete information). In real life, some of the cash flow components for the customer will be uncertain since they will be affected by random fluctuations in k, r, E_t and V_N. C_P^R is the riskiest cash flow element in C_p (mainly due to the uncertainty regarding technology obsolescence). In the presence of uncertainty, the modelling involves defining the probability distribution functions for the random variables representing some of the model parameters. The customer's objective function can be the expected NPV of cash flows or a combination of the mean and variance of this NPV.

In general, it is difficult to obtain the customer's optimal decision analytically and a simulation method would need to be used to decide whether to lease or buy.

10.3.2 Multiperiod Formulations (Leases with Options)

A multiperiod formulation can have two to four periods (as indicated in Fig. 10.2), and there can be several different scenarios, depending on the initial duration of the LC and the options (cancel, renew, extend, buy) available to the lessee.

Two-period formulation

This corresponds to Periods 1 and 2 of Fig. 10.2. The initial duration of the LC covers Periods 1 and 2. The lessee has the option to either terminate (cancel) or renew the lease at the end of Period 1, and there is a penalty for cancelling the LC.

Three-period formulation

This corresponds to Periods 1–3 of Fig. 10.2. The initial duration of the LC is Periods 1 and 2. The options available at the end of Period 1 include cancelling

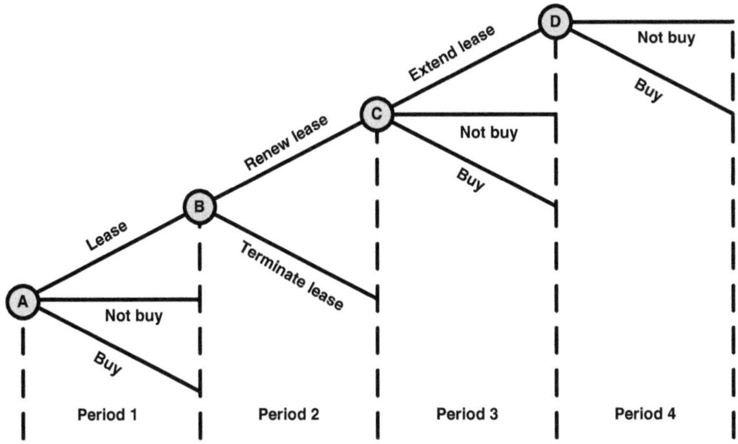

Fig. 10.2 Multiperiod characterisation of LCs

(with penalty) or renewing the lease. At the end of Period 2, the options available include (1) extending the lease, (2) buying the item and (3) not buying.

Four-period formulation

This corresponds to Periods 1–4 of Fig. 10.2. The initial duration of the LC is Periods 1 and 2. There is the option to renew or cancel at the end of Period 1, the option to extend at the end of Period 2 and the option to buy at the end of Periods 2 and 3.

Contingent Claims analysis (CCA)

With options included in the LC, the leased asset can be viewed as a traded security with options (to cancel the lease early, to extend its life or to purchase at some specified price) seen as claims whose value is contingent on the future value of the leased asset (or the LC). Contingent claims analysis has been used for the valuation of leases. It involves a tree-structured characterisation (as shown in Fig. 10.2).

We begin by defining the parameters and variables used.

i	Index for period ($i = 1, \ldots, 4$)
V_{i-1}	Value of leased item at the start of period i ($i = 1, \ldots, 4$)
S_{i-1}	Salvage value of leased item at the start of period i ($i = 2, 3$)
L_{i-1}	Rental payment at the start of period i ($i = 1, 2, 3$)
EX_{i-1}	Exercise price (to buy the item) at the start of period i ($i = 2, 3$)
D_i	Defined in Sect. 10.3.1
T	Defined in Sect. 10.3.1
I_{i-1}	Lease costs in period i $[I_{i-1} = L_{i-1}(1-T) + D_{i-1}T]$ ($i = 1, 2, 3$)
PEN	Penalty cost for the cancellation of lease contract.

There are many papers in the option theory literature that deal with valuation of leases under different options for deciding between leasing and buying.[4]

Model 10.3

The objective of the model is to determine the value of the LC at the start of Period 1 so as decide whether to lease or buy. The process involves a backward risk-neutral valuation during which the value of the LC with embedded options (C_t) is adjusted to $\left(C_t'\right)$ as discussed below.

The option to buy

The buy option is available at the start of Periods 2 and 3. At the start of Period 4, the value of contract if the lessee does not buy the item is V_3. If the item is bought, then the value is $V_3 + (S_3 - \mathrm{EX}_3)$. This implies that the lessee buys the item if $S_3 - \mathrm{EX}_3 \geq 0$. As a result, the adjusted value at the start of Period 3 is given by

$$C_3' = V_3 + \max\{S_3 - \mathrm{EX}_3, 0\} \tag{10.3}$$

Note that $C_3 = \max\{V_3, 0\}$.

A similar reasoning at the start of Period 3 between buy and terminate is based on comparing V_2 with $V_2 + (S_2 - \mathrm{EX}_2)$. This result is used later in looking at the renew option at the start of Period 2.

The option to cancel

The lessee may terminate a cancellable lease early, just before the rental payment is due. In this case, the adjustment is given by $C_t' = \max\{C_t - I_t, -\mathrm{PEN}\}$.

The option to extend

This occurs at the start of Period 3. The adjusted value is given by $C_t' = \max\{C_t - I_t, 0\}$.

Comment: This follows as there is no penalty incurred, the adjusted value of the lease is the unadjusted value minus the lease costs, and the decision to extend occurs only if this is greater than zero.

As a result, at the start of Period 3, the decision to extend the LC results in a valuation of the lease as $C_2 - I_2$. If the decision is to choose between buy and terminate, the valuation is $V_2 + \max\{S_2 - \mathrm{EX}_2, 0\}$. As a result, the adjusted value at the start of Period 3 is given by

$$C_2' = \max\{C_2 - I_2, V_2 + \max\{S_2 - \mathrm{EX}_2, 0\}\} \tag{10.4}$$

---

[4] Grenadier (1995), Kim et al. (1978), Myers et al. (1976), Miller and Upton (1976) and Trigeorgis (1996) is a small illustrative sample. Model 10.3 is based on material from Trigeorgis (1996).

Similarly, the adjusted value of the lease at the start of Period 2 is given by

$$C'_1 = \max\{C_1 - I_1, -\text{PEN}\} \tag{10.5}$$

Thus, one proceeds from the valuation at the start of Period 4 (given by Eq. 10.3) and then proceeds backwards to obtain the valuations at the start of Periods 3 and 2 (given by Eqs. 10.4 and 10.5, respectively). Note that the valuations are based on optimal decisions at the start of Periods 4, 3 and 2. At the start of Period 1, the decision to lease or not lease (buy or not buy) would involve comparing $C'_1 - I_0$ with the V_0 (value of asset under buy option) and the financing of it through borrowing funds.

10.3.3 Leasing in Different Industry Sectors

There are many papers dealing with choosing between buying and leasing in different industry sectors. A small sample is given below:

Aircraft leases: Rieple and Helm (2008), Hsu et al. (2011), Gavazzza (2010) and Bazargan and Hartman (2012).

Hospital equipment: Henry and Roenfeldt (1978), Nisbet and Ward (2001) and Roenfeldt and Henry (1979).

Government contracts: Mollaghasemi et al. (1995).

Industrial plants and products: Meier et al. (2010).

Retail leases: Lee (1995).

Finally, Sorensen and Johnson (1977) report on an empirical study of leasing practices and costs in industry.

10.4 Game-Theoretic Models

The models that have appeared in the literature fall into two categories:

- Microeconomics: The models deal with products[5] and try to explain (1) the actions of a monopolist for leasing as opposed to selling, (2) the effect of competition on the actions of a monopolist opting to both lease and sell and (3) the impact of intermediaries (such as retailers) in the supply chain in the leasing and selling of products.
- Marketing: The models deal with pricing and strategies to ensure some specified consumer behaviour in relation to the selling and leasing of products.

[5] The products are durable goods as opposed to non-durable goods.

In the industrial organisation literature, it is well established that the durability of a product can interfere with a monopolistic manufacturer's pricing of the product (Coase 1972). The reason for this is as follows. After the manufacturer sells its product to a subset of the market (in the first period), it has an incentive to continue production and selling its products at lower and lower prices in the subsequent periods. Consumers anticipate this opportunistic behaviour, and fewer are willing to buy at any given price. This issue is referred to as *time inconsistency* in reference to the fact that a monopolist's ability to sell the product at a price above the marginal cost is *inconsistent* with its own incentives to produce at a rate that causes the price to decrease. By leasing, the manufacturer internalises the effect of its future output and eliminates the problem of time inconsistency.[6]

Another issue (that has received attention in the marketing, operational research and economics literature) is channel coordination and the inefficiencies that occur with the presence of retailers. One of the primary sources of inefficiency is *double marginalisation*, which occurs when individual members of the channel add their own margins to the cost of the product and this leads to a final retail price that is higher than the one that would maximise the total channel profits. However, when the products from different manufacturers are highly substitutable, retailers benefit because double marginalisation from using the products mitigates the downward pressure on prices in competition.[7]

In this section, we discuss some of these GT models in context of the leasing of products. The models are highly stylised and differ significantly in terms of the variables included, the reasons for building the models and the notation. We discuss the main features of the model formulations—decision variables for the parties involved (such as manufacturers, retailers and customers) and other relevant variables; objective functions; and the game-theoretic structure. We omit any model analysis and its implications.[8]

10.4.1 One-Period Models

Assumptions: A manufacturer sells a product directly to customers at price p_p. The manufacturer is risk neutral, and the manufacturing cost per unit of product is c_m. The customers also have the option of leasing the product from the manufacturer for price p_l. The lease period has length L, and a customer who decides to lease always leases for the whole period. The product has a salvage value $S(L)$ at the end of the period.

[6] For more on this, see Stokey (1981) and Bulow (1982).

[7] For more on this and relevant references, see Bhaskaran and Gilbert (2009).

[8] This is done for a variety of reasons—(1) the need for deeper understanding of economics and marketing; (2) the models are very stylised (and unrealistic to some extent), and only some of the inferences are indirectly validated through real or test data; and (3) none deal with all the relevant issues. This is discussed further in Chap. 12. However, we do give relevant references where interested readers can get the details of model analysis and its implications.

Each time the product fails, a minimal repair is performed by an independent service agent and the cost of each repair is p_r. Repair costs are paid by the customer if the product is purchased and by the manufacturer if it is leased. The times taken to perform the repairs are very small compared with the mean time between product failures and so can be ignored in the failure modelling. If $N(t)$ denotes the number of failures that occur in the time interval [0, t), then, under the minimal repair assumption, $\{N(t);\ t \geq 0\}$ is an NHPP with intensity function $\lambda(t)$ and cumulative intensity function $\Lambda(t) = \int_0^t \lambda(t)\mathrm{d}t$.

The customers are homogeneous with respect to risk attitude and are all risk averse with utility function given by (8.11). Each customer has initial wealth ψ.

Key elements and decision variables: The cost of a minimal repair to the customer or the manufacturer is exogenous. The set of decision variables for the manufacturer is $y = \{p_p, p_l\}$, and the customer's decision variable is

$$x = \begin{cases} 1, & \text{if the product is purchased,} \\ 2, & \text{if the product is leased.} \end{cases}$$

Objective functions: For a given value of y chosen by the manufacturer, the customer's wealth at the end of the period [0, L) is given by

$$Y_C(x;y) = \begin{cases} \psi - p_p + S(L) - p_r N(L) & \text{if } x = 1, \\ \psi - p_l & \text{if } x = 2. \end{cases} \tag{10.6}$$

The customer's expected utility function for wealth $J_C(x;y)$ is derived by using (8.11), conditioning on the number of failures that will occur during the period [0, L) and then removing the conditioning. After some simple manipulation, we obtain

$$J_C(x;y) = \begin{cases} \frac{1}{\gamma}\left[1 - e^{-\gamma(\psi - p_p + S(L) - p_r' \Lambda(L))}\right] & \text{if } x = 1, \\ \frac{1}{\gamma}\left[1 - e^{-\gamma(\psi - p_l)}\right] & \text{if } x = 2, \end{cases} \tag{10.7}$$

with $p_r' = \left[e^{\gamma p_r} - 1\right]/\gamma$. Note that the parameter p_r' is increasing in γ and is always $>p_r$ for all $\gamma > 0$.

For a given value of x chosen by the customer, the manufacturer's expected profit is given by

$$J_M(y;x) = \begin{cases} p_p - c_m, & \text{if } x = 1, \\ p_l - c_m + S(L) - p_r \Lambda(L), & \text{if } x = 2. \end{cases} \tag{10.8}$$

Optimal decisions: The manufacturer is the leader, and the customer is the follower in the Stackelberg game between the two parties. For a given value of y chosen by the manufacturer, the customer selects the optimal $x^*(y)$ (purchase or lease) that maximises $J_C(x;\ y)$. Using (10, 7), the customer will decide to lease if $p_l < p_p - S(L) + p_r' \Lambda(L)$ and purchase if $p_l > p_p - S(L) + p_r' \Lambda(L)$.

The manufacturer then chooses the optimal purchase and lease prices $y^* = \{p_p^*, p_l^*\}$ for the product to maximise $J_M(y : x^*(y))$.

Model 10.5 (Lee 1995)

Lee (1995) investigates the structure of optimal retail lease contracts. The lessor is a landlord who owns retail space, and the lessee is a retailer (tenant) who will be able to sell goods by using the space and other inputs.

Assumptions: The amount of goods sold by the retailer is a random variable S with distribution function $F(s;\, m)$ and density function $f(s;\, m)$ where m is the effort the retailer uses to generate sales. The cost to the retailer to provide the sales effort is $C(m)$, a convex, increasing function of m.

The rental payment for the retail space $r(s)$ depends on the actual amount of sales s achieved by the retailer. It consists of a base rent plus a percentage of sales in excess of a threshold level. The exact form is given by

$$r(s) = \begin{cases} \beta, & \text{if } 0 \le s \le \hat{s}, \\ \beta + \alpha(s - \hat{s}), & \text{if } \hat{s} < s \le s_m, \end{cases} \tag{10.9}$$

with $\alpha \ge 0$ and $\beta \ge 0$, $\hat{s}$ the threshold sales level and s_m the maximum sales.

The landlord is risk neutral, and the retailer is risk averse with concave utility function for wealth Y given by $U(Y)$.

Decision variables: The set of decision variables for the landlord is $y = \{\beta,\, \alpha,\, \hat{s}\}$, and the retailer's decision variable is $x = m$.

Objective functions: For a given value of y chosen by the landlord, the retailer's expected utility function for wealth is

$$J_R(x; y) = \int_0^{\hat{s}} U(s - \beta) f(s; m) \mathrm{d}s + \int_{\hat{s}}^{s_m} U(s - \beta - \alpha(s - \hat{s})) f(s; m) \mathrm{d}s - C(m). \tag{10.10}$$

The retailer requires a certain minimum return on the investment, and this translates into the constraint

$$J_R(x; y) \ge u_m \tag{10.11}$$

where u_m is the retailer's reservation utility.

For a given value of x chosen by the retailer, the landlord's expected rent is given by

$$J_L(y; x) = \int_0^{\hat{s}} \beta f(s; m) \mathrm{d}s + \int_{\hat{s}}^{s_m} [\beta + \alpha(s - \hat{s})] f(s; m) \mathrm{d}s. \tag{10.12}$$

Optimal decisions: A Stackelberg game is played between the landlord (leader) and the retailer (follower). For a given value of y chosen by the landlord, the retailer selects the optimal sales effort $x^*(y)$ that maximises $J_R(x;y)$. The landlord then chooses the optimal rent structure $y^* = \{\beta^*, \alpha^*, \hat{s}^*\}$ to maximise $J_L(y;x^*)$, ensuring that the constraint $J_R(x^*;y^*) \geq u_m$ is satisfied.

10.4.2 Two-Period Models

There are several possible scenarios, depending on the number of different parties involved in the leasing market. In the two-party case, there are only customers and manufacturers present. We look at models involving both one and two manufacturers. In the three-party case, customers, manufacturers and dealer/renters are present and we again look at situations with both one and two manufacturers. Finally, there are models involving four parties (customers, manufacturers, dealer/renters and independent service agents).

10.4.2.1 Two Parties

Model 10.6 (Desai and Purohit 1998)

Desai and Purohit (1998) develop a model that deals with the strategic effects of simultaneous leasing and selling by a manufacturer.

Assumptions: The product is a car, made by a single (monopolistic) manufacturer and whose lifetime is two periods (Periods 1 and 2). In Period 1, only new cars are available in the market and these are either sold or leased by the manufacturer (with the lease duration being Period 1). In Period 2, there are new cars available for sale or lease as well as used and ex-leased cars available for sale. The used cars are those that have been bought in Period 1 and are now offered for sale by customers.

The marginal cost to the manufacturer to produce and market a car is assumed to be zero (this assumption does not alter the nature of the model results). The manufacturer is risk neutral. Customers are heterogeneous with respect to their valuation θ of the services provided a car during any period. θ is a random variable which is uniformly distributed on the interval [0, 1], and higher values of θ indicate a higher valuation for the service provided. A car depreciates as it ages, and in any period, a new car is more valuable than a not-new car.

Key elements and decision variables: A customer's gross utility per period from using either a new or not-new car is given by

$$\tilde{G}(\theta, n) = \theta(1 - n\tilde{\delta}_i) \tag{10.13}$$

where $n = 0[1]$ if the car is new [not-new] and $\tilde{\delta}_i$ is a random variable defined on the interval [0, 1] that represents a car's depreciation with age ($i = b$ when the car has been bought and $i = l$ when the car has been leased). $\tilde{\delta}_i$ has mean δ_i ($i = b, l$).

A customer's expected net utility per period is given by

$$U = G(\theta, n) - p \tag{10.14}$$

where $G(\theta, n) = E[\tilde{G}(\theta, n)]$ and p is the price paid the customer for either a new or not-new car.

There are seven different types of customer according to the decisions they make in Periods 1 and 2 (see Table 10.2).[9]

From these seven types, four customer groups can be identified: *Top* (those who get a new car in both periods), *Middle* (those who get a new car in Period 1 and then hold onto this car in Period 2), *Bottom* (those who do not use a car in Period 1 and then get a not-new car in Period 2) and *Inactive* (those who do not use a car in either period and so stay out of the market). Thus, *Top* ≡ {LL, BB}, *Middle* ≡ {BH, LX}, *Bottom* ≡ {IU, IX} and *Inactive* ≡ {II}.

Three different strategies for the manufacturer are analysed: *Pure Leasing* (all cars produced are offered for lease), *Pure Selling* (all cars produced are offered for purchase) and *Concurrent Leasing and Selling* (all cars produced are offered for purchase or lease). In the first case, the customer types involved in the market are {LL, LX, IX, II}. In the second case, the types of customers involved are {BB, BH, IU, II}, and in the third case, all seven customer types are present.

Customer decisions (in order to maximise expected utility) are not modelled explicitly. Instead, inverse demand functions are used to link prices to demand for cars. q_{ij} denotes the demand in Period $i = 1, 2$ with $j = n$ for a new car, $j = x$ for a leased car and $j = u$ for a used car. p_{ij} denotes the purchase price in Period $i = 1, 2$ with $j = n$ for a new car, $j = x$ for a leased car and $j = u$ for a used car. r_{ij} denotes the price of a leased car in Period $i = 1, 2$ with $j = n$ for new and $j = x$ for leased.

The decision variables for the manufacturer are the quantities to sell q_{ij}, and these determine the prices using the inverse demand functions. The manufacturer's objective function to be maximised is expected total discounted profit $J(\cdot)$ (= demand times price) for both periods.

Pure leasing

The manufacturer only leases the cars. In Period 1, there is only leasing on new cars, whereas in Period 2, there are both new and ex-leased cars available to customers. The system of inverse demand functions is given by

$$\begin{aligned} r_{2x} &= (1 - \delta_l)(1 - q_{1l} - q_{2l}) \\ r_{2n} &= l_{2x} + \delta_l(1 - q_{2l}) \\ r_{1n} &= 1 - q_{1l}. \end{aligned} \tag{10.15}$$

[9] The abbreviations used for customer type denote the following: B (buy new car), L (lease new car), H (hold onto car), X (buy ex-leased car), U (buy used car), I (inactive—do not buy car).

Table 10.2 Customer types

Period 1	Period 2	Type
Buy new	Sell and buy/lease new	BB
	Retain	BH
Lease new	Buy/lease new	LL
	Buy/lease off-lease	LX
Inactive	Buy/lease used	IU
	Buy/lease off-lease	IX
	Inactive	II

Objective function: The manufacturer's expected total profit is given by

$$J_M(q_{1l}, q_{2l}) = r_{1l}q_{1l} + \rho[r_{2n}q_{2n} + r_{2x}q_{1l}] \tag{10.16}$$

where $\rho\,(0 \le \rho \le 1)$ is the discount factor per period.

Optimal decisions: The optimal quantities that the manufacturer should offer for lease in the two periods are

$$q_{1l}^* = \frac{1}{2(1+\delta_l\rho - \delta_l^2\rho)} \quad \text{and} \quad q_{2l}^* = \frac{1}{2} - (1-\delta_l)q_{1l}^* \tag{10.17}$$

respectively.

Comment: This is an optimisation problem with no game structure.

Pure selling

The manufacturer only sells new cars, but there are also used cars available in the second-hand market. In this case, the manufacturer's sales of new cars in Period 2 involve competition with customers selling used cars bought in Period 1 from the manufacturer. The system of inverse demand functions for new and used cars is given by

$$\begin{aligned} p_{2u} &= (1-\delta_b)(1-q_{1b}-q_{2b}) \\ p_{2n} &= p_{2u} + \delta_b(1-q_{2b}) \\ p_{1n} &= (1-q_{1b}) + \rho p_{2u}. \end{aligned} \tag{10.18}$$

Objective function: The manufacturer's expected total discounted profit for both periods is given by

$$J_M(q_{1b}, q_{2b}) = p_{1n}q_{1b} + \rho p_{2n}q_{2b}. \tag{10.19}$$

where $\rho\,(0 \le \rho \le 1)$ is the discount factor per period.

Optimal decisions: The optimisation is done in two stages with Stage 2 being a Nash game (with customers) for a given q_{1b}.

Stage 1: The manufacturer maximises expected profits in Period 2 (given by $J_2(q_{2b}) = p_{2n}q_{2b}$) for a given q_{1b} which yields

$$q_{2b}^{*}(q_{1b}) = \frac{1 - (1 - \delta_b)q_{1b}}{2}. \tag{10.20}$$

Stage 2: Given q_{2b}^{*}, the manufacturer chooses the optimal quantity to sell in Period 1 by maximising $J(q_{1b}; q_{2b}^{*}) = p_{1n}q_{1b} + \rho[p_{2n}q_{2b}^{*}(q_{1b})]$. The optimal sales level for Period 1 is given by

$$q_{1b}^{*} = \frac{2}{4 + \rho + 2\delta_b\rho - 3\delta_b^2\rho}. \tag{10.21}$$

Concurrent leasing and selling

In this case, the objective function for the manufacturer is the total expected discounted profit and the decision variables are the quantities of cars sold in Periods 1 and 2 and the quantity leased in Period 1 taking into account the interaction between the markets for new and not-new cars.

The authors consider the following two marketing strategies:

Premium lease in which the manufacturer's quantity decisions make an LL strategy dominate a BB strategy for all top group consumers;
Value lease in which the manufacturer's quantity decisions make a BB strategy dominate an LL strategy for all top group consumers.

In both cases, a Nash game takes place in Period 2 so a two-stage optimisation approach needs to be used similar to that for the Pure Selling case. The effect of the values of the mean depreciation rates for sales and leases δ_b and δ_l is also investigated.

Model 10.7 (Bucovetsky and Chilton 1986)

Bucovetsky and Chilton (1986) deal with concurrent renting (leasing) and selling strategies for a manufacturer when there is a threat of a competing firm entering the market.

Assumptions: The "established" manufacturer (M1) produces a product in Periods 1 and 2, whereas the new or "entrant" manufacturer (M2) only produces the product in Period 2. Thus, M1 is a monopolist in Period 1 and there is competition between M1 and M2 in Period 2. Any product produced in Period 1 has a lifetime equal to Period 1 + Period 2, whereas any product produced in Period 2 only has a lifetime equal to Period 2.

Key elements and decision variables: Let X_i denote the number of items produced by M1 in Period i ($i = 1, 2$) and Y the number of items produced (and sold) by M2 in Period 2. The marginal cost of production for both manufacturers is c. All items produced in Period 1 are either sold or leased. s is the fraction of the Period 1 production that is sold, so sX_1 denotes the number of items sold and $(1 - s)X_1$ denotes the number of items leased (rented). The length of the lease for these $(1 - s)X_1$ items is Period 1, and they are available for sale in Period 2. Leased items from Period 1 and items produced in Period 2 are perfect substitutes.

Table 10.3 Decisions for M1 and M2 in Periods 1 and 2

	Period 1	Period 2	
	M1	M1	M2
Production	X_1	X_2	Y
Lease (rental)	$(1-s)X_1$	–	–
Sales	sX_1	S_2	Y

If S_2 denotes the number of items sold by M1 in Period 2, then it follows that $S_2 \le (1-s)X_1 + X_2$.

A breakdown of the number of items produced, leased and sold by each manufacturer in Periods 1 and 2 is shown in Table 10.3.

Decision variables: The sets of decision variables for manufacturer M1 in Periods 1 and 2 are $\{X_1, s\}$ and $\{X_2, S_2\}$, respectively. In Period 2, manufacturer M2 has the single decision variable Y.

Objective functions: The objective function for M1 in Period 1 is the sum of the profit earned in this period and the discounted profit for Period 2. In Period 2, the objective functions for M1 and M2 are the profits they earn in this period.

Optimal decisions: The optimal values of the decision variables for the two manufacturers are found by solving a two-stage game.

Stage 1: (Optimisation in Period 2)

M1's and M2's optimal decisions for Period 2 are based on a Nash game where M1's decisions in Period 1 are given. The price per item in Period 2 is $\phi(S_2 + Y + sX_1)$, where ϕ is decreasing. Thus, M1's objective function is given by

$$J^2_{M1}(X_2, S_2; X_1, s) = \phi(S_2 + Y + sX_1)S_2 - cX_2 \tag{10.22}$$

where $S_2 \ge 0$, $X_2 \ge 0$ and $S_2 \le (1-s)X_1 + X_2$. These conditions imply that $X_2 = \max\{0, S_2 - (1-s)X_1\}$, so M1 has only one decision variable S_2 that needs to be selected to maximise

$$J^2_{M1}(S_2; X_1, s) = \phi(S_2 + Y + sX_1)S_2 - c[\max\{0, S_2 - (1-s)X_1\}]. \tag{10.23}$$

M2's objective function is given by

$$J^2_{M2}(Y; X_1, s) = [\phi(S_2 + Y + sX_1) - c]Y \tag{10.24}$$

where $Y \ge 0$.

Let the optimal values of the decision variables for each manufacturer in Period 2 (from the solution of the Nash game) be denoted by $S_2^*(X_1, s)$ and $Y^*(X_1, s)$, respectively.

Stage 2 (Optimisation in Period 1)

The price per item in Period 1 is $\phi(X_1)$, so M1's optimal decisions for Period 1 are found by maximising the objective function

$$J_{M1}\left(X_1, s; S_2^*(X_1, s), Y^*(X_1, s)\right)$$
$$= [\phi(X_1) - c]X_1 + \delta\left\{\phi\left[sX_1 + S_2^*(X_1, s) + Y^*(X_1, s)\right]\left[sX_1 + S_2^*(X_1, s)\right] - X_2^*(X_1, s)\right\}. \tag{10.25}$$

where δ is the discount factor per period and $X_2^* = \max\{0, S_2^* - (1 - s)X_1\}$.

Let the optimal values of the decision variables be denoted by X_1^* and s^*, respectively. The optimal Period 2 decisions are then given by $S_2^*\left(X_1^*, s^*\right)$ and $Y^*\left(X_1^*, s^*\right)$.

Model 10.8 (Desai and Purohit 1999)

Desai and Purohit (1999) extend Model 10.6 by considering two competing manufacturers M1 and M2 who are both able to sell and lease their product (a car).

Assumptions: The assumptions regarding the lifetime of a car (Periods 1 and 2), zero marginal costs for production and marketing and deterioration characterised by a parameter δ are the same as in Model 10.6, but used cars are now just ex-lease returns made available for sale in Period 2 for cars which were leased in Period 1.

Key elements and decision variables: Each manufacturer needs to choose the optimal number of cars to produce in Periods 1 and 2 and the optimal mix of leases and sales in Period 1. Lower-case [upper-case] letters are used to denote the relevant variables for M1 [M2].

q_{in} and Q_{in} denote the quantities of new cars produced in Period $i = 1, 2$, by the two manufacturers with f and F the fraction of cars leased in Period 1. q_{2j} and Q_{2j} denote the number of cars the two manufacturers have available for sale in Period 2 with $j = n$ for a new car and $j = u$ for a used car. p_{ij} and P_{ij} denote the one-period selling prices of cars from the two manufacturers in Period i $(= 1, 2)$ with $j = n$ for a new car and $j = u$ for a used car.

For M1, the inverse demand functions for new cars (in Periods 1 and 2) and used cars (in Period 2) are given by

$$\begin{aligned} p_{1n} &= \alpha - q_{1n} - eQ_{1n} \\ p_{2n} &= \alpha - q_{2n} - \gamma q_{2u} - e\{Q_{2n} + \gamma Q_{2u}\} \\ p_{2u} &= \gamma\{\alpha - q_{2n} - q_{2u} - e(Q_{2n} + \gamma Q_{2u})\}. \end{aligned} \tag{10.26}$$

The corresponding inverse demand functions for M2 are given by

$$\begin{aligned} P_{1n} &= \alpha - Q_{1n} - eq_{1n} \\ P_{2n} &= \alpha - Q_{2n} - \gamma Q_{2u} - e\{q_{2n} + \gamma q_{2u}\} \\ P_{2u} &= \gamma\{\alpha - Q_{2n} - Q_{2u} - e(q_{2n} + \gamma q_{2u})\}. \end{aligned} \tag{10.27}$$

The parameters $\gamma\,(=1-\delta)$ and $e\,(0<e<1)$ represent the degree of substitutability between new and used cars and the degree of competition between the two manufacturers' cars, respectively.

The sets of decision variables for M1 and M2 are $\{q_{1n}, f, q_{2n}\}$ and $\{Q_{1n}, F, Q_{2n}\}$, respectively.

Objective functions: Each manufacturer wishes to maximise their total discounted profit for both periods.

Optimal Decisions: The optimal values of the decision variables for M1 and M2 are found by solving a two-stage game.

Stage 1: (Optimisation in Period 2)

M1 and M2 maximise their profits for Period 2 (by selecting q_{2n} and Q_{2n}) for given values of $\psi \equiv \{q_{1n}, f, Q_{1n}, F\}$—the decision variables of Period 1—as a Nash game. The objective functions for M1 and M2 are $J^2_{M1}(q_{2n};\psi) = p_{2n}q_{2n} + p_{2u}fq_{1n}$ and $J^2_{M2}(Q_{2n};\psi) = P_{2n}Q_{2n} + P_{2u}FQ_{2n}$, respectively. Using the inverse demand functions from (10.26) and (10.27) in these expressions and carrying out the optimisation for both manufacturers simultaneously, the optimal decisions are given by

$$q^*_{2n}(\psi) = \frac{\alpha(2-e) - \gamma q_{1n}(2f+2-e^2) - e\gamma Q_{1n}(1-F)}{4-e^2}, \tag{10.28}$$

and

$$Q^*_{2n}(\psi) = \frac{\alpha(2-e) - \gamma Q_{1n}(2F+2-e^2) - e\gamma q_{1n}(1-f)}{4-e^2}. \tag{10.29}$$

Stage 2: (Optimisation in Period 1)

Using the optimal responses from Stage 1 given in (10.28) and (10.29), M1 and M2 maximise their total discounted profits for both periods by selecting $\{q_{1n}, f\}$ and $\{Q_{2n}, F\}$ optimally.

We only give the expressions for M1 (the results for M2 are similar due to the symmetry of the problem).

The profit for Period 1 is $J^1_{M1}(q_{1n}, f) = (1-f)q_{1n} + fp_{in}q_{1n}$, and the total discounted profit for both periods is $J_{M1}(q_{1n}, f) = J^1_{M1}(q_{1n}, f) + \rho J^2_{M1}(q^*_{2n}(\psi);\psi)$, where $\rho\,(0 \le \rho \le 1)$ is the discount factor per period. The optimal first period decisions are given by

$$f^* = \mathrm{Max}\left\{0, 1 - \frac{e^2(2+e-2\gamma\rho+2\gamma^2\rho)}{\gamma(4+2e-e^2)}\right\} \tag{10.30}$$

and

$$q^*_{1n} = \begin{cases} \dfrac{\alpha}{2+e+2\gamma\rho-2\gamma^2\rho}, & \text{for } f^* > 0 \\[2ex] \dfrac{\alpha(8+4e-2e^2-e^3+e^2\gamma\rho)}{(2-e)(2+e)^3 + 2\gamma\rho(2-e)(2+e)^2 - \gamma^2\rho(12+6e-3e^2-2e^3)}, & \text{for } f^* = 0 \end{cases} \tag{10.31}$$

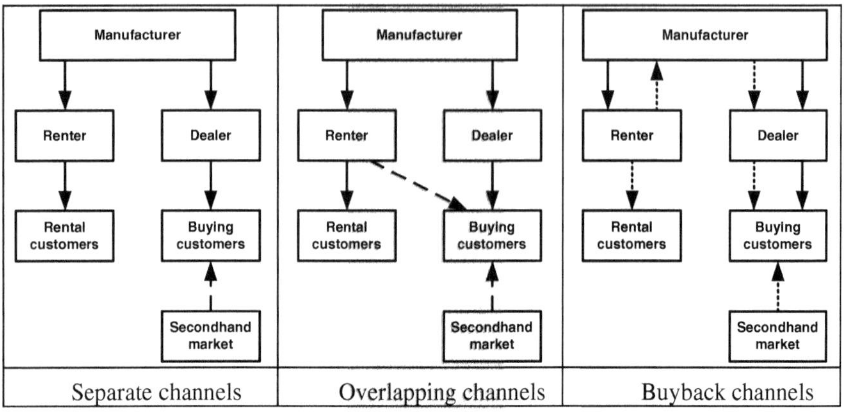

Fig. 10.3 Three different channels for marketing leases

The authors discuss the effect of the competitive intensity parameter (e) and the deterioration parameter (δ) on the optimal sell/lease decisions.

10.4.2.2 Three Parties

The three parties involved are manufacturers, intermediaries (dealers, renters or brokers) and customers. Manufacturers sell products to dealers who in turn sell them to customers. Alternatively, dealers may lease the products to customers. Dealers either sell the products to customers or lease them (acting as lessors or as intermediaries for manufacturer's leases). Renters only lease products to customers (acting as lessors or as intermediaries for manufacturers' leases). Brokers lease products using manufacturers' leases.

These options produce several different marketing channels for leasing. Figure 10.3 shows three of these channels in the case of a single manufacturer involved with a dealer and a renter.

Note that, in general, there can be one or more players in each party and the dominance structure between the players leads to many different types of GT model. We discuss some of these models.

Model 10.9 (Bhaskaran and Gilbert 2009)

Bhaskaran and Gilbert (2009) develop a two-period model that captures the interaction between a single manufacturer and a single dealer who sells and leases the product (acting as the lessor) to customers.

Assumptions: Customers decide to either buy or lease the product in Period 1. A lease lasts for one period, so in Period 2, both new and ex-leased items are sold by the dealer.

Key elements and decision variables: Let w_i denote the wholesale price at which the manufacturer sells the product to the dealer in Period i ($i = 1, 2$). q_{1n} and q_{1l} denote the quantities of the product that the dealer makes available to customers for sale and lease in Period 1. q_{2n} denotes the quantity made available for sale in Period 2, in addition to the ex-leased items from Period 1. Thus, the sets of decision variables for the manufacturer and dealer are $\{w_1, w_2\}$ and $\{q_{1n}, q_{1l}, q_{2n}\}$, respectively.

The dealer's price/unit to lease the product to customers in Period 1 is r_{1l}, and p_i is the product's selling price in Period $i = 1, 2$. As in Models 10.6 and 10.8, customer demands and prices are related by inverse demand functions. Marginal costs for manufacturer and dealer are both normalised to zero.

Objective functions: The manufacturer and dealer both want to maximise their total discounted profit for both periods.

Optimal Decisions: The optimal values of the decision variables for the manufacturer and dealer are found by solving a two-stage game.

Stage 1: (Optimisation in Period 2)

Total sales and price are related by the inverse demand function $p_2(q_{2n}, q_{1n}, q_{1l}) = a - q_{1l} - q_{1n} - q_{2n}$. The dealer's and manufacturer's objective functions are given by

$$J_D^2(q_{2n}; q_{1n}, q_{1l}, w_2) = (a - q_{1l} - q_{1n} - q_{2n})(q_{1l} + q_{2n}) - w_2 q_{2n} \tag{10.32}$$

and

$$J_M^2(w_2; q_{1n}, q_{1l}) = w_2 q_{2n} \tag{10.33}$$

respectively.

This is solved as a Stackelberg game where the dealer is the follower and decides the optimal $q_{2n}^*(w_2)$ by maximising the objective function in (10.32) for a given w_2 chosen by the manufacturer. The following optimal quantity is obtained:

$$q_{2n}^*(q_{1n}, q_{1l}, w_2) = \max\{\frac{a - w_2 - q_{1n} - 2q_{1l}}{2}, 0\}. \tag{10.34}$$

Using this optimal quantity in (10.32), the objective function for the manufacturer, the optimal wholesale price in Period 2 is given by

$$w_2^*(q_{1n}, q_{1l}) = \text{Max}\{\frac{a - q_{1n} - 2q_{1l}}{2}, 0\}. \tag{10.35}$$

Stage 2: (Optimisation in Period 1)

Here, the dealer first maximises total discounted profits for both periods for a given w_1 chosen by the manufacturer and also using the optimal decisions of both players in Period 2. The objective function for the dealer is given by

$$\begin{aligned} J_D(q_{1n}, q_{1l}; w_1) = {} & q_{1l}(a - w_1 - q_{1n} - q_{1l}) \\ & + q_{1n}(\{a - w_1 - q_{1n} - q_{1s} + \rho p_2(q_1, q_2, q_2^*(w_2^*))\} \\ & + \rho J_D^2(q_{2n}^*; q_{1n}, q_{1l}, w_2^*) \end{aligned} \tag{10.36}$$

where ρ $(0 \le \rho \le 1)$ is the discount factor per period.

The objective function for the manufacturer is given by

$$J_M(w_1) = w_1(q_{1n} + q_{1l}) + \rho J_M^2(w_2^*; q_{1n}, q_{1l}). \tag{10.37}$$

As in the previous stage, the optimal solution is obtained from a Stackelberg game and is as follows:

$$w_1^* = \frac{a(4+3\rho)^2}{16(2+\rho)}, \quad q_{1n}^* = 0 \quad \text{and} \quad q_{1l}^* = \frac{a(4+\rho)}{8(2+\rho)}. \tag{10.38}$$

Using these optimal values in (10.35) and (10.34) gives expressions $w_2^* = w_2^*(q_{1n}^*, q_{1l}^*)$ and $q_{2n}^* = q_{2n}^*(q_{1n}^*, q_{1l}^*, w_2^*)$.

Model 10.10 (Bhaskaran and Gilbert 2009)

In this model, the manufacturer uses a single broker to market leases and sells directly to customers in both periods. Note that, in this case, the manufacturer is the lessor.[10]

Key elements and decision variables: w_i denotes the manufacturer's lease margin in Period $i\,(i = 1, 2)$, and q_i denotes the quantity of leases that the broker makes available to customers in Period $i\,(i = 1, 2)$. The broker decides on the lease quantities to offer, depending on the lease margins.

Objective functions: The objective functions for the broker and manufacturer are their total discounted profits for the two periods and are given by

$$J_B(q_1, q_2; w_1, w_2) = q_1(a - q_1 - w_1) + \rho q_2(a - q_2 - w_2) \tag{10.39}$$

and

$$J_M(w_1, w_2; q_1, q_2) = q_1 w_1 + \rho q_2 w_2 \tag{10.40}$$

respectively.

(Note: The broker's profit is the lease price minus the lease margin.)

Optimal Decisions: The optimal values of the decision variables for the manufacturer and broker are found by solving a Stackelberg game with the manufacturer as the leader and the broker as the follower.

For given lease margins set by the manufacturer, the optimal lease quantities for the broker are given by

$$q_i^*(w_1, w_2) = \frac{a - w_i}{2}, \; i = 1, 2. \tag{10.41}$$

[10] This model is the same as the rental agency model of Purohit (1995).

The optimal lease margins are given by

$$w_1^* = w_2^* = a/2. \tag{10.42}$$

Using this result in (10.40) gives $q_1^* = q_2^* = a/4$.

Details of other models where there is competition between multiple dealers/brokers can also be found in Bhaskaran and Gilbert (2009). The implications of the manufacturer being forced to offer the same wholesale price to all dealers/brokers are investigated in these models.

Model 10.11 (Purohit and Staelin 1994)

Purohit and Staelin (1994) consider a single manufacturer who distributes cars through a dealer and a renter. In Period 1, the dealer sells the cars to customers and the renter leases them and only new cars are available. In Period 2, there is both a new and a second-hand market for cars (ex-leased cars and used cars that were bought in Period 1 and are now being resold). The three types of marketing channel shown in Fig. 10.3 are analysed for Period 2.

Assumptions: We focus on the separate channel structure. In Period 2, the manufacturer does not sell any cars to the renter and the renter does not sell any ex-leased cars. Thus, the only competition faced in this period by the dealer comes from the second-hand market arising from the dealer's sales in Period 1 (new cars sold in Period 1 become used cars available for sale in Period 2).

Key elements and decision variables: q_{in} denotes the quantities of new cars sold by the manufacturer to the dealer in Period $i = 1, 2$, and $q_{2u}(= q_{1n})$ denotes the quantity of used cars available in Period 2. p_{ij} denotes the one-period selling prices of the cars in Period $i = 1, 2$ with $j = n$ for a new car and $j = u$ for a used car.

The inverse demand functions for new cars (in Periods 1 and 2) and used cars (in Period 2) are given by

$$\begin{aligned} p_{1n} &= \alpha - \beta q_{1n} \\ p_{2u} &= \alpha - \beta(q_{1n} + q_{2n}) \\ p_{2n} &= \alpha - \beta(\gamma q_{1n} + q_{2n}) \end{aligned} \tag{10.43}$$

where γ $(0 \leq \gamma \leq 1)$ measures the degree of substitution of used cars for new cars.

The set of decision variables for the dealer is $\{q_{1n}, q_{2n}\}$, whereas the manufacturer's decision variable is w_n, the wholesale price for a new car.

Objective functions: The manufacturer and the dealer both wish to maximise their two-period profits.

Optimal decisions: The manufacturer is the leader, and the dealer is the follower in a Stackelberg game. For a given w_n from the manufacturer, the dealer chooses $q_{in}^*(w_n)$ and $q_{2n}^*(w_n)$. Knowing these optimal response quantities from the dealer, the manufacturer is then able to determine the optimal wholesale price w_n^*.

For a given w_n, the dealer's optimal quantities need to be chosen sequentially by first solving the Period 2 problem conditional on the quantity decision taken in

Period 1 and then solving the two-period problem. Given q_{1n} (and w_n), the dealer wishes to maximise Period 2 profit which is given by

$$J_D^2(q_{2n}; q_{1n}, w_n) = q_{2n}(p_{2n} - w_n). \tag{10.44}$$

The optimal quantity of new cars for the dealer in this period is then given by

$$q_{2n}^*(q_{1n}, w_n) = \frac{\alpha - w_n - \gamma\beta q_{1n}}{2\beta}. \tag{10.45}$$

The dealer now wishes to find the value of $q_{1n}(= q_{1n}(w_n))$ which maximises two-period profit given by

$$J_D(q_{1n}; q_{2n}^*(q_{1n}, w_n), w_n) = q_{1n}(p_{1n} - w_n) + q_{2n}^*(q_{1n}, w_n)(p_{2n} - w_n). \tag{10.46}$$

The optimal quantity of new cars for the dealer in Period 1 is then given by

$$q_{1n}^*(w_n) = \frac{\alpha(3 - \gamma) - w_n(1 - \gamma)}{\beta(8 - 2\gamma - \gamma^2)}. \tag{10.47}$$

Now, the manufacturer needs to determine the optimal wholesale price to charge the dealer. The manufacturer's two-period profit is given by

$$J_M(w_n; q_{1n}^*(w_n), q_{2n}^*(w_n)) = q_{1n}^*(w_n)(w_n - c) + q_{2n}^*(w_n)(w_n - c) \tag{10.48}$$

where c is the marginal cost of producing a new car. This objective function is maximised when

$$w_n^* = \frac{7\alpha + 5c}{10}. \tag{10.49}$$

The total number of new cars sold by the dealer is $Q_D^* = q_{1n}^* + q_{2n}^* = (7\alpha - 5c)/(4\beta(4 + \gamma))$, and the total number sold by the manufacturer is $Q_M^* = Q_D^* + q_{1r}$, where q_{1r} is the number of cars sold to the renter in Period 1.

Details of the analysis of the other two possible channel structures (overlapping and buyback) for Period 2 can be found in Purohit and Staelin (1994).

Model 10.12 (Xiong et al. 2012)

Xiong et al. (2012) develop a two-period dual-channel model for a manufacturer who sells a product directly to customers through an e-channel and also to an independent dealer who then sells and leases the product to the customers. A lease contract lasts exactly one period.

Assumptions: The product has a useful life of two periods. It is new when it is sold in Period 1 and is then classified as used in Period 2. Only, new products are available in Period 1, but both new and used products (those that were sold in Period 1) are available in Period 2. The product is assumed to be perfectly durable, so it does not deteriorate with time.

Key elements and decision variables: The marginal cost of production for the manufacturer is normalised to zero, and the marginal cost of selling through the e-channel is $c > 0$. The manufacturer needs to decide the wholesale price w_i and the number of units to sell through the e-channel q_{iM} in Period $i = 1, 2$. In Period 1, the dealer chooses the number of new units to sell q_s and lease q_l. The number of new units that the dealer sells in Period 2 is q_{2R}.

The model ends in Period 2, so selling a new product in this period is equivalent to leasing it. Thus, the number of units available for leasing in Period 2 is normalised to zero. The dealer is assumed to have zero marketing costs. The one-period lease prices in Periods 1 and 2 are given by

$$l_1 = a - (q_s + q_l) - q_{1M} \tag{10.50}$$

and

$$l_2 = a - (q_s + q_l) - q_{1M} - q_{2R} - q_{2M} \tag{10.51}$$

respectively, where a is the size of the potential market. If a customer purchases the product in Period 1, then the customer obtains the services it provides for both periods. Thus, the purchase price in Period 1 is $p_1 = l_1 + l_2$ (a zero discount rate is assumed, which implies that the discount factor per period $\rho = 1$).

The sets of decision variables for the manufacturer and dealer are $\{w_1, w_2, q_{1M}, q_{2M}\}$ and $\{q_s, q_1, q_{2R}\}$, respectively.

Objective functions: The manufacturer and the dealer both wish to maximise their two-period profits.

Optimal decisions: In each period, a two-stage game is played between the manufacturer and dealer. In Stage 1 of the game, the manufacturer announces the product's wholesale price to the dealer, and then, the dealer reacts by determining the quantities to sell and lease. In Stage 2, after the dealer determines these quantities to sell and lease, the manufacturer chooses the quantity to sell through the e-channel. Thus, the manufacturer is the leader, and the dealer is the follower in the Stackelberg game that takes place in Stage 1, whereas the roles are reversed for the Stage 2 Stackelberg game.

The optimal solution to the Period 2 problem is found first, and then, this information is used to find the optimal strategies for both parties in Period 1 (the method of backward induction).

Period 2 analysis

Stage 2: The manufacturer's and dealer's Period 2 profits are given by

$$J_M^2(q_{2M}, w_2; q_{2R}, q_s, q_l, q_{1M}) = w_2 q_{2R} + (l_2 - c) q_{2M} \tag{10.52}$$

and

$$J_D^2(q_{2R}; w_2, q_{2M}, q_s, q_l, q_{1M}) = (l_2 - w_2) q_{2R} + l_2 q_l \tag{10.53}$$

respectively.

For a given q_{2R} chosen by the dealer, the e-channel quantity that maximises the manufacturer's objective function in (10.52) (the manufacturer's optimal response) is denoted by $q_{2M}^*(q_{2R})$. The dealer's optimal quantity that maximises $J_D^2(q_{2R}; w_2, q_{2M}^*(q_{2R}), q_s, q_l, q_{1M})$ is denoted by q_{2R}^*. The manufacturer's optimal e-channel quantity is then $q_{2M}^* = q_{2M}^*(q_{2R}^*)$.

Stage 1: The manufacturer's optimal wholesale price is found by maximising $J_M^2(q_{2M}^*, w_2; q_{2R}^*, q_s, q_l, q_{1M})$ and is denoted by w_2^*.

Period 1 analysis

Stage 2: The manufacturer's and dealer's two-period profits are given by

$$\begin{aligned} J_M(q_{1M}, w_1; q_s, q_l) = {} & w_1(q_s + q_l) + (p_1 - c) q_{1M} \\ & + J_M^2(q_{2M}^*, w_2^*; q_{2R}^*, q_s, q_l, q_{1M}) \end{aligned} \tag{10.54}$$

and

$$\begin{aligned} J_D(q_s, q_l; q_{1M}, w_1) = {} & (p_1 - w_1) q_s + (l_1 - w_1) q_l \\ & + J_D^2(q_{2R}^*; w_2^*, q_{2M}^*, q_s, q_l, q_{1M}) \end{aligned} \tag{10.55}$$

respectively.

For given q_s and q_l chosen by the dealer, the e-channel quantity that maximises the manufacturer's objective function in (10.54) (the manufacturer's optimal response) is denoted by $q_{1M}^*(q_s, q_l)$. The dealer's optimal quantities that maximise $J_D(q_s, q_l; q_{1M}^*(q_s, q_l), w_1)$ are denoted by q_s^* and q_1^*. The manufacturer's optimal e-channel quantity is then $q_{1M}^* = q_{1M}^*(q_s^*, q_l^*)$.

Stage 1: The manufacturer's optimal wholesale price is found by maximising $J_M(q_{1M}^*, w_1; q_s^*, q_l^*)$ and is denoted by w_1^*.

Expressions for the optimal wholesale prices and optimal quantities in Periods 1 and 2 that are derived from the above procedure can be found in the paper.

10.4.3 Three-Period Models

Model 10.13 (Chemmanur et al. 2010)

Chemmanur et al. (2010) discuss a leasing problem with double-sided asymmetric information.[11] The lessor (manufacturer of capital equipment) has private information about the type of equipment being leased to the lessee/customer (who is also called the entrepreneur). The customer learns more about the equipment type over time as it is being used. Customers are also heterogeneous with respect to their maintenance costs for the leased equipment and have superior information about these costs compared to the manufacturer.

Assumptions: The manufacturer and the customers are risk neutral. At time 0 (start of Period 1), a customer requires one unit of the equipment to implement a project with positive NPV. A customer may buy or lease the equipment from the manufacturer, and more than one kind of leasing contract may be chosen.

The equipment is of two types G and B, with type G generating greater cash flows for customers than type B. At time 0, a customer cannot identify the exact type of equipment being purchased or leased but believes that it is type G [B] with probability $\theta\,[1-\theta]$.

There are two types of customer. A type L (low-cost) customer has a lower maintenance cost per period for the equipment c_L than a type H (high-cost) customer whose maintenance cost per period is c_H. $c = c_H - c_L$ denotes the difference in these maintenance costs. At time 0, before observing how a customer decides to acquire the equipment, the manufacturer cannot identify the exact customer type but believes that the customer is of type H [L] with probability $\phi\,[1-\phi]$.

The true type of equipment is revealed to a customer at the end of Period 1 after the equipment has been used. The customer then decides whether to perform maintenance for this period. A similar maintenance decision is made by the customer at the end of Period 2, and the use of maintenance by the customer affects the salvage value of the equipment. In addition to these maintenance decisions, a customer has other choices to make over time. If a short-term lease (lasting one period) has been chosen at time 0, then a choice has to be made whether or not to buy the equipment at time 1. Alternatively, if a customer initially chooses a long-term lease (lasting two periods), then there is an option to buy at time 2. The end of the useful life of the equipment occurs at time 3 (after three periods have elapsed). Figure 10.4 shows the sequence of events that takes place.

Key elements and decision variables: Type G equipment, if it is well maintained, returns a cash flow of x to a customer in each period for which it is used. The corresponding cash flow per period for type B equipment is fx ($f < 1$). The future cash flows generated by equipment (of either type) will be reduced by the fraction $1-\delta$ ($\delta < 1$) if it is not well maintained in any period. f is assumed to be sufficiently small such that $(1-\delta)fx < c_L$. This property implies that it is not

[11] Asymmetric information can lead to adverse selection. Adverse selection under leasing is discussed in Hendel and Lizzeri (2002).

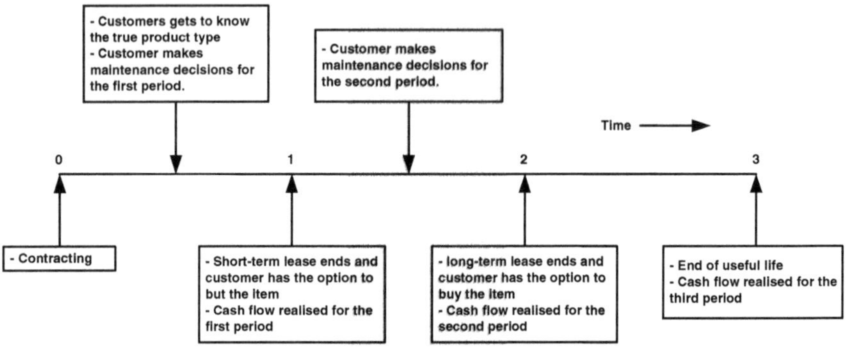

Fig. 10.4 Sequence of events over time

optimal for a type L (or a type H) customer to maintain type B equipment. It is also assumed that $c_{\mathrm{L}} < (1-\delta)x$, so it is optimal for a type L customer to maintain type G equipment. If $c_{\mathrm{H}} > (1-\delta)x$, then it will not be optimal for a type H customer to maintain this type of equipment.

If a customer returns the equipment to the manufacturer at the end of a lease, then no future cash flows will be received and the manufacturer will own the residual value of the returned equipment. If type G equipment has not been well maintained, then its residual value is assumed to be $\beta\,(\beta > 1)$ times the present value of future cash flows that the equipment would generate for a type H customer. The residual value to the manufacturer of type B equipment is assumed to be zero at any time.

If a customer decides to buy the equipment at time 0, then the sales price is S (for a sales contract) and the customer then owns the equipment until the end of its useful life (time 3). $\{M, R\}$ denotes a short-term LC with an option to buy. A customer pays an initial amount M at time 0 and then has the option to buy the equipment at time 1 for price R. A long-term LC with an option to buy is denoted $\{N, P\}$ where N is the initial lease payment and P is the purchase price at time 2.

The manufacturer first chooses the set of contracts to offer to a customer at time 0. The customer decides which contract to accept and then makes further decisions over time according to the terms of the initial contract.

Objective functions and optimal decisions: The manufacturer needs to decide on the menu of contracts to offer to customers plus the prices of these contracts in order to maximise the expected value of future cash flows. We only give the details of a small subset of the possible scenarios.

If the manufacturer provides type G equipment and only offers an $\{M, R\}$ LC to a customer, then the expected pay-off to the manufacturer is given by

$$\begin{aligned} J_M^{\mathrm{G}}(M,R) = M &+ \phi[I_S^{\mathrm{H}}R + (1-I_S^{\mathrm{H}})\beta(\delta x + \delta^2 x)] \\ &+ (1-\phi)[I_S^{\mathrm{L}}R + (1-I_S^{\mathrm{L}})\beta(\delta x + \delta^2 x)]. \end{aligned} \tag{10.56}$$

The indicator function is $I_S^{\mathrm{H}} = 1\ [0]$ if a type H customer purchases [does not purchase] the equipment when the short-term lease expires (at time 1). If the equipment is not purchased by the customer, no maintenance will be carried out and it will be returned to the manufacturer who then receives the residual value of the equipment. The indicator function I_S^{L} is defined similarly.

If the manufacturer provides type G equipment and only offers an $\{N, P\}$ LC to a customer, then the expected pay-off to the manufacturer is given by

$$J_{\mathrm{M}}^{\mathrm{G}}(N,P) = N + \phi[I_{\mathrm{L}}^{\mathrm{H}}P + (1 - I_{\mathrm{L}}^{\mathrm{H}})\beta\delta x] + (1-\phi)[I_{\mathrm{L}}^{\mathrm{L}}P + (1 - I_{\mathrm{L}}^{\mathrm{L}})\beta\delta x]. \quad (10.57)$$

The indicator functions $I_{\mathrm{L}}^{\mathrm{H}}$ and $I_{\mathrm{L}}^{\mathrm{L}}$ capture the purchase decisions for each type of customer when the long-term lease expires (at time 2). In this case, it is also assumed that the maintenance cost for a type H customer is sufficiently low so that both types of customer will perform maintenance in Period 1 but not in Period 2 if they decide not to purchase the equipment at the end of the lease.

If the manufacturer provides type G equipment and offers both an $\{M, R\}$ LC and an $\{N, P\}$ LC to a customer, then the expected pay-off to the manufacturer for the case where a type H (type L) customer accepts the long-term (short-term) lease is given by

$$\begin{aligned} J_{\mathrm{M}}^{\mathrm{G}}(M,R,N,P) = {} & \phi[N + I_{\mathrm{L}}^{\mathrm{H}}P + (1 - I_{\mathrm{L}}^{\mathrm{H}})\beta\delta x] \\ & + (1-\phi)[M + I_S^{\mathrm{L}}R + (1 - I_S^{\mathrm{L}})\beta(\delta x + \delta^2 x)]. \end{aligned} \quad (10.58)$$

The manufacturer of type G equipment finds the maximum of the expected pay-offs given in (10.56)–(10.58) to identify the best type(s) of LC to offer the customer.

We now consider the manufacturer of type B equipment. If this manufacturer offers an $\{S\}$ sales contract to a customer, then the expected pay-off is $J_{\mathrm{M}}^{\mathrm{B}}(S) = S$. If an $\{M, R\}$ LC is offered and the customer decides not to exercise the purchase option at time 1, the manufacturer's expected pay-off (assuming zero residual value for type B equipment) is $J_M^{\mathrm{B}}(M,R) = M$. If an $\{N, P\}$ LC is offered and the customer decides not to exercise the purchase option at time 2, the manufacturer's expected pay-off (assuming zero residual value for type B equipment) is $J_{\mathrm{M}}^{\mathrm{B}}(N, P) = N$. Finally, if both an $\{M, R\}$ LC and an $\{N, P\}$ LC are offered and the customer decides not to exercise the purchase option for both leases, the manufacturer's expected pay-off is $J_{\mathrm{M}}^{\mathrm{B}}(M, R, N, P) = \phi N + (1-\phi)M$. This type of manufacturer also compares expected pay-offs to identify the best type(s) of contract to offer the customer.

The objective of a customer at each time point is to maximise the expected value of future cash flows from using the equipment, net of any maintenance costs incurred. At time 0, the customer needs to choose a particular contract from the manufacturer and at subsequent times choose whether or not to perform maintenance and whether or not to purchase the equipment when the lease expires. Once again, we only give the details of a small subset of the possible scenarios.

As a first example, consider a type L customer who is offered a short-term $\{M, R\}$ LC and infers that the contract is from a type G manufacturer. The customer will accept the LC at time 0 only if

$$x - \mathrm{M} + I_S^{\mathrm{L}}(2x - 2c_{\mathrm{L}} - R) \geq 0, \tag{10.59}$$

and, at time 1, the customer will purchase the equipment ($I_S^{\mathrm{L}} = 1$) only if $2x - 2c_{\mathrm{L}} - R \geq 0$. The corresponding conditions for a type H customer to accept the LC at time 0 and purchase the equipment at time 1 are

$$x - \mathrm{M} + I_S^{\mathrm{H}}[\max(2x - 2c_{\mathrm{H}}, \delta x + \delta^2 x) - R] \geq 0, \tag{10.60}$$

and $\max(2x - 2c_{\mathrm{H}}, \delta x + \delta^2 x) - R \geq 0$, respectively.

As a second example, consider a type L customer who is offered a long-term $\{N, P\}$ LC and infers that the contract is from a type G manufacturer. The customer will accept the LC at time 0 only if

$$2x - c_{\mathrm{L}} - N + I_{\mathrm{L}}^{\mathrm{L}}(x - c_{\mathrm{L}} - P) \geq 0, \tag{10.61}$$

and will purchase the equipment $\left(I_{\mathrm{L}}^{\mathrm{L}} = 1\right)$ at time 2 only if $x - c_{\mathrm{L}} - P \geq 0$. The corresponding conditions for a type H customer to accept the LC at time 0 and purchase the equipment at time 2 are

$$\begin{aligned} &x + \max(x - c_{\mathrm{H}}, \delta x) - N + I_{\mathrm{L}}^{\mathrm{H}}[I_{MT}^{\mathrm{H}}(x - c_{\mathrm{H}}) + (1 - I_{MT}^{\mathrm{H}})\delta^2 x - P] \geq 0, \text{ and} \\ &I_{MT}^{\mathrm{H}}(x - c_{\mathrm{H}}) + (1 - I_{MT}^{\mathrm{H}})\delta^2 x - P \geq 0, \end{aligned} \tag{10.62}$$

respectively. The indicator function is $I_{MT}^{\mathrm{H}} = 1\,[0]$ if the customer maintains [does not maintain] the equipment at time 1.

As a third example, consider a type L customer who is offered both a short-term $\{M, R\}$ LC and a long-term $\{N, P\}$ LC and infers that these contracts are from a type G manufacturer. The customer will choose the short-term lease rather than the long-term lease at time 0 if the break-even constraint (10.59) and the constraint

$$x - \mathrm{M} + I_S^{\mathrm{L}}(2x - 2c_{\mathrm{L}} - R) \geq 2x - c_{\mathrm{L}} - N + I_{\mathrm{L}}^{\mathrm{L}}(x - c_{\mathrm{L}} - P) \tag{10.63}$$

are both satisfied. The customer will make the opposite decision if the break-even constraint (10.61) and the opposite of constraint (10.63) are both satisfied. The conditions for a type H customer to choose the long-term lease rather than the short-term lease at time 0 are that the break-even constraint (10.62) and the constraint

$$\begin{aligned} &x + \max(x - c_{\mathrm{H}}, \delta x) - N + I_{\mathrm{L}}^{\mathrm{H}}[I_{MT}^{\mathrm{H}}(x - c_{\mathrm{H}}) + (1 - I_{MT}^{\mathrm{H}})\delta^2 x - P] \\ &\quad \geq x - M + I_S^{\mathrm{H}}[\max(2x - 2c_{\mathrm{H}}, \delta x + \delta^2 x) - R] \end{aligned} \tag{10.64}$$

are both satisfied. The customer will choose the short-term lease if the break-even constraint (10.60) and the opposite of constraint (10.64) are both satisfied. Both types of customer will make their purchase decisions when the leases expire by using similar conditions to those used where only one lease is offered.

As a final example, consider a type L or a type H customer who is offered an $\{S\}$ sales contract and both infer that the contract is from a type B manufacturer. In this case, each type of customer will accept the sales contract at time 0 only if

$$-S + fx + f\delta x + f\delta^2 x \geq 0. \tag{10.65}$$

The technique that Chemmanur et al. (2010) use to solve the dynamic game (with imperfect information) between the manufacturer and the customers is perfect Bayes equilibrium (PBE).[12] The equilibrium of the game can be summarised as follows: If the quality factor f is large enough ($f \geq \underline{f}$) and the maintenance cost difference $c = c_{\mathrm{H}} - c_{\mathrm{L}}$ is sufficiently small ($\underline{c} \leq c < \bar{c}$), then

1. A type G manufacturer will offer both a short-term and a long-term LC, whereas a type B manufacturer will only offer a sales contract.
2. If a manufacturer offers both types of LC, then both types of customer believe with certainty that it is a type G manufacturer. A type L customer will accept the short-term LC, purchase the equipment at time 1 and perform maintenance in both the first and the second period. A type H customer will accept the long-term LC, perform maintenance only in the first period and not purchase the equipment at time 2.
3. If a manufacturer offers only a sales contract, then both types of customer believe with certainty that it is a type B manufacturer and so will accept the contract, use the equipment for three periods and not perform any maintenance.

10.4.4 Other Leasing Models

We now consider another GT leasing model involving two parties (a manufacturer and customers). The model uses a discrete time formulation (Periods 1, 2,…) where decisions are taken in each period over an infinite time horizon.

Model 10.14 (Tilson et al. 2006)

Tilson et al. (2006) consider a monopolist manufacturer who produces a single product. The manufacturer leases the product to a single corporate customer and also sells and leases the product to a population of individual customers (see Fig. 9.2 for details on this type of market structure). The interaction between the manufacturer and both types of customer is modelled as an infinite horizon

[12] See Fudenberg and Tirole (1991).

dynamic game, and the players' optimal strategies are found using the concept of Markov perfect equilibrium (MPE).[13]

Assumptions: Individual customers own or lease at most one unit of the product during each time period, whereas the corporate customer may require to use multiple units. The lifetime of the product is two periods: it is new in Period 1 and used in Period 2. The retail and corporate LCs both last one period, and the off-lease items are then returned to the manufacturer who resells them in the second-hand market.

In the dynamic game between the manufacturer and the customers, decisions are made sequentially in each period. The manufacturer moves first setting the prices for new and used products and placing the off-lease products from the previous period on sale in the second-hand market. This market is competitive, and price for these used items is determined such that the market clears in this period. In response to these prices for new and used products, each customer decides to buy, lease or do nothing. The actions of all the individual customers in the population and the corporate customer then determine the number of new products that will purchased and leased and the price for used products.

Individual customers (key elements, decision variables, objective function and optimal decisions): Individual customers are heterogeneous in their valuation of the product. A customer of type θ places a value $u_\theta(k)$ on the per-period usage of a product where $k = 0$ indicates a new product and $k = 1$ indicates a used product. Customers of different types are distributed in the population with probability density function $f(\theta)$ with $\theta \in [0, 1]$.

At the beginning of each period, an individual customer may be in one of two states: Does not own the product ($s = 0$) or owns a used product of age one period ($s = 1$). If A_s denotes the set of feasible actions for the customer who is in state s, then $A_0 = \{I, L, N, U\}$ where I = do nothing, L = lease a new product (for one period), N = buy a new product and U = buy a used product; and $A_1 = \{K, S, SL, SN\}$ where K = keep the product, S = sell it and do not replace it, SL = sell it and lease a new product, and SN = sell it and buy a new product.

$\underline{p} = \{p_0, p_l, p_1\}$ denotes the set of retail prices (p_0 is the sale price of a new product, p_l is the lease price, and p_1 is the price of a used product) that an individual customer has to pay, and these prices do not vary between periods. α denotes the transaction cost for a customer to sell a product in the second-hand market.

Each action $a \in A_s$ that a type θ individual customer takes results in an immediate net reward $r_\theta(s, a; \underline{p})$ and a state transition function $T(s, a)$ which specifies the customer's next period state given the current state and the action taken. If $V_\theta(s; \underline{p})$ denotes a type θ customer's expected total discounted reward

[13] See Maskin and Tirole (2001).

Table 10.4 Immediate rewards for type θ individual customers

State ($s = 0$)		State ($s = 1$)	
Action (a)	Reward ($r_\theta(s, a; \underline{p})$)	Action (a)	Reward ($r_\theta(s, a; \underline{p})$)
I	0	K	$u_\theta(1)$
L	$u_\theta(0) - p_l$	S	$p_1 - \alpha$
N	$u_\theta(0) - p_0$	SL	$p_1 - \alpha + u_\theta(0) - p_l$
U	$u_\theta(1) - p_1$	SN	$p_1 - \alpha + u_\theta(0) - p_0$

over an infinite time horizon by following an optimal strategy starting in state s, then $V_\theta(s; \bar{p})$ satisfies the optimality equations

$$V_\theta(s; \bar{p}) = \max_{a \in A_s} \{ r_\theta(s, a; \underline{p}) + \gamma V_\theta(T(s, a); \underline{p}) \} \quad (s = 0,\ 1) \tag{10.66}$$

where $\gamma\,(0 < \gamma \leq 1)$ is the discount factor per period.

The immediate rewards are shown in Table 10.4, and the state transitions are given by $T(0, a) = 0$ if $a = L,\ U,\ I$; $T(0, a) = 1$ if $a = N$; $T(1, a) = 0$ if $a = SL,\ K,\ S$; and $T(1, a) = 1$ if $a = SN$.

The rewards, transition functions and product prices do not vary with time, and the sets of feasible actions are finite, so only stationary customer strategies need to be examined. An analysis of the optimality equations shows that a type θ individual customer should always make decisions that cover two periods and should choose from (1) using a new product in each period (denoted by $\{00\}$), (2) using a new product in the first period and a used one in the second period (denoted by $\{01\}$), (3) using a used product in each period (denoted by $\{11\}$) or (4) doing nothing in each period (denoted by $\{22\}$). The optimal choice for the customer is made by comparing the value placed on using the particular product "bundle" over the two-period interval with the associated price that has to be paid which is calculated easily in terms of p_0, p_1 and p_l. The manufacturer uses this information about individual customer behaviour to choose optimal pricing policies.

Corporate customer (key elements, decision variables, objective function and optimal decisions): The corporate customer leases new products from the manufacturer. The lease price per product, set by the manufacturer, is $\bar{p}_l$, and at this price, the customer's leasing quantity is $\bar{D}_{\mathrm{L}}$ products. The corporate customer does not play strategically against either the manufacturer or the individual customers. All leases last for one period, so a lease quantity decision made in the current period does not affect any future decisions or profits. The corporate customer's objective is to find the optimal lease quantity $\bar{D}^*_{\mathrm{L}}(\bar{p}_l)$ that maximises single-period profit $J_{\mathrm{CC}}(\bar{D}_{\mathrm{L}}; \bar{p}_l)$. This optimal response function is assumed to be known to the manufacturer, and the demand is always satisfied since the manufacturer has no capacity constraints. The manufacturer's price and quantity decisions in the retail market are affected by the supply of used products coming from the corporate customer when leases expire.

Manufacturer (key elements, decision variables, objective function and optimal decisions): The manufacturer's decision variables are p_0, p_l and $\bar{p}_l$. The values of these prices are chosen simultaneously by the manufacturer, taking into account the dynamic interactions among the different sales channels. They affect the supply and demand of used products and thus the price p_1 for items in the second-hand market. The manufacturer's problem is complicated by the fact that there is a heterogeneous population of individual customers. Each customer knows their type, but the manufacturer only knows the distribution of types in the population.

Due to the strategies used by individual customers discussed above, the retail market does not contain two types of item (new and used products) but rather four types of two-period "bundles" labelled {00},{01},{11}and {22}. Depending on the prices of the bundles, there is a separation of customer types who prefer using one particular bundle over any of the others and this generates the two-period demand for each bundle $D_{\{00\}}$, $D_{\{01\}}$, $D_{\{11\}}$ and $D_{\{22\}}$.

If $D_{\mathrm{L}}, D_{\mathrm{N}}$ and D_{U} denote the quantities of products that the manufacturer leases, sells new and sells used per period to individual customers and D_I denotes the number of customers per period who do not participate in the market (choose to be idle), then it follows that

$$D_{\mathrm{L}} = D_{\{00\}}, D_{\mathrm{N}} = D_{\{01\}}/2, D_{\mathrm{U}} = D_{\{11\}} \quad \text{and} \quad D_I = D_{\{22\}}. \tag{10.67}$$

The manufacturer's profit per period is given by

$$\begin{aligned} J_{\mathrm{M}}(p_0, p_l, \bar{p}_l; p_1) = (p_0 - c)D_{\mathrm{N}}(p_0, p_l, p_1) + (p_l - c)D_{\mathrm{L}}(p_0, p_l, p_1) \\ + (p_1 - \beta)D_{\mathrm{U}}(p_0, p_l, p_1) + (\bar{p}_l - c)\bar{D}^*_{\mathrm{L}}(\bar{p}_l), \end{aligned} \tag{10.68}$$

where c is the constant marginal cost of producing and marketing the product and β is the cost to dispose of a used product in the second-hand market. The manufacturer wishes to maximise total discounted profit over an infinite horizon, but in the steady state, this is equivalent to maximising the profit per period given in (10.68). The second-hand market clearing condition that determines the used product price p_1 is given by

$$D_{\mathrm{U}}(p_0, p_l, p_1) = D_{\mathrm{L}}(p_0, p_l, p_1) + \bar{D}_{\mathrm{L}}(\bar{p}_l). \tag{10.69}$$

The manufacturer needs to find the values of p_0, p_l and $\bar{p}_l$ that maximise one-period profits. This optimisation problem is solved sequentially in two stages. In Stage 1, for a given value of $\bar{p}_l$ or, equivalently $\bar{D}_{\mathrm{L}}(\bar{p}_l)$, the optimal values $p_0^*(\bar{D}_{\mathrm{L}})$ and $p_l^*(\bar{D}_{\mathrm{L}})$ are found by maximising (10.68). In Stage 2, these conditional optimal values are substituted back into (10.68), and then using the known expression for $\bar{D}_{\mathrm{L}}(\bar{p}_l)$, the global optimal value $\bar{p}_l^*$ (and hence $\bar{D}_{\mathrm{L}}^*$) is determined. The remaining two global optimal selling prices are then $p_0^* = p_0^*(\bar{D}_{\mathrm{L}}^*)$ and $p_l^* = p_l^*(\bar{D}_{\mathrm{L}}^*)$, respectively.

By assuming simple forms for a type θ customer's product usage values $u_\theta(k)$ and for the density function of the customer-type distribution $f(\theta)$ and by making simplifying assumptions about other model parameters, Tilson et al. (2006) are able to obtain an analytical solution to the dynamic game. Useful insights are then gained about the impact of corporate leasing on the retail market. Further details can be found in the paper.

10.5 Maintenance Decision Models

In the case of an operational lease, the maintenance of the equipment is the responsibility of the lessor. The lessor has to decide on an effective maintenance policy, and this will depend on the following factors:

- The duration of the lease.
- The penalty terms in the LC.
- The equipment's usage intensity (which is under the control of the lessee) and the environment in which the equipment is used (which might or might not be under the control of the lessee).
- The state of the equipment at the beginning of the lease (this applies in the case of used equipment).

The lessor has to choose the type of maintenance policy to use and then determine the optimal values for the parameters of this policy. In order to do this, both equipment failures and the effect of maintenance actions (CM and PM) on these failures need to be modelled. We assume that all equipment failures are rectified through minimal repair. The times needed to rectify the failed equipment are small compared to the mean time between failures and so can be ignored in the failure modelling. Although insignificant for the purpose of modelling failures, these repair times need to be considered when assessing penalties in the LC. The minimal repair assumption implies that the number of equipment failures over time with no PM actions follows an NHPP. The failure intensity function then has the same form as the hazard function for time to first failure.

The effect of PM actions can be modelled in various ways. We assume that the actions are imperfect and consider two modelling methods. In the first case, each PM action reduces the failure intensity function, and in the second case, a PM action reduces the equipment's age. These PM modelling techniques are described in Chap. 3.

For each maintenance model that we discuss, we state the model assumptions, describe the model formulation and then perform the model analysis and optimisation. The following notation is used:

L	Duration of lease period
$F(t)$	Distribution function for the time to first failure of new equipment
$f(t)$, $h(t)$	Density and hazard functions associated with $F(t)$

$H(t)$	Cumulative hazard function $[= \int_0^t h(x)\mathrm{d}x]$
$\lambda_0(t)$	Failure intensity function with only CM actions $[=h(t)]$
$\Lambda_0(t)$	Cumulative failure intensity function with only CM actions $[= \int_0^t \lambda_0(x)\mathrm{d}x = H(t)]$
$\lambda(t)$	Failure intensity function with both CM and PM actions
$\Lambda(t)$	Cumulative failure intensity function with both CM and PM actions $[= \int_0^t \lambda(x)\mathrm{d}x]$
$N(L)$	Number of equipment failures over the lease period
C_f	Average cost of performing a minimal repair (CM action)
T	Time to perform a minimal repair (random variable)
$G(t)$	Distribution function for T
t_j	Time instant of the jth PM action
δ_j	Reduction in failure intensity due to the jth PM action
$C_p(\delta_j)$	Cost of the jth PM action
γ, τ	Parameters of penalty costs
C_n	Penalty cost per failure (when the number of failures exceeds γ)
C_t	Penalty cost per unit time (when repair time exceeds τ)
A	Age of used equipment at the beginning of lease
x	Reduction in age at an overhaul
$C_u(x)$	Cost of an overhaul with reduction in age x.

10.5.1 New Equipment Lease

Many different types of PM policy may be defined for use by the lessor. We consider a few of these policies in the case of a new equipment lease.

Policy 1 (Jaturonnatee et al. 2006)

Model formulation: The equipment is new and is leased for a period of length L. According to the LC, two types of penalty may be incurred by the lessor. Penalty 1 occurs if the equipment fails and is not restored to its operating condition within a specified period of time. If the random variable T denotes the time to restore the equipment from the failed state to the operating state, the penalty cost is given by $C_t\max[0, T - \tau]$. Penalty 2 occurs if there are any failures during the lease period, and the penalty cost is given by $C_n\max[0, N(L)]$.

The time to first failure of the equipment has distribution function $F(t)$, and the associated hazard function $h(t)$ is strictly increasing with $h(0) = 0$. The lessor rectifies all equipment failures during the lease period by performing minimal repairs, so the number of failures that occur up to time $t \in [0, L)$ with no PM actions is an NHPP with intensity function $\lambda_0(t) = h(t)$.

The equipment is subjected to k imperfect PM actions by the lessor during the lease period. The time instants at which these actions are carried out are given by $\{t_j, 1 \le j \le k\}$ with $t_i < t_j$ for $i < j$. The reduction in the failure intensity function due to the jth PM action is δ_j. Thus, the $2k + 1$ decision variables for the policy are $k, \underline{t} = \{t_j, 1 \le j \le k\}$ and $\underline{\delta} = \{\delta_j, 1 \le j \le k\}$.

Under this PM policy, the equipment's failure intensity function is given by

$$\lambda(t) = \lambda_0(t) - \sum_{i=0}^{j} \delta_i \quad \text{for} \quad t_j \le t < t_{j+1} \tag{10.70}$$

with $t_0 = \delta_0 = 0$. The δ_j must satisfy the constraints

$$0 \le \sum_{i=1}^{j} \delta_i \le \lambda_0(t_j), \quad 1 \le j \le k. \tag{10.71}$$

Model analysis and optimisation: The cost of the jth PM action is $C_p(\delta_j) = a + b\delta_j, \quad a > 0,\ b \ge 0$, so the total cost of the PM actions during the lease period [0, L) is given by

$$\sum_{j=1}^{k} C_p(\delta_j) = ka + b\sum_{j=1}^{k} \delta_j. \tag{10.72}$$

The expected total cost of the CM actions during the lease period $[0, L) - [t_0, t_{k+1})$ is given by

$$C_f \sum_{j=0}^{k} \left\{ \int_{t_j}^{t_{j+1}} \lambda(u)\mathrm{d}u \right\} = C_f \left[\Lambda_0(L) - \sum_{j=1}^{k} (L - t_j)\delta_j \right]. \tag{10.73}$$

The expected total Penalty 1 cost is given by

$$C_t \int_{\tau}^{\infty} \bar{G}(t)\mathrm{d}t \left[\Lambda_0(L) - \sum_{j=1}^{k} (L - t_j)\delta_j \right] \tag{10.74}$$

where $\bar{G}(t) = 1 - G(t)$, and the expected total Penalty 2 cost is given by

$$C_n \left[\Lambda_0(L) - \sum_{j=1}^{k} (L - t_j)\delta_j \right]. \tag{10.75}$$

Combining the costs in (10.72)–(10.75) gives the lessor's expected total cost using this PM policy as

$$J(k, \underline{t}, \underline{\delta}) = C\left[\Lambda_0(L) - \sum_{j=1}^{k} (\tilde{L} - t_j)\delta_j\right] + ka \tag{10.76}$$

where $C = C_f + C_t \int_{\tau}^{\infty} \bar{G}(t)\mathrm{d}t + C_n$ and $\tilde{L} = L - (b/C)$.

The optimal values of the decision variables are obtained by minimising the objective function given in (10.76) subject to the constraints given in (10.71). The optimisation is carried out using the following three-stage process.

Stage 1: For fixed k and $\underline{t}$, the optimal values $\underline{\delta}^*(k, \underline{t}) = \{\delta_1^*, \delta_2^*, \ldots, \delta_k^*\}$ are those that minimise $J(k, \underline{t}, \underline{\delta})$.

Stage 2: For fixed k, the optimal values $\underline{t}^*(k) = \{t_1^*, t_2^*, \ldots, t_k^*\}$ are obtained by minimising $J(k, \underline{t}, \underline{\delta}^*(k, \underline{t}))$.

Stage 3: The optimal k^* is obtained by minimising $J(k, \underline{t}^*(k), \underline{\delta}^*(k, \underline{t}^*(k)))$.

In Stage 1, for fixed k and $\underline{t}$, the optimal intensity reductions at the PM actions are given by

$$\delta_j^*(k, \bar{t}) = \lambda_o(t_j) - \lambda(t_{j-1}), \quad 1 \le j \le k. \tag{10.77}$$

The optimal PM action at time t_j is to reduce the failure intensity by the maximum possible amount if $t_j < \tilde{L}$ and carry out no PM action if $t_j \ge \tilde{L}$.

In Stage 2, for fixed k, the optimal times for PM actions are obtained by minimising

$$J(k, \underline{t}, \underline{\delta}^*(k, \underline{t})) = C\left[\Lambda_0(L) - \sum_{j=1}^{k} (t_{j+1} - t_j)\lambda_0(t_j)\right] + ka \tag{10.78}$$

where $0 < t_1 < t_2 < \cdots < t_k < \tilde{L}$ and $t_{k+1} = \tilde{L}$.

If the time to first failure of the equipment has a Weibull distribution with scale parameter $\alpha = 1$ and shape parameter $\beta > 1$, then an analytical solution for $\underline{t}^*(k) = \{t_1^*, t_2^*, \ldots, t_k^*\}$ can be obtained by solving the first-order conditions $\partial J(k, \underline{t}, \underline{\delta}^*(k, \underline{t})/\partial t_j = 0$, $1 \le j \le k$. In this case, the failure intensity function with no PM actions is given by $\lambda_0(t) = \beta t^{\beta-1}$. The optimal PM times are defined recursively by

$$t_k^* = v_{k+1}\tilde{L}, \tag{10.79}$$

$$t_{j-1}^* = v_j t_j^*, \quad j = k, k-1, \ldots, 2 \tag{10.80}$$

where

$$v_{j+1} = \frac{\beta - 1}{\beta - v_j^{\beta-1}}, \quad j = 1, 2, \ldots, k, \ (v_1 = 0). \tag{10.81}$$

In Stage 3, a numerical search procedure is needed to obtain the optimal number of PM actions k^*.

Policy 2 (Pongpech and Murthy 2006)

A periodic PM policy is considered where the PM actions are carried out at times $j\tau$, $j = 1, 2, \ldots, k$, over the lease period. k is the largest integer less than L/τ. The reduction in the failure intensity function due to the jth PM action is δ_j, so the $k + 1$ decision variables for this policy are τ and $\underline{\delta} = \{\delta_j, 1 \le j \le k\}$.

The lessor's expected total cost using the policy is given by

$$J(\tau, \underline{\delta}) = C\left[\Lambda_0(L) - \sum_{j=1}^{k} (\tilde{L} - j\tau)\delta_j\right] + ka. \tag{10.82}$$

The optimal values of the decision variables are obtained by minimising this objective function subject to the constraints

$$0 \le k < L/\tau \quad \text{and} \quad 0 \le \sum_{i-1}^{j} \delta_i \le \lambda_0(j\tau), \quad 1 \le j \le k. \tag{10.83}$$

The optimisation is carried out using the following two-stage process.

Stage 1: For fixed τ, the optimal values $\underline{\delta}^*(\tau) = \{\delta_1^*, \delta_2^*, \ldots, \delta_{k(\tau)}^*\}$ where $k(\tau)\tau < L \le (k(\tau) + 1)\tau$ are those that minimise $J(\tau, \underline{\delta})$. The optimal intensity reductions are given by

$$\delta_j^*(\tau) = \lambda_o(j\tau) - \lambda((j-1)\tau), \quad 1 \le j \le k(\tau). \tag{10.84}$$

Thus, the optimal PM action at time $j\tau$ is to reduce the failure intensity by the maximum possible amount if $j\tau < L$ and carry out no PM action if $j\tau \ge \tilde{L}$.

Stage 2: The optimal τ^* is obtained by minimising

$$J(\tau, \underline{\delta}^*(\tau)) = C\left[\Lambda_0(L) - \sum_{j=1}^{k(\tau)-1} \tau\lambda_0(j\tau) - (\tilde{L} - k(\tau)\tau\right] + k(\tau)a. \tag{10.85}$$

It is impossible to derive any analytical results for this optimisation, so a numerical search procedure must be used.

Policy 3 (Yeh et al. 2009)

k imperfect PM actions are performed by the lessor during the lease period $[0, L)$ at times $t_1, t_2, \ldots, t_k$. At each PM action, the failure intensity function is reduced by the constant amount δ. Thus, the $k + 2$ decision variables for this policy are k, δ and $\underline{t} = \{t_j, 1 \leq j \leq k\}$.

The lessor's expected total cost using this PM policy is given by

$$J(k, \delta, \underline{t}) = C\left[\Lambda_0(L) - \delta \sum_{j=1}^{k} (\tilde{L} - t_j)\right] + ka, \tag{10.86}$$

and the optimal values of the decision variables are obtained by minimising this objective function subject to the constraints

$$\lambda_0(t_j) - j\delta \geq 0, \quad 1 \leq j \leq k. \tag{10.87}$$

The optimisation is carried out using a three-stage process. In Stage 1, for fixed $k > 0$ and $\delta > 0$, it is easy to show that the optimal time instant to perform the jth PM action is when $\lambda_0(t_j) = j\delta$, so

$$t_j^*(k, \delta) = \lambda_0^{-1}(j\delta) = h^{-1}(j\delta). \tag{10.88}$$

Using this result, the objection function then becomes

$$J(k, \delta, \underline{t}^*(k, \delta)) = C\left[H(L) - \delta \sum_{j=1}^{k} (\tilde{L} - h^{-1}(j\delta))\right] + ka. \tag{10.89}$$

For Stages 2 and 3, Yeh et al. (2009) provide a numerical algorithm that can be used to search for k^* and δ^*, the optimal values for the number of PM actions and the constant reduction in the failure intensity function at each action. If the time to first failure of the equipment has a Weibull distribution with scale parameter $\alpha > 0$ and shape parameter $\beta > 1$, then the failure intensity function with no PM actions is given by $\lambda_0(t) = \alpha\beta(\alpha t)^{\beta-1}$. In this case, an analytical solution for $\delta^*(k)$ can be obtained, and then, a simple numerical search procedure is needed to identify k^*.

In an example for the Weibull case, Yeh et al. (2009) compare the optimal values of the decision variables and the minimum expected total costs for Policies 1, 2 and 3. When $\beta = 2$, each policy has approximately the same optimal expected cost. As β increases, Policy 1 has the smallest cost. The performance of Policy 3 is better than that of Policy 2 and is very close to that of Policy 1. The optimal value for the number of PM actions k^* is found to be approximately the same for the three policies.

10.5.2 Used Equipment Lease

We now discuss a lessor's maintenance policy for used equipment.

Policy 4 (Pongpech et al. 2006)

The age of the equipment at the beginning of the lease period is $A > 0$. In order to reduce CM costs and penalties for equipment failures, the lessor carries out an overhaul (upgrade) which reduces the age by an amount $x \, (0 \leq x \leq A)$ before the equipment is leased out. The cost of the upgrade is given by the increasing function

$$C_u(x) = \frac{wx}{1 - e^{-\varphi(A-x)}} \tag{10.90}$$

with parameters $w > 0$ and $\varphi > 0$. PM actions are carried out during the lease period as in Policy 1. The lessor's expected total cost using this used equipment maintenance policy is given by

$$J(x, k, \underline{t}, \underline{\delta}) = C\left[\Lambda_0(A + L - x) - \Lambda_0(A - x) - \sum_{j=1}^{k} (\tilde{L} - t_j)\delta_j\right] + ka + \frac{wx}{1 - e^{-\varphi(A-x)}}. \tag{10.91}$$

The problem is to find the optimal age reduction for the upgrade x^*, the optimal number of PM actions k^*, the optimal time instants for the PM actions $\underline{t}^* = \{t_1^*, t_2^*, \ldots, t_k^*\}$ and the optimal failure intensity reduction values $\underline{\delta}^* = \{\delta_1^*, \delta_2^*, \ldots, \delta_k^*\}$ that minimise the objective function given in (10.91) subject to the constraints $0 \leq x \leq A$, $0 < t_1 < t_2 < \cdots < t_k < \tilde{L}$ and

$$0 \leq \sum_{i=1}^{j} \delta_i \leq \lambda_0(A + t_j - x) - \lambda_0(A - x), \quad 1 \leq j \leq k. \tag{10.92}$$

The optimisation is carried out using the following four-stage process.

Stage 1: For fixed x, k and $\underline{t}$, the optimal values $\underline{\delta}^*(x, k, \underline{t})$ are those that minimise $J(x, k, \underline{t}, \underline{\delta})$.

Stage 2: For fixed x and k, the optimal values $\underline{t}^*(x, k)$ are obtained by minimising $J(x, k, \underline{t}, \underline{\delta}^*(x, k, \underline{t}))$.

Stage 3: For fixed x, the optimal $k^*(x)$ is obtained by minimising $J(x, k, \underline{t}^*(x, k), \underline{\delta}^*(x, k, \underline{t}^*(x, k)))$.

Stage 4: Find x^*, the value of x that minimises $J(x, k^*(x), \underline{t}^*(x, k^*(x)), \underline{\delta}^*(x, k^*(x), \underline{t}^*(x, k^*(x))))$.

Once x^* is obtained, then proceeding backwards gives $k^* = k^*(x^*)$, $\underline{t}^* = \underline{t}^*(x^*, k^*)$ and $\underline{\delta}^* = \underline{\delta}^*(x^*, k^*, \underline{t}^*)$. Pongpech et al. (2006) give details of the optimisation procedure when the equipment's failure intensity has the Weibull form $\lambda_0(t) = \beta t^{\beta-1}$.

10.5.3 Other Maintenance Decision Models

We now give a brief description of some other maintenance models for equipment under lease. In these models, all failures are rectified by minimal repairs and each PM action reduces the equipment's age. The details of this type of PM modelling can be found in Chap. 3.

Yeh and Chen (2006) derive the optimal periodic PM policy when the time to first failure of the equipment has a Weibull distribution and each PM action reduces the age by a constant amount. In Yeh et al. (2011a), a more general model is analysed where the PM actions are non-periodic with non-constant age reductions. It is shown that in the case of a Weibull time to first failure distribution with shape parameter $\beta > 1$, the optimal PM policy is in fact periodic with constant age reduction where the interval between PM actions is equal to the amount of age reduction. Optimal policies are discussed for various forms of PM costs.

In Chang and Lo (2011), PM actions are scheduled when the equipment's age reaches a specified control limit and each action reduces the age by a constant amount. The optimal PM policy and length of lease period are determined taking into account the equipment's residual value at the end of the lease. It is shown that the equipment should be restored to its original condition after each PM action and the PM actions are periodic. Yeh et al. (2011b) extend the Chang and Lo (2011) model by allowing multiple lease periods for the equipment and treating the number of lease periods as an extra decision variable.

References

Bazargan M, Hartman J (2012) Aircraft replacement strategy: model and analysis. J Air Transp Manag 25:26–29

Bhaskaran SR, Gilbert SM (2009) Implication of channel structure for leasing or selling durable goods. Mark Sci 28:918–934

Bucovetsky S, Chilton J (1986) Concurrent renting and selling in a durable goods monopoly under threat of entry. Rand J Econ 17:261–278

Bulow J (1982) Durable goods monopolists. J Polit Econ 90:314–332

Chang WL, Lo HC (2011) Joint determination of lease period and preventive maintenance policy for leased equipment with residual value. Comput Ind Eng 61:489–496

Chemmanuar T, Jiao Y, Yan A (2010) A theory of contractual provisions in leasing. J Financ Intermed 19:116–142

Coase R (1972) Durability and monopoly. J Law Econ 15:143–149
Desai P, Purohit D (1998) Leasing and selling: optimal marketing strategies for a durable goods firm. Manage Sci 44(11):19–34
Desai P, Purohit D (1999) Competition in durable goods markets: the strategic consequences of leasing and selling. Mark Sci 18:42–58
Fudenberg D, Tirole J (1991) Perfect Bayesian equilibrium and sequential equilibrium. J Econ Theory 53:236–260
Gamba A, Rigon R (2008) The value of embedded real options: evidence from consumer automobile lease contracts—A note. Finance Res Lett 5:213–220
Gavazzza A (2010) Asset liquidity and financial contracts: evidence from aircraft leases. J Financ Econ 95:62–84
Grenadier S (1995) Valuing lease contracts a real-options approach. J Financ Econ 38:297–331
Hendel I, Lizzeri A (2002) The role of leasing under adverse selection. J Polit Econ 110:113–143
Henry JB, Roenfeldt RL (1978) Cost analysis of leasing hospital equipment. Inquiry 33–37
Hsu CI, Li HC, Liu SM, Chao CC (2011) Aircraft replacement scheduling: a dynamic programming approach. Transp Res Part E Logistics Transp Rev 47:41–60
Jaturonnatee J, Murthy DNP, Boondiskulchok R (2006) Optimal preventive maintenance of leased equipment with corrective minimal repairs. Eur J Oper Res 174:201–215
Kim EH, Lweellen WG, McConnell JJ (1978) Sale-and-leaseback agreements and enterprise valuation. J Financ Quant Anal 13:871–881
Lee K (1995) Optimal retail lease contracts: the principal-agent approach. Reg Sci Urban Econ 25:727–738
Maskin E, Tirole J (2001) Markov perfect equilibrium: I. observable actions. J Econ Theory 100:191–219
Meier H, Roy R, Seliger G (2010) Industrial product-service systems—IPS2. CIRP Ann Manufact Technol 59:607–627
Miller M, Upton C (1976) Leasing, buying, and the cost of capital services. J Finance 31:761–786
Mollaghasemi M, Pet-Edwards J, Gupta U (1995) A Multiple criteria buy versus lease analysis for government contracts. IEEE Trans Eng Manage 42:278–287
Myers SC, Dill DA, Bautista AJ (1976) Valuation of financial lease contracts. J Finance 31:799–819
Nisbet A, Ward A (2001) Radiotherapy equipment—purchase or lease? Br J Radiol 74:735–744
Pongpech J, Murthy DNP (2006) Optimal periodic preventive maintenance policy for leased equipment. Reliab Eng Syst Saf 91:772–777
Pongpech J, Murthy DNP, Boondiskulchok R (2006) Maintenance strategies for used equipment under lease. J Qual Maintenance Eng 12:52–67
Purohit D (1995) Marketing channels and the durable goods monopolist: renting versus selling reconsidered. J Econ Manage Strategy 4:69–84
Purohit D, Staelin R (1994) Rental, sales and buybacks: managing secondary distribution channels. J Mark Res 31:325–338
Rieple A, Helm C (2008) Outsourcing for competitive advantage: an examination of seven legacy airlines. J Transp Manage 44:280285
Roenfeldt RL, Henry JB (1979) Lease-cost measurement of hospital equipment under cost-based reimbursement. Financ Manage 8:24–35
Saaty TL (1980) The analytic hierarchy process. McGraw Hill, New York
Sorensen IW, Johnson RE (1977) Equipment financial leasing practices and costs: an empirical study. Financ Manage 6:33–40
Stokey N (1981) Rational expectation and durable good pricing. Bell J Econ 12:112–128
Tilson V, Wang Y, Yang W (2006) Channel strategies for durable goods: co-existence of selling and leasing to individual and corporate consumers. Technical Memorandum 813, Department of Operations, Weatherhead School of Management, Case Western Reserve University, Cleveland, Ohio
Trigeorgis L (1996) Evaluating leases with complex options. Eur J Oper Res 91:315–329

Vargas LG, Saaty TL (1981) Financial and intangible factors in fleet lease or buy decisions. Ind Mark Manage 10:1–10

Xiong Y, Yan W, Fernandes K, Xiopng ZK, Guo N (2012) "Bricks vs. Clicks": the impact of manufacturer encroachment with a dealer leasing and selling of durable goods. Eur J Oper Res 217:75–83

Yeh RH, Chen CK (2006) Periodical preventive-maintenance contract for a leased facility with Weibull life-time. Qual Quant 40:303–313

Yeh RH, Kao KC, Chang WL (2009) Optimal preventive maintenance policy for leased equipment using failure rate reduction. Comput Ind Eng 57:304–309

Yeh RH, Kao KC, Chang WL (2011a) Preventive-maintenance policy for leased products under various maintenance costs. Expert Syst Appl 38:3558–3562

Yeh RH, Chang WL, Lo HC (2011b) Optimal length of lease period and maintenance policy for leased equipment with a control-limit on age. Math Comput Modell 54:2014–2019

Part IV
Management Issues

Chapter 11
Management of EWs/MSCs and LCs

11.1 Introduction

Assets (products, plants and infrastructure) are either bought, built or leased to meet some specified performance objectives over a specified time interval. Asset performance depends on the characteristics of the asset (such as reliability and quality), the usage mode and intensity, the operating environment, and on the support service (such as maintenance and maintenance logistics). If the asset is bought (or built), the owner of the asset can either do the maintenance in-house or outsource some or all of the maintenance through an EW or MSC. If the asset is leased, depending on the terms of the lease contract, the responsibility for maintenance either rests with the lessor or rests with the lessee.

EWs, MSCs and LCs are contracts that involve two or more parties with the most two important ones (in the context of maintenance service) being (1) customers who purchase the contracts and (2) providers who sell the contracts. Customers can be households buying or leasing consumer products or businesses/government agencies buying or leasing products, plants or infrastructures under an EW, MSC or LC.[1] Contracts deal with support services in the case of EWs and MSCs and with the asset and support services in the case of LCs. The customers and providers for EWs/MSCs and for LCs are as indicated below, and we will use this terminology in the rest of the chapter.

	EW/MSC	Lease
Customer	Owner of asset	Lessee
Provider	Service agent	Lessor

EWs, MSCs and LCs are complex processes, and both parties (customers and the providers) need to manage them properly. Failure to do this can lead to serious consequences. Management deals with issues such as the information needed for decision-making, implementation of the decisions and monitoring the outcomes.

[1] Often terms such as companies, corporations, firms, organisations, etc. are used instead of businesses.

D. N. P. Murthy and N. Jack, *Extended Warranties, Maintenance Service and Lease Contracts*, Springer Series in Reliability Engineering,
DOI: 10.1007/978-1-4471-6440-1_11,

Some of the issues are common to both parties and others different. As such, the framework needed for effective management is different for the two parties. Maintenance of an asset is the responsibility of the provider in the case of EWs and of the provider and/or customer in the case of MSCs and LCs. In this chapter, we look at the issues and the framework needed for effective management of EWs, MSCs and LCs focussing on the maintenance aspect from both the customer and provider perspectives.

The outline of the chapter is as follows. We start with a general discussion of management in Sect. 11.2. Sections 11.3 and 11.4 deal with the framework and issues from the customer and provider perspectives, respectively. Customer service is a critical issue that providers need to take into account, and this is discussed in Sect. 11.5. Maintenance logistics plays a very important role in ensuring effective customer service and is the focus of Sect. 11.6. Information is critical for proper management, and this is discussed in Sect. 11.7 where we also discuss information management systems. Since uncertainty is a very significant factor in the EW, MSC and LC processes, effective management requires proper risk analysis. This issue is discussed in Sect. 11.8.

11.2 Management

Management involves making decisions and coordinating the efforts of people to accomplish desired goals and objectives of a business using available resources efficiently and effectively. The tasks involved in management include the following:

- Planning: Deciding what needs to happen in the future (short to long term) and generating plans for action.
- Structuring and coordinating: This deals with the relationships among workers and making optimum use of the resources required to enable the successful carrying out of plans.
- Staffing: Recruiting and hiring of people with appropriate skills for executing the different tasks.
- Leading/directing: Determining what must be done in a situation and getting people to do it.
- Controlling/monitoring: Checking progress against plans.
- The decision-making is done at three different levels:
    - The *Strategic level* deals with decisions that have long-lasting effect on the business.
    - The *Tactical level* typically includes decisions that are updated anywhere between once every quarter and once every year.
    - The *Operational level* refers to short-term (day-to-day or week-to-week) decisions.

Most organisations have a three-level management structure—top, middle and junior—in a hierarchy of authority to perform different tasks. The top level (assisted by the middle level) deals with strategic issues, the middle level (assisted by the junior level) deals with tactical issues, and the junior level deals with operational issues.

11.2.1 Maintenance Management

Maintenance management (MM) of an asset deals with activities required for (1) maintenance planning (philosophy, maintenance workload forecast, capacity and scheduling), (2) maintenance organisation (work design, standards, work measurement and project administration) and (3) maintenance control (of works, materials, inventories, costs and quality-oriented management). It involves a simple input–output relationship—the inputs are manpower, management, tools, equipment, etc., and the output is the asset performing satisfactorily.[2]

Marquez and Gupta (2006) discuss MM in terms of the *framework* (the supporting structure needed to manage maintenance effectively) and the *process* (the course of actions and the series of stages to follow). MM needs to be aligned with actions at three different levels of business activities—strategic (business priorities transformed into maintenance priorities), tactical (assignment of maintenance resources—equipment, material and human—to execute the maintenance plans) and operational (maintenance tasks are executed properly). They propose a framework for MM involving three pillars: (1) the information technology (IT) pillar, (2) the maintenance engineering (ME) pillar and (3) the organisational (or behavioural) pillar, and they discuss the elements and issues for each of the three pillars. An illustrative sample of the maintenance-related problems that need to be solved at these three different levels in the case of in-house maintenance is indicated in Fig. 11.1. When maintenance of an asset is outsourced or the asset is leased, then the management of maintenance is more complex and involves both the provider and customer and is discussed in the remainder of the chapter. The responsibility for the different tasks can be with either party depending on the contract.

11.3 Customer Perspective

A framework for effective management of EWs, MSCs and LCs from the customer perspective involves three stages as indicated in Fig. 11.2. A brief discussion of the stages is given below.

---

[2] For more on this, see Duffuaa et al. (1998).

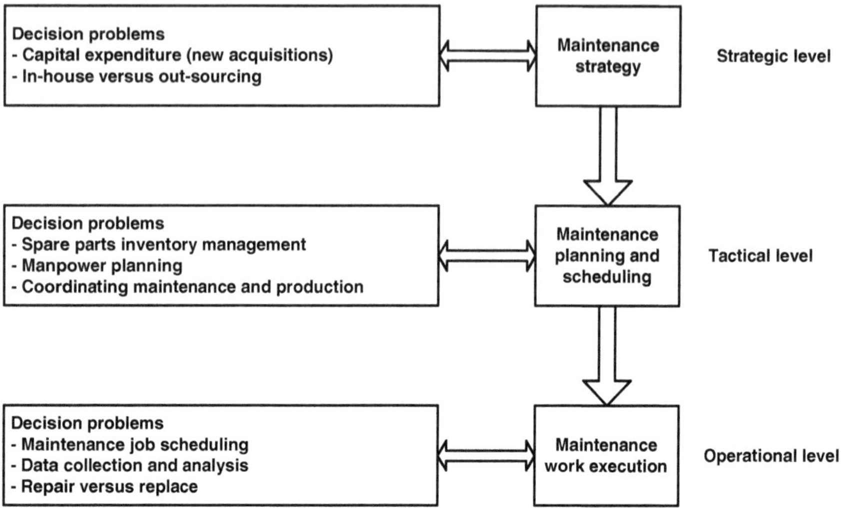

Fig. 11.1 Maintenance management at the strategic, tactical and operational levels

- Stage 1 [Pre-purchase]: Searching for information regarding alternative providers and details of contracts. The contract can be standard or custom-designed depending on the asset and the power relationship between the customer and the provider.
- Stage 2 [Purchase]: Evaluating the contract options (terms, period, exclusions, etc.) and selecting the best option to purchase.
- Stage 3 [Post-purchase]: Monitoring the performance (of maintenance service in the case of EWs/MSCs and of both asset and maintenance service in the case of LCs) over the contract period. The performance depends on asset usage (under the control of customer) and the maintenance from the provider.

We focus mainly on Stages 1 and 2 in the remainder of the section. Issues of interest in Stage 3 are customer satisfaction and dispute resolution, and these are discussed in the next section.

11.3.1 EWs

Most EW customers are households buying a single consumer product at a time. They have very limited or no technical background to evaluate the long-term performance of the product being purchased. In Stage 1, customers obtain information from retailer regarding the alternative EW options (provided by the OEM and/or other independent providers) at the time of purchase and they need to decide whether to buy an EW or not. If the customer does not buy an EW at the time of purchase, in some cases, the manufacturer might contact the customer to offer the option to buy an EW before the BW expires. The decision to purchase or

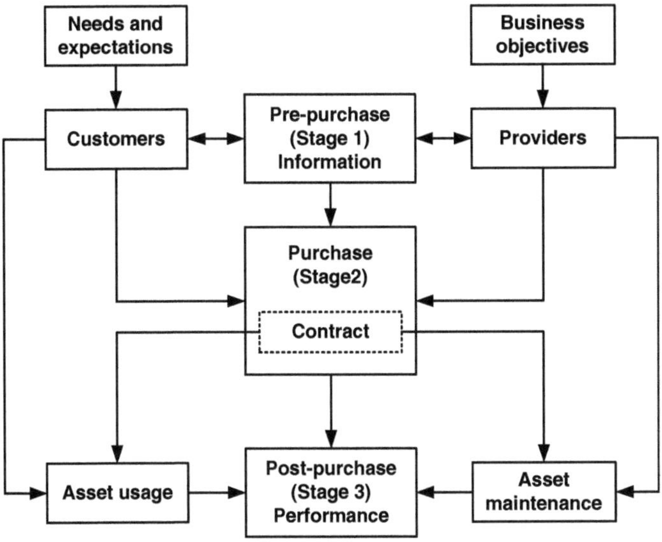

Fig. 11.2 Framework for management (customer perspective)

not is influenced by the customer's attitude to risk and the "feeling of security" that the EW provides.

Businesses and government agencies usually buy a batch of standard products (such as computers) and often negotiate the terms of the BW so that an EW is not an important issue. In this case, the strategic management deals with the choice of EW. With a dominant customer, the terms would be decided jointly by the customer and the provider.

In Stage 3, it is essentially the execution of claims under the EW and ensuring that the conditions of the EW are not violated as this would result in the EW becoming null and void.

11.3.2 MSCs

Depending on the customer (businesses and government agencies) and the asset (product, plant or infrastructure), there is a wide variation in the types of MSC contract. The customer has a degree of technical competence, and as such, Stage 1 usually involves the providers being asked to provide a lot of information relating to the technical aspects of maintenance being outsourced and their capability to carry out the activities involved. At this stage, a decision is made regarding which maintenance activities are to be outsourced.

Stage 2 is a complex process as the evaluation can involve several criteria—some hard and objectively quantifiable and others soft which can only be characterised subjectively in a qualitative sense. These include

- Evaluation of alternative options. These could include several criteria (such as cost and risk). The cost aspects have been discussed in earlier chapters, and the risk issue is discussed later in the chapter.
- Choosing among the alternative options. This is a multicriteria decision problem.[3] The *Analytical Hierarchy Process* (AHP) provides a method to assist in the decision-making process, and this is discussed later in the section.
- Contract negotiations in the case of a customised MSC.[4]

Stage 3 involves monitoring, collecting relevant data and initiating actions if the provider is not carrying out the tasks as per the contract terms. The data management issue is discussed later in the chapter.

11.3.3 LCs

This case is very similar to a MSC. In Stage 1, the customer collects information regarding the different lessors and the other parties involved in the lease process. Stage 2 involves choosing the best lease contract. For complex plants and infrastructure, this is a multicriteria decision-making process and the risk factors become more critical. Stage 3 involves monitoring the asset as well as the support service of the provider and proper data collection to ensure that the provider meets the terms of the contract.

11.3.4 The Analytical Hierarchy Process

The AHP hierarchy is a structured method of modelling the decision at hand. It consists of an overall *goal*, a group of options or *alternatives* for reaching the goal and a group of factors or *criteria* that relate the alternatives to the goal as indicated in Fig. 11.3 where there are M options and K different criteria. The criteria can be further broken down into subcriteria, sub-subcriteria and so on, to produce as many levels as the problem requires.

The first step is to decompose the decision problem into a hierarchy of subproblems, each of which can be analysed independently. The elements of the hierarchy can relate to any aspect of the decision problem—tangible or intangible, carefully measured or roughly estimated, well or poorly understood—anything at all that applies to the decision at hand.

---

[3] The literature on multicriteria decision-making is vast, see for example, Roy (1996) and Belton and Stewart (2002). In the context of maintenance, see de Almeida (2005, 2007) and de Melo Brito et al. (2010).

[4] For more on contract negotiations, see for example, Kumar et al. (2004), and Jackson and Pascual (2008).

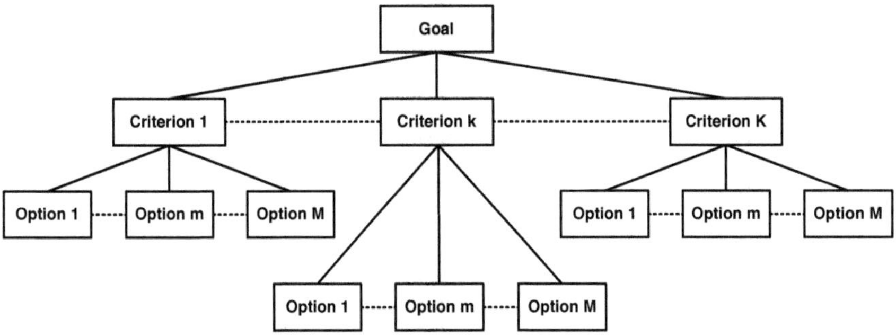

Fig. 11.3 The AHP process

Once the hierarchy is built, one evaluates the various elements by comparing them to one another two at a time, with respect to their impact on an element above them in the hierarchy. In making the comparisons, one uses data about the elements as well as subjective judgments about the elements' relative meaning and importance. This essence of the AHP is that it combines human judgments with the underlying information in performing the evaluations.

The AHP converts these evaluations to numerical values that can be processed and compared over the entire range of the problem. A numerical weight (or priority) is derived for each element of the hierarchy, allowing diverse and often incommensurable elements to be compared with one another in a rational and consistent way. This capability distinguishes the AHP from other decision-making techniques.

In the final step of the process, numerical priorities are calculated for each of the decision alternatives. These numbers represent the alternatives' relative ability to achieve the decision goal, so they allow a straightforward consideration of the various courses of action.

The AHP methodology for decision-making was conceived and developed by Thomas Saaty, and his book on the topic (Saaty 1980) is a classic. The literature on the AHP and its application is vast, and overviews of the applications can be found in Vargas (1990) and Vaidya and Kumar (2006). Sundarraj (2004) deals with a Web-based AHP approach to standardise the process of managing service contracts. The use of the AHP in the context of maintenance and leasing has received some attention in the literature. Triantaphyllou et al. (1997) suggest the use of the AHP for deciding on the best maintenance strategy using four criteria—cost, reparability, reliability and availability. Bevilacquaa and Braglia (2000) describe an application of the AHP for selecting the best maintenance strategy from a set of five alternatives in the context of the maintenance of an oil refinery. Bertolini et al. (2004) look at maintenance outsourcing service selection using the AHP. Vargas and Saaty (1981) use the AHP process to decide between leasing and buying a fleet taking into account financial and intangible factors. Yang and Lee (1997) deal with the use of the AHP in the selection of facility location.

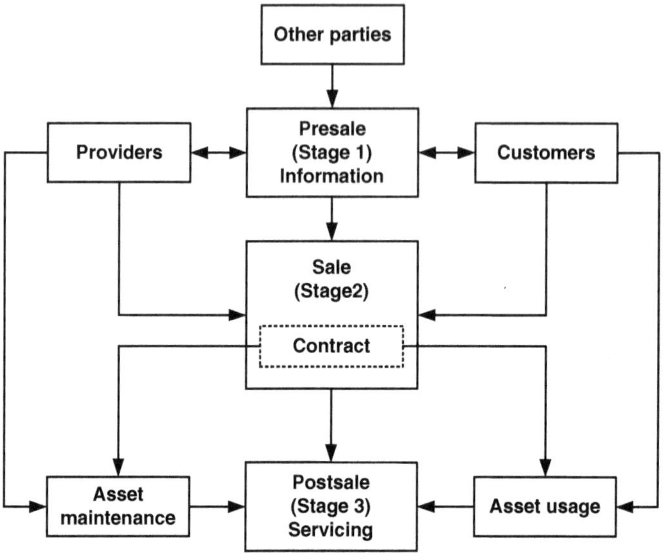

Fig. 11.4 Framework for management (provider perspective)

11.4 Provider Perspective

A framework for effective management of EWs, MSCs and LCs from the provider perspective also involves three stages as indicated in Fig. 11.4. A brief discussion of the stages is given below.

Stage 1 [Presale]: Searching for information relating to customers, competitors, other parties, assets, etc., so as to decide on various management decisions from an overall business perspective at the strategic level. This includes the following:

- Number of customers to service (size of business, regions to service, etc.)
- Marketing strategies—channels of distribution, etc.[5]
- EW/MSC/LC offerings (terms, price, etc.)
- Technology acquisition—for example, e-maintenance (of ship engines or assets in remote locations)
- Partnership with other parties
- Facilities needed to deliver the service such as service centres and warehouses (number, location, capacities, etc.),
- Risk management (discussed in Sect. 11.8).

Stage 2 [Sale]: Marketing of contracts for standard contracts—pricing, promotions, etc.

[5] For more on marketing channels see, Lewis (1968) and on the linkage between product distribution and service support channels see, Loomba (1996).

Stage 3 [Post-sale]: The management decisions are mainly at the operational level. These include the following:

- Customer service and satisfaction.
- Maintenance logistics—inventory of spares, scheduling of maintenance activities, decisions regarding repair versus replace of a failed component, etc.

We focus mainly on Stages 1 and 2 in the remainder of the section. Issues of interest in Stage 3 are discussed in the next two sections.

11.4.1 EWs

EW contracts are standard contracts, and their management is very similar to that for BWs. Murthy and Blischke (2006) deal with the management issues at all three stages.

11.4.2 MSCs

For a standard MSC, the terms are defined by the provider and as such the management issues are very similar to that for an EW. However, for a customised MSC, the process in Stage 1 is different. The information flow is very critical and is discussed further in a later section of the chapter.

11.4.3 LCs

LCs for consumer and most industrial and commercial products are standard contracts. For the lease of complex plants and infrastructure, the contracts are usually customised contracts. The management issues at Stages 1–3 (for leasing of products) and at Stages 1 and 3 (for plants and infrastructure) are very similar to the standard and customised MSCs.

11.5 Customer Service

11.5.1 Customer Satisfaction/Dissatisfaction

Customer dissatisfaction can arise due to poor performance of the leased item in the case of LCs and of the maintenance service in the case of EWs, MSCs and LCs. This can impact the customer business due to the quality of goods and services produced decreasing and operating costs increasing.

Customer dissatisfaction impacts the provider's business performance in a similar manner. The direct impact is higher costs and damage to the provider's reputation. The indirect impact is the loss of existing customers (with customers switching provider) and the negative word-of-mouth effect resulting in loss of new potential customers. The consequences are difficult and costly to rectify, and hence, it is very important that the providers avoid this occurring in the first instance. Hence, ensuring customer satisfaction is very critical for providers.

There are several dimensions to maintenance service quality, and many of these are intangible and can vary significantly from customer to customer. For example, customers can have undue expectations regarding performance for a variety of reasons (exaggerated statements made during promotion, customer being not fully informed, etc.). However, other dimensions are more tangible and can be objectively assessed. These include response time to attend to a failure, the time taken to rectify the failed item, delays resulting from lack of spares, workshop resources, etc. Through effective service logistics, the negative impacts resulting from these can be minimised.[6]

11.5.2 Service Recovery

While providers cannot eliminate complaints, they can learn to respond effectively to them. This response, termed *service recovery*, is defined as the process by which the provider attempts to rectify a service- or asset-related failure. Service recoveries are critical because customers perceiving poor recovery efforts may not renew the contract (or in the worst case terminate the contract prematurely) and purchase elsewhere.

Complaint handling is an important element of service recovery. It is often the manufacturer's (or service agent's) response to a failure, rather than a failure itself, that triggers discontent which in turn leads to dissatisfaction. It is important that the response be perceived as fair, as this has a significant impact on satisfaction with the asset performance and maintenance service.

11.5.3 Dispute Resolution

Disputes in the context of EWs, MSCs and LCs arise when the provider refuses to admit a complaint from customer as being valid under the terms of the contract. There are several paths that a customer may use in seeking redress, as shown in Fig. 11.5. The first course of action is to complain to the provider. This is a resolution process

---

[6] There is extensive literature on customer satisfaction and improving service quality. See, for example, Maxham and Netemeyer (2002), Haugen and Hill (1999), Ehinlanwo and Zairi (1996) and Kurata and Nam (2010).

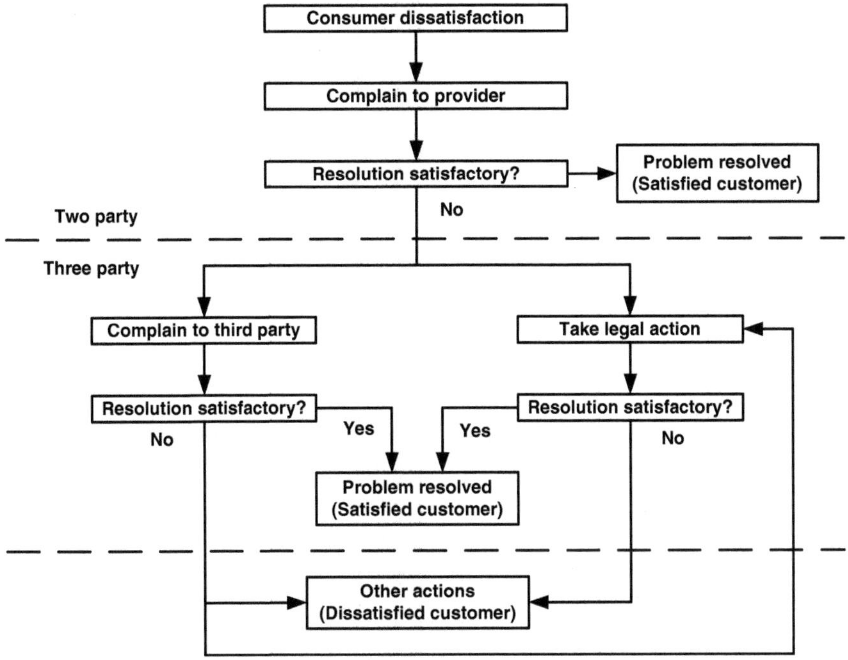

Fig. 11.5 Complaint resolution process

involving only two parties—provider and dissatisfied customer. If the resolution is satisfactory, then the problem is resolved. If not, the customer might either complain to a third party (such as a consumer protection agency or a media channel—especially in the case of EWs) and then seek legal action should the problem remain unresolved or go directly for resolution through legal action. If this leads to a resolution, no further action is necessary. If not, the customer might pursue other actions.[7]

11.6 Maintenance Logistics

The Society of Logistic Engineers (SOLE 1996) defines logistics as "integrated design, management and operation of human, physical, financial and information resources, during product, system or service life time". Carrasqueira and Machado (2007) view logistics as consisting of two phases: logistics engineering and operational logistics. Logistics engineering includes planning activities and

[7] Palfrey and Romer (1983) deal with dispute resolution in the context of BWs and this is also applicable to EWs and MSCs.

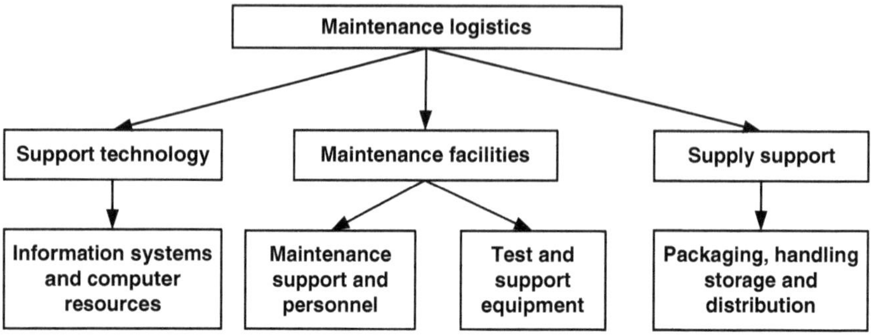

Fig. 11.6 Key elements of maintenance logistics

obtaining the necessary resources before their use, and operational logistics deals with activities needed to keep systems operative.

In the context of EWs, MSCs and LCs, the overall purpose of maintenance logistics is to guarantee proper support to ensure that the asset under consideration is performing as required over the contract period.[8] The main tasks of maintenance logistics are the following:

1. To assess asset condition,
2. To plan maintenance requirements to ensure the desired asset performance, and
3. To execute the maintenance actions.

These need to be done in a manner that takes into account cost and asset availability. The three key elements of maintenance logistics for maintaining a product or plant under an EW, MSC and LC are as shown in Fig. 11.6. The logistics for the maintenance of infrastructure is more complex due to the spatial dimension of the asset.

11.6.1 Maintenance Logistics Management

Maintenance logistics management deals with decision-making at *strategic*, *tactical* and *operational* levels. The strategic level deals with decisions that include the number and the location and capacities of warehouses. The tactical level typically includes decisions that are updated anywhere between once every quarter and once every year. This embraces purchasing decisions, inventory policies and transportation strategies, including the frequency with which the retailers are visited. The operational level refers to day-to-day decisions such as scheduling, routing trucks and measuring performance.

Some of the key issues, all of which have been studied extensively, are the following:

[8] Murthy et al. (2004) deal with issues and challenges in product warranty logistics.

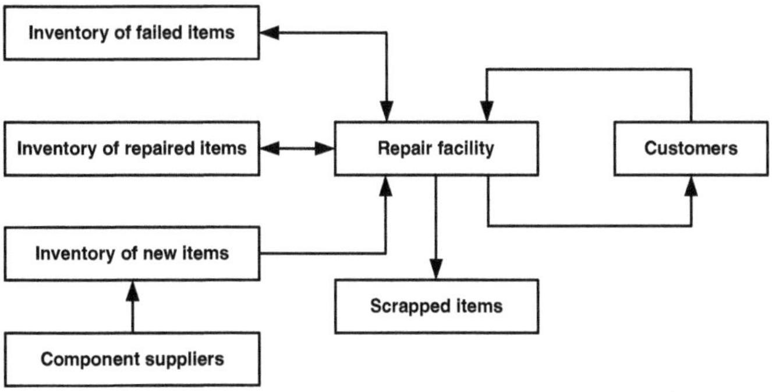

Fig. 11.7 Elements of maintenance logistics management

- Allocation of maintenance resources, including location of repair facilities, level of repair, capacity of repair facilities, size of repair equipment, etc.
- Inventory management
- Scheduling of maintenance actions
- Transportation and distribution of maintenance material and repair personnel
- Information management and decision support systems.

Figure 11.7 shows the key elements of maintenance logistics management.

Failed items are brought to repair facility by customers. Some are scrapped due to a variety of reasons and others get repaired. As a result, there are two kinds of spares: repaired and new items. The inventory level of repaired items increases with each repair, and the inventory level of new items increases with each purchase. These levels decrease as repaired and new items get used.

Maintenance tasks include planned and unplanned activities. These need to be performed in a sequential manner taking into account the various constraints (e.g. production or operation requirements, priorities and the times needed for different tasks). This is referred to as maintenance scheduling.

Some maintenance tasks have to be performed on site (e.g. lifts and air conditioners). In this case, a repair crew has to visit the site and carry the maintenance resources needed (e.g. spares, material and equipment). This leads to two problems—the repairman problem and the knapsack problem.

There is extensive literature dealing with the topics mentioned above. A small illustrative sample where interested readers can get more details is given below.

- *Logistics management*: Bowersox and Closs (1996), Christopher (1998), Coyle et al. (1992) and Aras et al. (2011).
- *Location problem*: Daskin (1995), Dresner (1992), Handler and Mirchandani (1979), Jayaraman and Srivastava (1995) and Owen and Daskin (1998) and Tersine (1994).
- *Inventory management*: Hadley and Whitin (1963), Gupta and Korugan (1998), Nozick and Turnquist (2001), Alfredsson (1997) and Sherbrooke (1992).
- *Repair analysis*: Barros and Riley (2001) and Cassady et al. (2001).
- *Transportation*: Qu et al. (1999) and Evers (2001).

- *Job scheduling*: Hajri et al. (2000), Jianer and Miranda (2001) and Ponnambalam et al. (2001).
- *Travelling repairman*: Afrati et al. (1986), Yang (1989) and Agnihothri (1998).
- *Repair versus replace*: Jack and Van der Duyn Schouten (2000), Jack and Murthy (2001), Iskandar and Murthy (2003) and Iskandar et al. (2005).

11.7 Information Flow and Management

Many different kinds of information are needed in the effective management of MSCs and LCs. They can be broadly grouped into the following categories:

- Technical(relating to asset and services needed for maintenance)
- Operations (servicing-related data)
- Financial (relating to various types of costs)
- Legal (relating to contract details)
- Commercial (relating to customers, marketing, etc.).

The flow, management and use of information are very critical topics.[9] McFarlane and Cuthbert (2012) propose a model for information requirements in complex engineering services. MSCs and LCs for plants and infrastructure involve both parties negotiating to formulate the contract, and information flow plays a very critical role in this process. We use the model proposed by McFarlane and Cuthbert to characterise the information flow for MSCs and LCs.

11.7.1 Information Flow for MSC Management

The model for information flow is shown in Fig. 11.8.[10] It involves six elements (shown as six boxes in the figure), and there are four different information flows as listed below.

- Design information (from customer to provider)
- Delivery information (from provider to customer)
- Assessment/evaluation information (from customer to provider)
- Information flow between provider and external suppliers such as vendors supplying material and spares, equipment manufacturers providing the equipment needed for carrying out maintenance service, specialists (for example for oil analysis), etc.

We discuss briefly each of these.

---

[9] There are many papers dealing with information in the context of service delivery, see for example Berkley and Gupta (1995).

[10] The information flow model of McFarlane and Cuthbert contains only the middle four boxes.

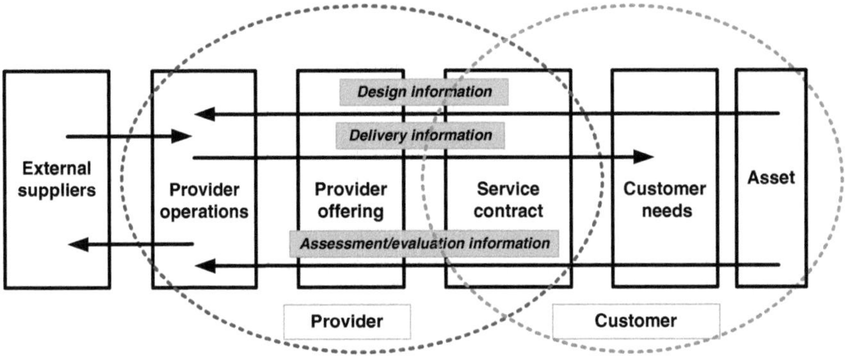

Fig. 11.8 Information flows for MSC management

11.7.1.1 Design Information

- *Asset*: Asset condition at the start of the contract (past history of operation and maintenance)
- *Customer needs*: Conceptual information about the customer's requirements for asset performance (reliability and financial related)
- *Service contract*: Information to formalise the MSC
- *Provider offering*: Alternative MSCs, asset performance implications, etc.
- *Provider operations*: Technical information to plan and develop the delivery of the MSC offered, cost information, resources (organisation, equipment needed, skill base), etc.
- *External Suppliers*: Technical information of equipment and material needed by the provider for delivery of the maintenance service, cost information, etc.

11.7.1.2 Delivery Information

- *Customer needs*: Information from the provider to enable the customer to achieve better coordination between maintenance service and asset operation
- *Service contract*: Information regarding the details of maintenance services to be delivered by the provider
- *Provider offering*: Information regarding the delivery details of alternative maintenance service offerings
- *Provider operations*: Technical information on how the provider can deliver the agreed maintenance services (relating to logistics of maintenance service delivery)
- *External Suppliers*: Delivery logistics for spares, material, etc.

11.7.1.3 Assessment and Evaluation Information

- *Asset*: Asset condition over the contract period
- *Customer needs*: Information to determine fulfilment of customer need
- *Service contract*: Performance requirements defined through various metrics (reliability, financial, operations, penalties, etc.), customers' responsibilities, etc.
- *Provider offering*: Information relating to the effectiveness of the maintenance service offerings defined through suitable performance measures
- *Provider operations*: Operational information on performance of service infrastructure and operations
- *External suppliers*: Component suppliers, transport services, etc.

11.7.2 Information Flow for LC Management

This depends on the type of asset (infrastructure or plant) and the type of lease contract. Also, depending on the lease, it can involve other parties (such as creditors and regulators). The information flow model would have a structure similar to that for a MSC.

11.7.3 Information Management System

An effective information management system (IMS) for both providers (and customers) needs to have the four interlinked elements shown in Fig. 11.9.

11.7.3.1 Database

The database must be maintained and managed for it to be of value to users. This involves file maintenance, database administration and information retrieval.

File maintenance involves adding records to tables, updating data in tables and deleting records from tables. Data from many different sources (maintenance, operations, suppliers, etc.) are needed for decision-making at strategic, tactical and operational levels. In the context of the IMS, file maintenance is carried out by staff from the maintenance and operations units. The maintenance staff would enter data relating to fault types and repair activities and so on.

Database administration involves the creation, deletion and restrictions on the use of files. As time progresses, some files may become obsolete and require deletion, while additional files may be required in order to support changes to the asset configuration. It is important that only certain personnel have the ability to make such changes.

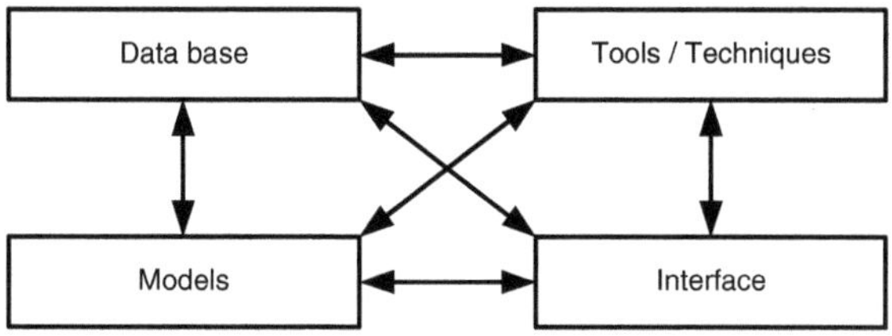

Fig. 11.9 Components of an IMS

Information retrieval involves the input of search criteria and the extraction of useful information based on the data within the database. Operational users will require the ability to access information based on their inputs. Strategic users (the managerial decision-makers) will need to retrieve strategic information linking multiple modules so that trade-off decisions can be made.

11.7.3.2 Models

Many different kinds of models are needed to assist in the decision-making at the strategic, tactical and operational levels. This element of the IMS is a library of the different models that have been developed either internally or externally.

The operational user should be able to select models and estimate the parameter values from data stored in the system. Conservative default values (and probability profiles) should be offered if appropriate data are not available (or judgement values solicited).

Often, several models need to be linked to find a solution to a specific decision problem. Also, the IMS must have the flexibility that allows for upgrading of models and the addition of new models to the system.

11.7.3.3 Tools and Techniques

A variety of tools are needed for data analysis, model building and model analysis and for determining the optimal decisions. Some of the packages are standard commercial packages, whereas others might be specialised packages. A large number of statistical packages are available for various tasks of model building (such as model selection, parameter estimation and model validation). Similarly, a large number of software packages are available for model analysis and optimisation.

11.7.3.4 Interface

Two interface requirements are that the IMS should have a user interface and an application interface. The user interface facilitates the flow of information from the user to the IMS and back, while the application interface provides the link between a variety of external programs and databases that may be called upon to solve problems or upload data to and/or download data from the IMS.

11.7.4 IMS for Rail Infrastructure Maintenance Outsourcing

Railway transport is a complex system with two subsystems: (1) rolling stock and (2) railway infrastructure. Railway infrastructure is a complex and distributed system, technically divided into substructures, namely bridges, tunnels, rails,[11] turnouts, sleepers, electrical assets (both low and high voltage), signalling systems including systems for traffic control, telecom systems such as systems for radio communication, telecommunications and detectors. Most of these are discrete elements except for rails and power network (in the case of electric trains) which are distributed.

The IMS for rail infrastructure maintenance is a complex system and involves several modules. When some or all of the maintenance is outsourced to one or several maintenance service providers, the owner needs to ensure that the providers' IMS is compatible with the owner's IMS. The IMS modules of Israel Rail (owner of the rail infrastructure in Israel) are described below.

1. Inventory Manager: This module stores and manages the entire railway network infrastructure and item inventory. The data are to be stored and managed using asset type pre-configured attribute templates and values. User interface and master data records include an interface to SAP (see ERP/SAP module) and support GIS functionality. All maintenance and inspection activities are documented in reference with the inventory listed in this module.
2. Inspections Manager: This module manages all track geometric inspections data and additional inspections, measurements and data. This module also handles media recordings of all railway inspections.
3. Inspections Analysing: This module analyses all the data from the inspections, including visualisation of the results. This module serves as the main decision support tool for infrastructure engineers.
4. Work Plan Optimisation: This module generates and clusters both automatically and manually recommended treatments based on parameters such as defect types, treatment types and location and will allow the maintenance planners to control and balance approval of activities based on cost–benefit ratio and expected performance level of the network as a whole (i.e. via "What if" scenarios).

[11] Also referred to as "*permanent ways*".

5. Bridge Management: This module stores and analyses inspection results of the bridge structural condition and subelement condition; these results trigger maintenance activity based on defect repair methods and cost optimisation.
6. ERP/SAP Interface: This module allows receiving and updating data records and attributes from SAP to the proposed system and vice versa, such as budgets, maintenance works and projects.
7. GIS Interface: This module allows viewing all spatial and linear data on the GIS such as inventory, inspection data, work plans and treatments. Also, the module enables controlling and creating inventory or work plans from the GIS.
8. Reporting: This module enables pre-defined reports and queries with export procedures for text and excel formats: inventory (rails, switches, etc.), intermediate reports for capacity of present inspection, defects reports, current condition reports, condition predictions reports, list of sections needing M&R, budgeting reports, etc.
9. Data Integration and Uploading: This module integrates and controls data from different sources: GIS, ERP (SAP), track measurement vehicle, trolley and manually inspected data, and any other system that interfaces with the system proposed.

11.8 Risk Management

We start with a brief discussion of some basic concepts and then look at various types of risks involved in EWs, MSCs and LCs from both the customer and provider perspectives.

11.8.1 Some Basic Concepts

11.8.1.1 Definitions of Risk[12]

A simple definition of risk from Johansen (2010) is the following:

> Risk is a characteristic of the future concerning the uncertain consequences decisions and contingencies

ISO 31000 (2009) defines risk as

> Risk is the effect of uncertainty on objectives

[12] There is a lot of controversy on defining risk. Aven (2010) discusses a vast number of risk definitions.

Table 11.1 Risk table

Scenario	Likelihood	Consequence
s_1	p_1	x_1
s_2	p_2	x_2
$\vdots$	$\vdots$	$\vdots$
s_n	p_n	x_n

A quantitative definition of risk proposed by Kaplan and Garrick (1981) aims to answer the following three questions:

1. What can happen? (i.e. what can go wrong?)
2. How likely is it that it will happen?
3. If it does happen, what are the consequences?

Kaplan and Garrick suggest making a list shown in Table 11.1. Here, line i corresponds to a scenario description s_i, the probability of the scenario occurring p_i and the consequence x_i should the scenario occur. The table characterises all possible scenarios (so that $\sum_{i=1}^{n} p_i = 1$), and the x_i's are arranged as a non-decreasing sequence. As a result, a formal quantitative definition of risk is given by a set of triplets $\Re \equiv \{<s_i, p_i, x_i>\ ,\ 1 \leq i \leq n\}$.

11.8.1.2 Risk Assessment

Risk assessment can be defined as the process and procedures for identifying, analysing and evaluating risks and their significance.

- *Risk identification* is the process of identifying different scenarios for risk.
- *Risk analysis* is the process of quantifying the probabilities and consequence values for an identified risk.
- *Risk evaluation* is the complex process of developing acceptable levels of risk.

Two other associated concepts are the following:

- *Risk aversion* is the act of avoiding and/or reducing risk.
- *Risk acceptance* is the willingness of a decision-maker to accept a specific risk to obtain some gain or benefit.

11.8.1.3 Risk Management

Risk management can be defined as any techniques used either to minimise the probability of an accident (or bad event occurring) or to mitigate its consequences with, for instance, good operating practices, preventive maintenance and evaluation plans.[13]

There are many different ways of managing risk. These include taking out insurance to cover the risk, different kinds of options, etc.[14]

11.8.1.4 Types of Risks

There are many different types of risks.[15] We give a brief description of some of the risks of relevance in the context of EWs, MSCs and LCs.

- *Commodity (material/spares) risk*: Risk that commodity prices and/or their implied volatility will change.
- *Credit risk*: Risk that a borrower will default on any type of debt by failing to make payments which it is obligated to do. The risk is primarily that of the lender and includes lost principal and interest, disruption to cash flows, and increased collection costs.
- *Demand risk*: Risk associated with variation of the demand for a public infrastructure service from initial expectations.
- *Financing risk*: Risk associated with variation of the financing costs from initial expectations. This includes the following:
  - *Interest rate risk*: Risk that interest rates and/or their implied volatility will change.
  - *Currency risk*: Risk that foreign exchange rates and/or their implied volatility will change.
- *Legal risk*: Risk that a business may incur losses due to violation of laws and regulations, breach of contract, entering into improper contracts or other legal factors
- *Liability risk*: Risk to a business arising from the possibility of liability for damages resulting from the purchase, ownership or use of a good or service offered by that business.
- *Market risk*: Risk of losses due movements in the market. This includes interest rate risk, currency risk, material/spare parts risk and demand risk.
- *Operational risk*: Risk of loss resulting from inadequate or failed internal processes, people and systems or from external events to carry out the activities in the normal manner.

---

[13] In *risk-based* decision-making risk analysis is the sole input whereas *risk-informed* decision-making it is one of many other inputs (such as decision analysis, cost-benefit analysis, etc.).

[14] There are many papers that deal with these topics, see for example, Hogart and Kunreuther (1989) for insurance and Chen and Shen (2012) for options contracts.

[15] In the sociocultural, political and economic context of western societies, Lupton (1999) defines six pressing risk domains. These are environmental risk, lifestyle risk, medical risk, interpersonal risk, economic risk and criminal risk.

Fig. 11.10 Provider's risks

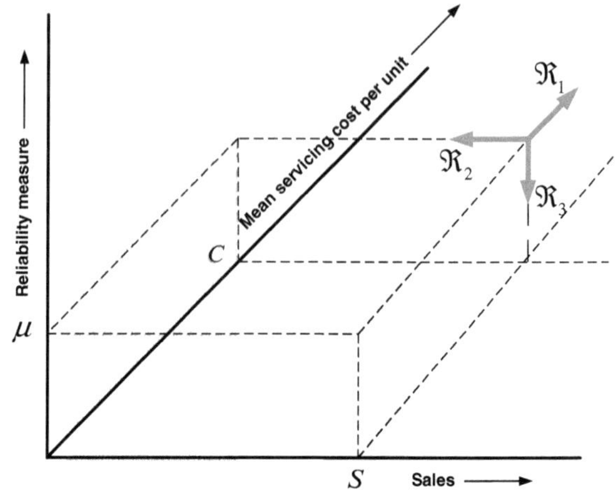

- *Regulatory change risk*: Risk that may incur in losses due to changes in various regulations or systems, such as those related to law, taxation and accounting.[16]
- *Technological risk*: Risk associated with technological change that could render the existing asset obsolete or the losses incurred due to asset not performing as per expectation.

11.8.2 Risks in EWs/MSCs

The risks from the provider's perspective are different from those of the customer.

11.8.2.1 Provider Perspective

In the case of products and plants, at the strategic level, the service provider needs to decide on the price (P) and sales (S) based on product reliability (characterised by a parameter μ). These can be viewed as the nominal values to ensure profit. Based on sales, servicing logistics and reliability, the servicing cost per unit is a random variable with mean C. Risk from a provider's point of view is making a loss instead of profit. The three important risks from the provider perspective are (1) operational risk $\Re_1$, (2) market risk $\Re_2$ and (3) technological risk $\Re_3$. These occur due to changes from the nominal values as indicated in Fig. 11.10.

[16] Losses due to violation of laws and regulation, breach of contract, entering into improper contracts or other legal factors are not part of the risk as these are deliberate actions.

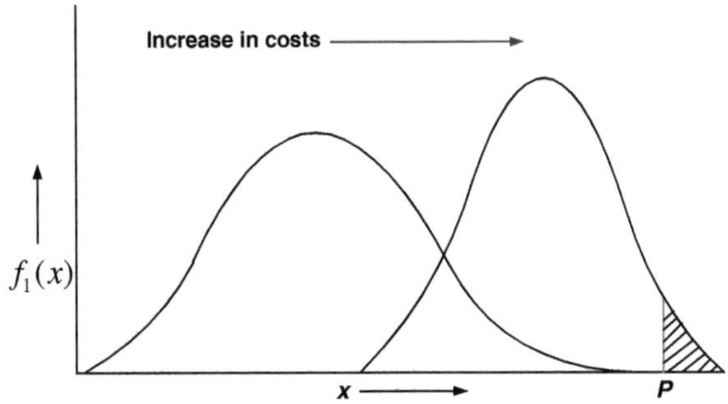

Fig. 11.11 Operational risk

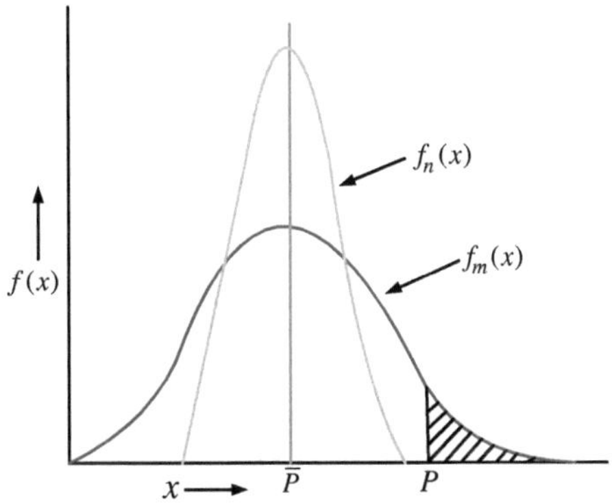

Fig. 11.12 Market risk

Operational risk

As discussed in Chaps. 7 and 8, the per unit EW/MSC cost is a random variable with density function $f_1(x)$ and mean *C*. The costs can increase—for several reasons (new legislation, cost of spare parts increasing, labour costs going up, etc.) so that the density function for cost moves to the right as shown in Fig. 11.11. Let *P* be the sale price of each EW/MSC. Then, under nominal conditions, the cost per unit is less than the sale price so that provider is always making a positive profit. If the cost of servicing increases significantly, then the provider makes a loss instead of a profit. The probability of incurring a loss is given by the shaded area in Fig. 11.11.

Market risk

As the number of EW/MSC customers increases, the variability in per unit servicing cost decreases. Let σ^2 denote the variance of the cost per unit. With n customers, the servicing cost per unit has the same mean, but variance is σ^2/n. With sale price *P*, the provider's profit is positive with probability close to one as shown in Fig. 11.12. Should the actual sales drop significantly, to say $m<n$, then the unit servicing cost can exceed *P* resulting in a loss. The shaded area in the figure shows the probability of this happening, and the probability increases as m decreases.

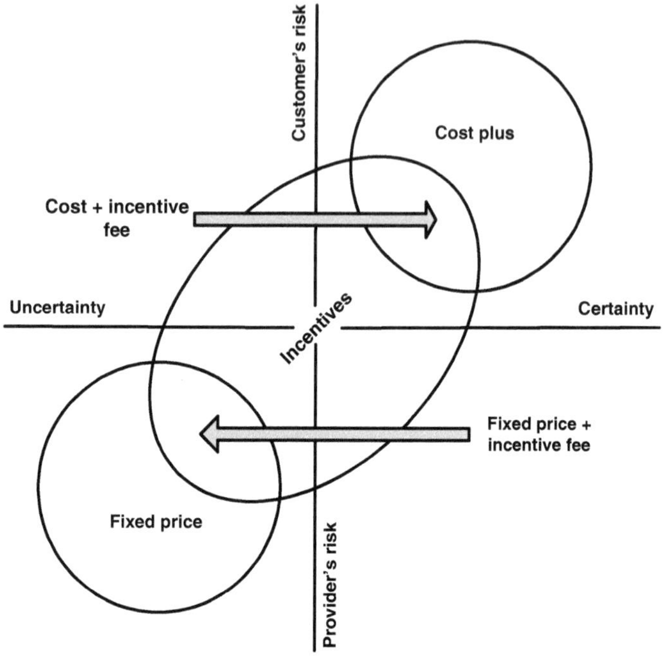

Fig. 11.13 Provider and customer risks

Technological risk
The expected number of failures increases as product/plant reliability decreases. If the provider overestimates the reliability and decides on the pricing based on this, then the EW/MSC servicing cost (per unit) can exceed the sale price so that the provider makes a loss instead of profit. This risk is high with MSCs since the provider is uncertain of the asset condition (as it depends on past usage and maintenance history) and the customer does not reveal this information.

11.8.2.2 Customer Perspective

For complex plants, the two main risks from the customer perspective are (1) quality of service risk and (2) residual value risk.

Quality of service risk
Poor quality of service results when the provider is not capable of providing the expected service. This can be due to several reasons—provider cutting costs, lacking technical competence, not having adequate resources, etc. This in turn affects the operations (production of goods and services) of the customer and the customer incurring significant losses.

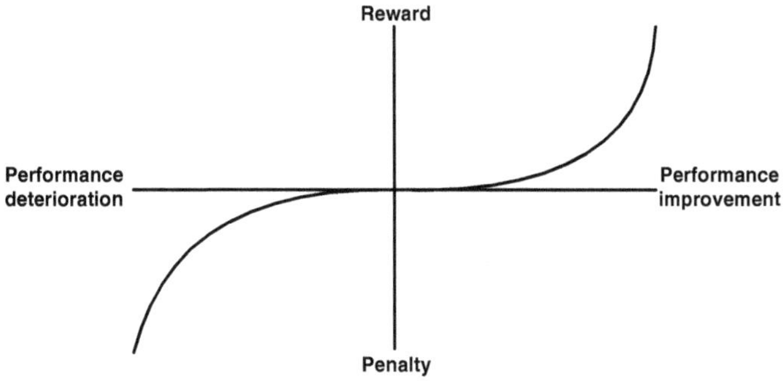

Fig. 11.14 Incentive structure

Residual value risk
For plants and infrastructures, the residual value at the end of the contact is very important. Poor maintenance by the provider can decrease the residual value, and this in turn can have a negative impact on the 'bottom line' of the customer's balance sheet.

11.8.2.3 Managing Risks

Managing risk is important for both providers and customers. The terms of the contract play a critical role in minimising the risks. As an illustrative case, consider a MSC where the uncertainty is due to operational risks. Figure 11.13 shows the customer's and provider's risk for two different terms of the contract. In the "cost plus" contract, the provider is not facing any risk, whereas the customer is taking all the risk. In the case of the "fixed price" contract, it is the reverse. By introducing an incentive component to the contract then sharing of risks occurs as shown in the figure.

The incentives involve rewards and penalties based on asset performance, and Fig. 11.14 shows a typical structure of incentive.

11.8.3 Risks in Leasing

Most leasing involves three players—lessor, lessee and a financial agent (bank or some other financial institution)—which is different from the lessor. The risks faced by the lessor and the lessee are similar to that faced by the provider and the customer in a MSC. The additional risk to the lessee is that the payment can increase due to interest rates increasing. The risk to the financial agent is the credit risk.

11.8.4 Infrastructures

There is considerable literature on managing risks in MSCs and LCs for infrastructures—Reichelsten (1992), Hirano (2004), Vickerman (2004), Scandizzo and Ventura (2010), Seyedshohadaie et al. (2010) is an illustrative sample.

References

Afrati F, Papadimitriou CH, Papageorgiou G, Papakostantinou N (1986) The complexity of the travelling repairman problem. Informatique-Theorique-et-Appl 20:79–87

Agnihothri SR (1998) Mean value analysis of the travelling repairman problem. IIE Trans 20:223–229

Alfredsson P (1997) Optimization of multi-echelon repairable item inventory systems with simultaneous location of repair facilities. Eur J Oper Res 99:584–595

Aras N, Gullu R, Yurulmez S (2011) Optimal inventory and pricing policies for remanufactured leased products. Int J Prod Econ 133:262–271

Aven T (2010) How to define, understand and describe risk. Reliab Eng Syst Saf 95:623–631

Barros L, Riley M (2001) A combinatorial approach to level of repair analysis. Eur J Oper Res 129:242–251

Belton V, Stewart J (2002) Multiple criteria decision analysis—an integrated approach. Kluwer Academic, London

Berkley B, Gupta A (1995) Identifying the information requirements to deliver quality service. Int J Serv Ind Manag 6:16–35

Bertolini M, Bevilacqua M, Braglia M, Frosolini M (2004) An analytical method for maintenance outsourcing service selection. Int J Qual Reliab Manage 21:772–788

Bevilacqua M, Braglia M (2000) The analytic hierarchy process applied to maintenance strategy selection. Reliab Eng Syst Saf 70:71–83

Bowersox DJ, Closs DJ (1996) Logistical management: the integrated supply chain process. McGraw-Hill Companies Inc, New York

Carrasqueira M, Machado VC (2007) Strategic logistics, a way to evolve towards a lean organization. In: Proceedings of the 2007 IEEE IEEM, pp 1709–1712

Cassady CR, Murdock WP, Pohl EA (2001) Selective maintenance of support equipment involving multiple maintenance actions. Eur J Oper Res 129:252–258

Chen X, Shen ZS (2012) An analysis of a supply chain with options contracts and service requirements. IIE Trans 44:805–819

Christopher M (1998) Logistics and supply chain management, strategies for reducing cost and improving service, 2nd edn. Financial Times Professional Ltd, London

Coyle JJ, Bardi EJ, Langley CJ (1992) The management of business logistics. West Publishing Company, St. Paul, Minnesota

Daskin MS (1995) Network and discrete location: models, algorithms, and applications. Wiley, New York

de Almeida AT (2005) Multicriteria modelling of repair contract based on utility and ELECTRE method with dependability and service quality criteria. Ann Oper Res 138:113–126

de Almeida AT (2007) Multicriteria decision model for outsourcing maintenance contracts selection based on utility and ELECTRE method with dependability and service quality criteria. Comput Oper Res 34:3569–3574

de Melo Brito AJ, de Almeida Filho AT, de Almeida AT (2010) Multicriteria decisionb models for selecting repair contracts by applying utility theory and variable independent parameters IMA. J Manage Math 21:349–361

Dresner Z (ed) (1992) Facility location: a survey of applications and methods. Springer, New York
Duffuaa SO, Raouf A, Campbell JD (1998) Planning and control of maintenance systems: modelling and analysis. Wiley, New York
Ehinlanwo OO, Zairi M (1996) Best practice in the car after-sales service—an empirical study of Ford, Toyota, Nissan and Fiat in Germany—Part II. Bus Process Re-Eng Manage J 2:39–53
Evers PT (2001) Heuristics for assessing emergency transhipments. Eur J Oper Res 129:311–316
Gupta SM, Korugan A (1998) A multi-echelon inventory system with returns. Comput Ind Eng 35:145–148
Hadley G, Whitin TM (1963) Analysis of inventory systems. Prenctice Hall Englewood Cliffs, New Jersey
Hajri S, Liouane N, Hammadi S, Borne P (2000) A controlled genetic algorithm by fuzzy logic and belief functions for job-shop scheduling. IEEE Trans Syst, Man Cybern Part B (Cybern) 30:812–818
Handler G, Mirchandani P (1979) Locations on network: theory and algorithms. MIT Press, Cambridge, Mass
Haugen DL, Hill AV (1999) Scheduling to improve field service quality. Decis Sci 30:783–804
Hirano J (2004) Outsourcing transport infrastructure maintenance: a theoretical approach with applications. Unpublished Doctoral Thesis, MIT, Cambridge, Mass
Hogart RM, Kunreuther H (1989) Risk, ambiguity, and insurance. J Risk Insur 2:5–35
Iskandar BP, Murthy DNP (2003) Repair-replace strategies for two-dimensional warranties. Math Comput Modell 38:1233–1241
Iskandar BP, Murthy DNP, Jack N (2005) A new repair-replace strategy for items sold with a two-dimensional warranty. Comput Oper Res 32:669–682
ISO 31000 (2009) Risk management—principles and guidelines. ISO, Geneva
Jack N, Van der Duyn Schouten F (2000) Optimal repair-replace strategies for a warranted product. Int J Prod Econ 67:95–100
Jack N, Murthy DNP (2001) Servicing strategies for items sold with warranty. J Oper Res 52:1284–1288
Jackson C, Pascual R (2008) Optimal service contract negotiation with aging equipment. Eur J Oper Res 189:387–398
Jayaraman V, Srivastava R (1995) A service logistics model for simultaneous siting of facilities and multiple levels of equipment. Comput Oper Res 22:191–204
Jianer C, Miranda A (2001) A polynomial time approximation scheme for general multiprocessor job scheduling. SIAM J Comput 31:1–17
Johansen IL (2010) Foundations of risk assessment. ROSS report # 20102, Department of production and quality engineering, Nowegian University of Science and Technology, Trondheim, Norway
Kaplan S, Garrick J (1981) On the quantitative definition of risk. Risk Anal 1:11–27
Kumar R, Markeset T, Kumar U (2004) Maintenance of machinery: negotiating service contracts in business-to-business marketing. Int J Serv Ind Manag 15:400–413
Kurata H, Nam SH (2010) After-sales service competition in a supply chain: optimization of customer satisfaction level or profit or both? Int J Prod Econ 127:136–146
Lewis EH (1968) Marketing channels: structure and strategy. McGraw Hill, New York
Loomba APS (1996) Linkages between product distribution and service support functions. Int J Phys Distrib Logistics Manag 26:4–22
Lupton D (1999) Risk. Routledge, London
Marquez AC, Gupta JND (2006) Contemporary maintenance management: process, framework and supporting pillars. Omega 34:313–326
Maxham JG III, Netemeyer RG (2002) Modeling customer perceptions of complaint handling over time: the effect of perceived justice on satisfaction and intent. J Retail 78:239–252
McFarlane D, Cuthbert R (2012) Modelling information requirements in complex engineering services. Comput Ind 63:349–360

Murthy DNP, Blischke WR (2006) Warranty Management and Product Manufacture. Springer Verlag, London
Murthy DNP, Solem O, Roren T (2004) Product warranty logistics: issues and challenges. Eur J Oper Res 156:110–126
Nozick LA, Turnquist MA (2001) Inventory, transportation, service quality and the location of distribution centers. Eur J Oper Res 129:362–371
Owen SH, Daskin MS (1998) Strategic facility location. Eur J Oper Res 111:423–447
Palfrey T, Romer T (1983) Warranties, performance, and the resolution of buyer-seller disputes. Bell J Econ 14:97–117
Ponnambalam SG, Ramkumar V, Jawahar N (2001) A multiobjective genetic algorithm for job shop scheduling. Prod Plann Control 12:764–774
Qu WW, Bookbinder JH, Iyogun P (1999) An integrated inventory-transportation system with modified periodic policy for multiple products. Eur J Oper Res 115:254–269
Reichelsten S (1992) Constructing incentive schemes for government contracts: An application of agency theory. Account Rev 67:712–731
Roy B (1996) Multicriteria methodology for decision aiding. Kluwer Academic Publishers, Dordrecht, Netherlands
Saaty T (1980) The analytical heirarchy process. McGraw Hill, New York
Scandizzo PL, Ventura M (2010) Sharing risks through concession contracts. Eur J Oper Res 207:363–370
Seyedshohadaie SR, Dammjanovic I, Butenko S (2010) Risk based maintenance and rehabilitation for transportation infrastructure network. Transp Res Part A 44:236–248
Sherbrooke CC (1992) Optimal inventory modeling of systems: multi-echelon techniques. Wiley, New York
SOLE (1996) SOLE Members' handbook. Society of logistic engineers
Sundarraj RP (2004) A web-based AHP approach to standardize the process of manageing service contracts. Decis Support Syst 37:343–365
Tersine RJ (1994) Principles of inventory and materials management. Prentice Hall, Englewood Cliffs, NJ
Triantaphyllou E, Kovalerchuk B, Mann L, Knapp GM (1997) Determining the most important criteria in maintenance decision making. J Quality Maintenance Eng 3:16–24
Vaidya OS, Kumar S (2006) Analytical hierarchy process: an overview of applications. Eur J Oper Res 169:1–29
Vargas LG (1990) An overview of the analytical hierarchy process and its applications. Eur J Oper Res 48:2–8
Vargas LG, Saaty TL (1981) Financial and intangible factors in fleet lease or buy decision. Ind Mark Manage 10:1–10
Vickerman R (2004) Maintenance incentives under different infrastructure regimes. Utilities Policy 12:315–322
Yang CE (1989) A dynamic programming algorithm for the travelling repairman problem. Asia–Pacific J Oper Res 6:192–206
Yang J, Lee H (1997) An AHP decision model for facility location selection. Facilities 15:241–254

Chapter 12
Epilogue

12.1 Introduction

Assets (products, plants and infrastructures) are getting more complex, expensive and require specialised service support. As a result, businesses need to critically evaluate four different options before making any decisions regarding a new asset. These options are as indicated in Fig. 12.1.

Option A: In the traditional approach the customer decides to purchase the asset and then carries out all the maintenance (preventive and corrective) in-house.
Option B: The customer purchases the asset but outsources some or all of the maintenance through an EW/MSC.
Option C: The customer leases the asset with the maintenance being carried out in-house.
Option D: The customer leases the asset with the maintenance carried out either by the lessor or outsourced to some other external agent.

Choosing between these options requires a proper methodology. For Options B–D, a game-theoretic approach is needed since there are two or more decision-makers with different objective functions. Mathematical models play a very important role in the decision-making. Firstly, they allow one to focus on issues to gain a better insight into the dynamics of the interactions between the decision-makers. Secondly, they characterise the optimal strategies for the different players. In this book, we have reviewed the game-theoretic models for EWs/MSCs/LCs that have appeared in the economics, management and operational research literature. In this chapter, we first comment on the current status of the modelling that has taken place and then, we suggest topics for further research in the future.

D. N. P. Murthy and N. Jack, *Extended Warranties, Maintenance Service and Lease Contracts*, Springer Series in Reliability Engineering,
DOI: 10.1007/978-1-4471-6440-1_12, © Springer-Verlag London 2014

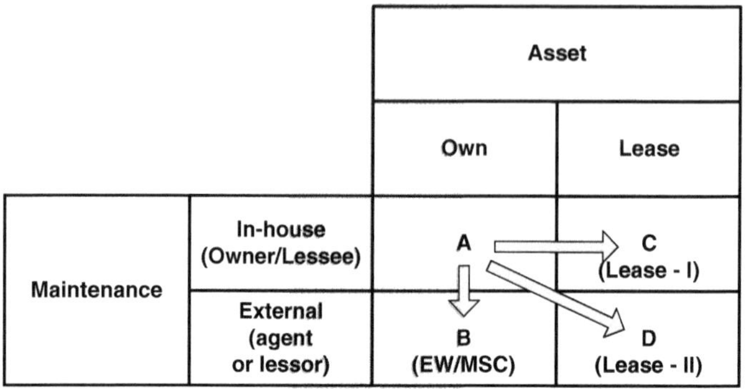

Fig. 12.1 Different options regarding a new asset

12.2 Current Status

The bulk of the models relating to EWs, MSCs and LCs involve two decision-makers. The models in the economics and marketing literature are highly stylised and mainly static in nature. In contrast, the models in the operational research literature are dynamic with the possibility of multiple asset failures occurring over time. Previously, the focus has been on narrow specific issues and mostly on cost analysis and decision-making to minimise costs. There has been very little interaction between researchers from different disciplines. This situation is very similar to the story of six blind men trying to describe an elephant by standing at different locations around it—each is partially correct but the complete description is missing!

Very few businesses and industry sectors now follow Option A above. Most have made the transition from A to B, C or D, or combinations of these. This implies that more realistic models are needed to help the decision-making process.

The literature dealing with applications is rather limited and mainly qualitative in nature with very few papers dealing with data issues. An exhaustive search of the literature provided the following—Albaum and Wiley (2010), Anderson and Bird (1980), Chen and Huang (2005), Chu and Chintagunta (2009), Fakhoury and Alhamed (2008), Huysentruyt and Read (2010), Mont et al. (2006), Ng et al. (2009), Ng and Nudurupati (2009), Oum et al. (2000), Stenbeck (2004) and Sturgeon (2005). The gap between theory and application is very wide.

12.3 Future Research in EWs/MSCs

Topic 1: The literature on BWs deals with a range of issues that include the following: Legal, Accounting, Economics, Finance, Marketing, Legislative, Consumerist, Logistics, Operational Research, Management, Historical, etc. Blischke and Murthy (1996) integrate the warranty literature until 1995, and Murthy and Djamaludin (2002) cover the literature from 1995 to 2011. The EW literature is discussed briefly in both of these publications. Thus, there is a need to carry out something similar for EWs and MSCs—to integrate the diverse literature.

Topic 2: As mentioned earlier, the bulk of the literature deals with two (owner and EW/MSC provider) or three (owner, EW/MSC provider and independent service agent) parties with one player in each party and with the dominance structure implying the use of Stackelberg game models. Both EW and MSC markets are characterised by several EW/MSC providers competing against each other. In this case, both Stackelberg game and Nash game concepts are required to build models to characterise the optimal decisions and the market outcomes.

Topic 3: The models in the economic and management literature are mainly static. The extension of these in a dynamic context will result in more realistic modelling to understand the different issues. This topic offers considerable scope for new research.

Topic 4: The bulk of the models assume complete and symmetric information. Issues about informational aspects (such as uncertainty, asymmetry) need to be addressed in a more effective manner. In a multistage context, the various players acquire new information about other players over time and this implies the need for a Bayesian framework (for updating information). A variety of models can be constructed based on different information structures, and then, the analyses of these models need to be carried out.

Topic 5: EWs and MSCs are contracts which require all the parties acting as per the terms of the contract. When one or more of the parties violate the contract (for personal advantage) then a dispute can arise. Very little work has been done on the modelling of dispute resolution—Lai et al. (2004) deal with this in the context of building service contracts, and so this is an open topic for research.[1]

[1] Palfrey and Romer (1983) deal with dispute resolution in the context of BWs.

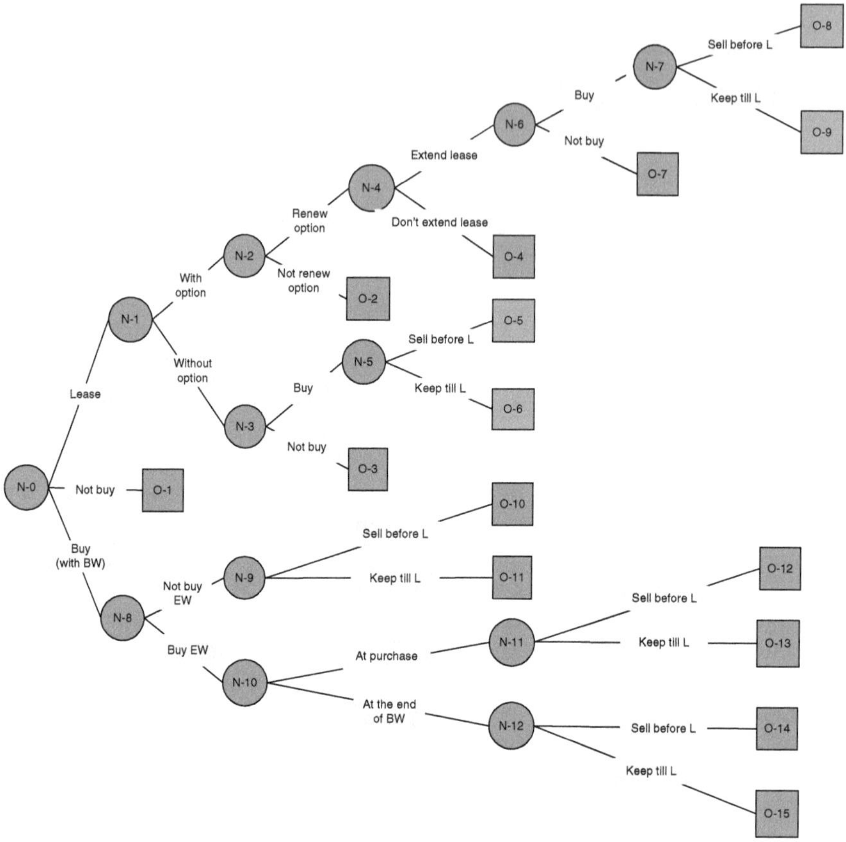

Fig. 12.2 Decision tree (customer perspective)

12.4 Future Research in LCs

Topics 1–5 are also applicable for LCs and so will not be repeated. A LC study involves additional issues such as upgrades, interaction between new products and second-hand products, remanufacturing and sustainability. These add extra dimensions to LC problems compared to those involving EWs/MSCs.

12.5 Integrated Approach to EWs, LCs and MSCs

In the past, EWs, MSCs and LCs have been studied in isolation. Choosing between Options A–D requires a unified framework. From a customer perspective, decision-making can be viewed as a multistage process best characterised by a decision tree as indicated in Fig. 12.2.

The tree has thirteen decision nodes (denoted by numbered circles), and the decisions needing to be made at each node are based on data and the information available. The terminating branches of the tree lead to 15 outcomes (denoted by numbered squares). Each of these represents the outcome of a particular decision history over the period [0, *L*) and the associated pay-off. The Bayesian approach combined with the use of dynamic programming is needed to identify the optimal decisions. However, this requires modelling both the data and the information aspects.

12.6 Conclusion

EWs, MSCs and LCs are of great interest to both practitioners and researchers. For practitioners, the benefits are higher availability of the asset, lower costs and higher business profits. For researchers, there is scope for many new researches in the future in the areas of EWs, MSCs and LCs. A closer relationship between practitioners and researchers would help both groups as well as providing interesting case studies for use in education programmes.

References

Albaum G, Wiley J (2010) Consumer perceptions of extended warranties and service providers. J Consum Mark 27:516–523

Anderson PF, Bird MM (1980) Marketing to the industrial lease buyer. Ind Mark Manag 9:111–116

Blischke WR, Murthy DNP (1996) Product Warranty Handbook. Marcel Dekker, New York

Chen KY, Huang S (2005) Durable goods lease contracts and used-goods market behavior: an experimental study. Exp Bus Res 2:1–19

Chu J, Chintagunta PK (2009) Quantifying the economic value of warranties in the US sever market. Mark Sci 28:99–121

Fakhoury B, Alhamed H (2008) Life cycle cost based model for successful maintenance outsourcing process—case study. Thesis No. TD025/2008, Department of Terotechnology, Vaxjo University, Sweden

Huysentruyt M, Read D (2010) How do people value extended warranties? Evidence from two field surveys. J Risk Uncertainty 40:197–218

Lai JHK, Yik FWH, Jones P (2004) Disputes arising from vaguely defined contractual responsibilities in building service contracts. Facilities 22:44–52

Mont O, Dalhammar C, Jacobsson N (2006) A new business model for baby prams based on leasing and product remanufacturing. J Cleaner Prod 14:1509–1518

Murthy DNP, Djamaludin I (2002) Product warranty—A review. Int J of Prod Econ 79:231–260

Ng ICL, Nudurupati SS (2009) Outcome-based service contracts in the defence industry—mitigating the challenges. J Serv Manag 21:656–674

Ng ICL, Maull R, Yip N (2009) Outcome-based contracts as a driver for systems thinking and service-dominant logic in service science: evidence from the defence industry. Eur Manag J 27:377–387

Oum TH, Zhang A, Zhang Y (2000) Optimal demand for operating leases of aircraft. Transp Res Part B 34:17–29

Palfrey T, Romer T (1983) Warranties, performance, and the resolution of buyer-seller disputes. Bell J Econ 14:97–117

Stenbeck T (2004) Incentives to innovations in road and rail maintenance and operations. Licentiate thesis, Department of Infrastructure, Royal Institute of Technology, Stockholm

Sturgeon J (2005) New incentives heat up auto leasing. http://www.bankrate.com/brm/news/auto/20050713a1.asp

Appendix A
Basic Concepts from Probability Theory

In this Appendix, we give a brief introduction to elementary probability theory, which is the basis of the mathematical approach to modelling failures. The presentation is non-rigorous. The objective is to develop an intuitive feel for the topic that forms the foundation for most models used in solving reliability-related problems.

A.1 Scalar Random Variables

Consider an experiment whose outcome is not known in advance but is such that the set of all outcomes (called the "sample space" $\Im$) is known. Any subset of the sample space $\Im$ is called an event. A random variable is a function, which maps outcomes from the sample space $\Im$ to $\Re$ the space of real numbers. In other words, for every outcome ω in the sample space $\Im$, $X(\omega)$ assigns a real number to ω. It can be either discrete or continuous. A discrete random variable takes on at most a countable number of values [for example the set of non-negative integers), and a continuous random variable can take on values from a set of possible values, which is uncountable (for example values in the interval $(-\infty, \infty)$].

Because the outcomes are uncertain, the value assumed by X is uncertain before the event occurs. Once the event occurs, X assumes a certain value. The standard convention used is as follows: X (upper case) represents the random variable before the event, and the value it assumes after the event is represented by x (lower case).

A.1.1 Distribution and Density Functions

The distribution function $F(x;\, \theta)$ is defined as the probability that $X \le x$ and is given by

$$F(x; \theta) = P\{X \le x\} \tag{A.1}$$

The domain of $F(x;\, \theta)$ is $(-\infty,\, \infty)$, the range is $[0,\, 1]$, and θ denotes the set of parameters of the distribution function. Often the parameters are omitted for

D. N. P. Murthy and N. Jack, *Extended Warranties, Maintenance Service and Lease Contracts*, Springer Series in Reliability Engineering,
DOI: 10.1007/978-1-4471-6440-1,

notational ease, so that one uses $F(x)$ instead of $F(x, \theta)$. We will do this in the remainder of the Appendix.

$F(x)$ has the following properties:

- $F(x)$ is a non-decreasing function of x.
- $F(-\infty) = 0$ and $F(\infty) = 1$
- For $x_1 < x_2$, $P\{x_1 < X \leq x_2\} = F(x_2) - F(x_1)$

When X is continuous valued and $F(x)$ is differentiable, the density function $f(x)$ is given by

$$f(x) = \frac{dF(x)}{dx} \tag{A.2}$$

$f(x)$ may be interpreted as

$$P\{x < X \leq x + \delta x\} \approx f(x)\delta x + O(\delta x^2). \tag{A.3}$$

When X takes on only values in a set $(x_1, x_2, \ldots, x_n)$, with n being finite or infinite, the probability that $X = x_i$ is given by

$$p_i = P\{X = x_i\}, \quad i = 1,\ 2,\ \ldots,\ n \tag{A.4}$$

In this case, X is called a *discrete* random variable, and the CDF is a step function with steps of height p_i at each of the possible values x_i. The probabilities p_i have the following properties: (i) $p_i \geq 0$ and (ii) $\sum_{i=1}^{n} p_i = 1$

Moments of Random Variables

The jth *moment* of the random variable X, $M_j(\theta)$, is given by[1]

$$M_j(\theta) = E[X^j] = \begin{cases} \int_0^{\infty} x^j f(x)\mathrm{d}x, & \text{if } X \text{ is continuous} \\ \sum_x x^j P\{X = x\}, & \text{if } X \text{ is discrete} \end{cases} \tag{A.5}$$

The first moment of X is called the *mean* and is usually denoted by μ, so that

$$\mu = E[X] \tag{A.6}$$

The jth *central moment* of X, μ_j, is given by

$$\mu_j = E[(X - \mu)^j] \tag{A.7}$$

The second central moment of X is called the *variance* and is usually denoted by σ^2, so that

[1] The parameters are omitted for notational ease, so that one uses M_j instead of $M_j(\theta)$.

$$\sigma^2 = E[(X - \mu)^2] \tag{A.8}$$

σ is called the *standard deviation.*

A.1.2 Discrete Distributions

The following are some well-known discrete distributions that are useful in failure modelling[2]:

Bernoulli Distribution X assumes two possible values, 0 and 1, with probabilities given by

$$p_0 = p \quad \text{and} \quad p_1 = (1 - p) \tag{A.9}$$

The parameter set is $\theta = \{p\}$, with $0 \le p \le 1$.

Binomial DistributionX assumes integer values from 0 to n, where n is a positive integer and p_i, $0 \le i \le n$, is given by

$$p_i = \frac{n!}{i!(n-i)!} p^i (1 - p)^{(n-i)} \tag{A.10}$$

The parameter set is $\theta = \{n, p\}$ with $0 \le p \le 1$ and $0 < n < \infty$.

Geometric Distribution X assumes integer values from 0 to ∞, with probabilities p_i, $0 \le i < \infty$, given by

$$p_i = (1 - p)^i p \tag{A.11}$$

The parameter set is $\theta = \{p\}$ with $0 \le p \le 1$.

Poisson DistributionX assumes integer values from 0 to ∞. p_i, $0 \le i < \infty$, is given by

$$p_i = \frac{e^{-\lambda} \lambda^i}{i!} \tag{A.12}$$

The parameter set is $\theta = \{\lambda\}$, with $\lambda > 0$.

A.1.3 Continuous Distributions

Continuous distribution functions useful in failure modelling can be grouped into three categories—(1) basic, (2) those derived from basic and (3) those involving two or more basic/derived distributions[3]:

---

2 Most basic books on statistics and probability discuss some of the well-known distributions. (Johnson and Kotz 1969a, b) gives a more comprehensive coverage of many discrete distributions.

3 Most basic books on statistics and probability discuss some of the well-known distributions. (Johnson and Kotz 1970a, b) give a more comprehensive coverage of many continuous distributions.

A.1.3.1 Basic Distributions and Density Functions

Exponential Distribution The distribution function for the exponential distribution is given by

$$F(x;\theta) = 1 - e^{-\lambda x}, \quad x \geq 0. \tag{A.13}$$

The parameter set is $\theta = \{\lambda\}$, with $\lambda > 0$.

Gamma Distribution The gamma density function is given by

$$f(x;\theta) = \frac{x^{\alpha-1}e^{-x/\beta}}{\beta^{\alpha}\Gamma(\alpha)}, \quad x \geq 0 \tag{A.14}$$

The parameter set is $\theta = \{\alpha, \beta\}$, with $\alpha > 0$ and $\beta > 0$. Here, $\Gamma(\cdot)$ is the gamma function. Extensive tables can be found in Abramowitz and Stegun (1964).

Weibull Distribution The two-parameter Weibull distribution function is given by

$$F(x;\theta) = 1 - e^{-(x/\alpha)^{\beta}}, \quad x \geq 0. \tag{A.15}$$

The parameter set is $\theta = \{\alpha, \beta\}$, with $\alpha > 0$ and $\beta > 0$.

A.1.3.2 Derived Distributions and Density Functions

The derived distributions given below are obtained by (1) transformation of the random variable from a basic distribution, (2) modification of the form of a basic distribution by introducing additional parameters (for example the exponentiated Weibull distribution) and (3) devising forms that involve two or more basic distribution functions (for example mixtures of distributions, competing risk models). We present some of each form of derived distribution.[4]

Three-Parameter Weibull Distribution This is an extension of the two-parameter Weibull distribution (A.16), and the distribution function is given by

$$F(x;\theta) = 1 - e^{-(\{x-\tau\}/\alpha)^{\beta}} x \geq \tau. \tag{A.16}$$

The additional parameter is the location parameter $\tau > 0$.

Exponentiated Weibull Distribution The distribution function is given by

$$F(x) = [1 - \exp\{-(x/\alpha)^{\beta}\}]^{\nu}, \quad x \geq 0, \tag{A.17}$$

with $\nu \geq 0$. The distribution reduces to the two-parameter Weibull (A.15) when $\nu = 1$.

[4] For additional details with regard to the three types, see (Blischke and Murthy 2000) and (Murthy et al. 2003).

A.1.3.3 Distributions Involving Two or More Basic/Derived Distributions

Mixtures of Distributions A *finite mixture* of distributions is a weighted average of distribution functions given by

$$F(x) = \sum_{i=1}^{K} p_i F_i(x) \tag{A.18}$$

with $p_i \geq 0,\ i = 1, 2, \ldots, K,\ \sum_{i=1}^{K} p_i = 1$ and $F_i(x) \geq 0,\ i = 1, 2, \ldots, K$ distribution functions (called the *components* of the mixture).

Competing Risks The distribution function is given by

$$F(x) = 1 - \prod_{i=1}^{K} (1 - F_i(x)) \tag{A.19}$$

Multiplicative The distribution function is given by

$$F(x) = \prod_{i=1}^{K} F_i(x),\ x \geq 0 \tag{A.20}$$

A.2 Vector Random Variables

We now give important probability results for the case where two or more random variables are needed to represent the outcomes of an uncertain event.

A.2.1 Two Random Variables

Let the two continuous random variables be denoted as X and Y.

A.2.1.1 Joint, Marginal and Conditional Distributions and Density Functions

The *joint distribution function* $F(x, y)$ is given by

$$F(x, y) = P\{X \leq x, Y \leq y\} \tag{A.21}$$

The random variables are said to be jointly continuous if there exists a function $f(x, y)$, called the *joint probability density function*, such that

$$f(x, y) = \frac{\partial^2 F(x, y)}{\partial x \partial y} \tag{A.22}$$

The *marginal distribution functions* $F_x(x)$ and $F_y(y)$ are given by

$$F_X(x) = F(x, \infty) \quad \text{and} \quad F_Y(y) = F(\infty, y) \tag{A.23}$$

The marginal density functions are given by

$$f_X(x) = \frac{\mathrm{d}F_X(x)}{\mathrm{d}x} \quad \text{and} \quad f_Y(y) = \frac{\mathrm{d}F_Y(y)}{\mathrm{d}y}. \tag{A.24}$$

The *conditional distribution* of X given that $Y = y$ is denoted $F(x|y)$ and is given by

$$F(x|y) = P\{X \le x | Y = y\} \tag{A.25}$$

The conditional distribution of Y given that $X = x$, $F(y|x)$, is defined similarly.

For jointly continuous random variables with a joint density function $f(x, y)$, the *conditional probability density* function of X, given $Y = y$, is given by

$$f(x|y) = \frac{f(x, y)}{f_Y(y)} \tag{A.26}$$

Similarly,

$$f(y|x) = \frac{f(x, y)}{f_x(x)} \tag{A.27}$$

The random variables X and Y are said to be *independent* (or *statistically independent*) if and only if

$$F(x, y) = F_X(x)\, F_Y(y) \tag{A.28}$$

for all x and y.

The results are similar for discrete random variables, with summation replacing integration.

A.2.1.2 Moments of Two Random Variables

The covariance of X and Y is defined as

$$\mathrm{Cov}(X, Y) = E[\{X - E[X]\}\{Y - E[Y]\}] = E[X\,Y] - E[X]\,E[Y] \tag{A.29}$$

The correlation ρ_{XY} is defined as

$$\rho_{XY} = \frac{\mathrm{Cov}(X, Y)}{\sigma_X \sigma_Y}, \tag{A.30}$$

where σ_x and σ_y are the standard deviations of X and Y, respectively. The random variables X and Y are said to be *uncorrelated* if $\rho_{XY} = 0$. Note that independent random variables are uncorrelated but that the converse is not necessarily true.

A.2.1.3 Conditional Expectation

$E[X|Y=y]$ is called the conditional expectation of X given that $Y = y$. The unconditional expectation of X, given by

$$E[X] = \int_{-\infty}^{\infty} x\, f_X(x)\, \mathrm{d}x, \tag{A.31}$$

is related to the conditional expectation by the relation

$$E[X] = \int_{-\infty}^{\infty} E[X|Y=y]\, f_Y(y)\, \mathrm{d}y. \tag{A.32}$$

This is written symbolically as

$$E[X] = E[E[X|Y]] \tag{A.33}$$

A.2.1.4 Sum of Two Independent Random Variables

Let X and Y be two independent random variables with density functions $f_X(x)$ and $f_Y(y)$, respectively, and let $Z = X + Y$. Then, the density function for Z, $f_Z(Z)$, is given by

$$f_Z(z) = \int_{-\infty}^{\infty} f_Y(t)\, f_X(z-t)\, \mathrm{d}t \tag{A.34}$$

or

$$f_Z(z) = \int_{-\infty}^{\infty} f_X(t)\, f_Y(z-t)\, \mathrm{d}t \tag{A.35}$$

This operation is called the convolution operation indicated by the symbol "*".

Thus,

$$f_Z(z) = f_X(z) * f_Y(z) = f_Y(z) * f_X(z) \tag{A.36}$$

A.2.2 The General Case

The k (>2) random variables may be represented by the vector $(X_1, X_2, \ldots, X_k)$. The approach is similar to the two random variable case, but involving an k-dimensional distribution function $F(x_1, x_2, \ldots, x_k)$. We have k marginal distributions and several different conditional distributions, depending on how the k-variables are divided into two sets, with the distribution of the first-set conditioned on the values of the variables in the second. Similarly, there are many different correlation coefficients. Details can be found in Johnson and Kotz (1972).

A.2.2.1 Sums of Independent Random Variables

When Z is the sum of n independent variables, X_i $(i = 1, 2, \ldots, n)$, with respective density functions $f_i(x)$, then the density function for Z is given by

$$f_Z(z) = f_1(z) * f_2(z) * \cdots * f_n(z) \tag{A.37}$$

References

Abramowitz M, Stegun IA (1964) Handbook of mathematical functions. Applied Mathematics Series No. 55. National Bureau of Standards, Washington, DC

Blischke WR, Murthy DNP (2000) Reliability. Wiley, New York

Johnson NL, Kotz S (1969a) Discrete distributions. Houghton Mifflin Co., Boston

Johnson NL, Kotz S (1969b) Distributions in statistics: discrete distributions. Wiley, New York

Johnson NL, Kotz S (1970a) Distributions in statistics: continuous univariate distributions—I. Wiley, New York

Johnson NL, Kotz S (1970b) Distributions in statistics: continuous univariate distributions—II. Wiley, New York

Johnson NL, Kotz S (1972) Distributions in statistics: continuous multivariate distributions. Wiley, New York

Murthy DNP, Xie M, Jiang R (2003) Weibull models. Wiley, New York

Appendix B
Introduction to Stochastic Processes

In this Appendix, we give a brief introduction to stochastic processes and discuss some of the processes that are used in the book. Our presentation will be intuitive and non-rigorous and will highlight the important concepts. Readers interested in a deeper understanding of the underlying theory should consult the references given at the end.

B.1 Stochastic Processes

In Appendix A, we defined a random variable, $X(\omega)$, as function that map outcomes from the sample space to real numbers. A stochastic process $X(t, \omega), \quad t \in T$, where T is a set of non-negative numbers, can be viewed as an extension of $X(\omega)$ in the following sense—t represents a time instant in the set T, which may be either finite or infinite. For a fixed $t \in T$, $X(t, \omega)$ is a random variable in the usual sense. For a fixed ω (outcome), $X(t, \omega)$ can be viewed as a function of t and $X(t, \omega)$ denotes the state of the process at time t. If T is countable, then $X(t, \omega)$ is called a *discrete-time stochastic process*. If T is a continuum, then it is called a *continuous-time stochastic process*. Henceforth, we omit ω and represent $X(t, \omega)$ as simply $X(t)$.

Let $t_i, i = 1, 2, \ldots, n$, denote n different time instants. The probabilistic characterisation of the process $X(t)$ at these n points can be done through the joint probability distribution

$$F(t_1, x_1;\ t_2, x_2;\ \ldots;t_n, x_n) = P\{X(t_1) \le x_1;\ X(t_2) \le x_2;\ \ldots;X(t_n) \le x_n\} \quad \text{(B.1)}$$

As n increases, this function becomes cumbersome and is of limited use in modelling real world problems.

D. N. P. Murthy and N. Jack, *Extended Warranties, Maintenance Service and Lease Contracts*, Springer Series in Reliability Engineering,
DOI: 10.1007/978-1-4471-6440-1, © Springer-Verlag London 2014

B.2 Markov Property

A stochastic process $X(t)$ is said to have the *Markov property* if

$$P\{X(t+\tau)\leq x|X(u)=x(u),\ -\infty<u\leq t\}=P\{X(t+\tau)\leq x|X(t)=x(t)\} \quad \text{(B.2)}$$

In other words, the probabilistic characterisation of $X(t+\tau)$ (a future event) given $\{X(u)=x(u),\ -\infty<u\leq t\}$ (past history and present value of the process) depends only on the present value $X(t)$ and not its past values. This simplifies the mathematical characterisation of the process considerably. Using conditional probability, we have, for an increasing sequence in t_i,

$$\begin{aligned}P\{X(t_1)\leq x_1;\ X(t_2)\leq x_2;\ \ldots;\ X(t_n)\leq x_n\} =& P\{X(t_n)\leq x_n|X(t_{n-1})\leq x_{n-1}\}\ldots \\ & P\{X(t_2)\leq x_2|X(t_1)\leq x_1\}P\{X(t_1)\leq x_1\}\end{aligned} \quad \text{(B.3)}$$

Thus, the joint probability distribution for $X(t)$ at n different points along the time axis can be obtained in terms of the conditional distribution of $X(t)$ involving two different values of t. In other words, the probabilistic characterisation of the process can be done as a function of sets of four variables $F(t_i,x_i;t_j,x_j)$ with

$$F(t_i,\ x_i;\ t_j,\ x_j)=P\{X(t_i)\leq x_i \text{ and } X(t_j)\leq x_j\} \quad \text{(B.4)}$$

for all t_i and t_j over the interval T and all x_i and x_j over the real line.

B.3 Classification of Stochastic Processes

Stochastic processes can be divided into four categories depending on whether:

1. the values assumed by the process $X(t)$ are discrete or continuous, and
2. the values assumed by the time variable t are discrete or continuous.

We briefly discuss each of these four categories.

Discrete State/Discrete Time Process

Here, both $X(t)$ and t assume only discrete values. Let the values assumed by $X(t)$ be denoted by $s_i,\ i=1,\ 2,\ \ldots,\ r$. r may be either finite or infinite. The values assumed by $t_i,\ i=1,\ 2,\ \ldots,$ form an increasing sequence. If the process is Markovian, then it is called a discrete-time Markov chain (DTMC).

Discrete State/Continuous-Time Process

Here, $X(t)$ assumes only discrete values with r either finite or infinite, and t assumes a continuous range of values in the interval $(-\infty,\ \infty)$. If the process is Markovian, it is called a continuous-time Markov chain (CTMC).

Continuous State/Discrete-Time Process

In this case, $X(t)$ assumes a continuous range of values and t assumes discrete values. If the process is Markovian, it is called a discrete-time Markov process.

Continuous State/Continuous-Time Process

In this process, both $X(t)$ and t assume continuous ranges of values. If the process is Markovian, it is called a continuous-time Markov process (or simply a Markov process).

A further subclassification is *stationary* and *non-stationary* stochastic processes. A stochastic process is said to be stationary if the joint distribution function is invariant under a shift in t, i.e. if

$$F(t_1', x_1;\ t_2', x_2;\ \ldots; t_n', x_n) = F(t_1, x_1;\ t_2, x_2;\ \ldots; t_n, x_n) \tag{B.5}$$

with $t_i' = t_i + \tau,\ (1 = 1, 2, \ldots, n)$ for all τ and n.

B.4 Point Processes

A *point process* is a continuous-time stochastic process characterised by events that occur randomly along the time continuum. An example, in the context of reliability, is an item being put into operation or an item failing. The theory of point processes is very rich, as a variety of such processes have been formulated and studied. Of particular interest to reliability modelling is the counting process.

B.4.1 Counting Processes

A point process $\{N(t),\ t \geq 0\}$ is a *counting process* if it represents the number of events that have occurred until time t. It must satisfy:

1. $N(t) \geq 0$.
2. $N(t)$ is integer valued.
3. If $s < t$, then $N(s) \leq N(t)$.
4. For $s < t$, $\{N(t) - N(s)\}$ is the number of events in the interval $(s, t]$.

We shall confine ourselves to $t \geq 0$. The behaviour of $N(t)$, for $t \geq 0$, depends on whether or not $t = 0$ corresponds to the occurrence of an event. The analysis of the case with $t = 0$ corresponding to the occurrence of an event is simpler than the alternate case. Also, we assume that $N(0) = 0$.

A counting process $\{N(t),\ t \geq 0\}$ is said to have *independent increments* if, for all choices $0 \leq t_1 < t_2 < \cdots < t_n$, the $(n - 1)$ random variables

$\{N(t_2) - N(t_1)\},\ \{N(t_3) - N(t_2)\}, \ldots, \{N(t_n) - N(t_{n-1})\}$ are independent. A counting process $\{N(t),\ t \geq 0\}$ is said to have *stationary independent increments* if, for each $s > 0$, $\{N(t_2 + s) - N(t_2)\}$ and $\{N(t_1 + s) - N(t_1)\}$ have the same distribution function, i.e. if the distribution function of $\{N(t + s) - N(t)\}$ does not depend on t.

Two special counting processes of particular importance to reliability modelling are (1) the Poisson process and (2) the renewal process.

B.4.2 Poisson Processes

We first consider the stationary Poisson process and later discuss some extensions.

B.4.2.1 Stationary Poisson Process

Definition 1 A counting process, $N(t)$, $t \geq 0$, is a *stationary* or *homogeneous* Poisson process (HPP) if

1. $N(0) = 0$.
2. The process has independent increments.
3. The number of events in any interval of length t is distributed according to Poisson distribution with parameter λt, i.e.

$$P\{N(t+s) - N(s)\} = \frac{e^{-\lambda t}(\lambda t)^n}{n!} \tag{B.6}$$

$n = 0, 1, 2, \ldots$, and for all s and $t \geq 0$.

It can be shown through simple analysis [see, e.g. Ross (1970)] that for an HPP, the times between events (also called inter-event times) are independent and identically distributed exponential random variables with mean $(1/\lambda)$. This is the basis of a second definition for an HPP.

Definition 2 Consider a counting process. Let X_1 denote the time instant of the first event occurrence, and for $j \geq 2$, let X_j denote the time interval between the $(j-1)$st and jth events. The counting process is an HPP with parameter λ if the sequence X_j, $j \geq 1$, are independent and identically distributed exponential random variables with mean $(1/\lambda)$.

We also have a third definition for an HPP [see Ross (1970)].

Definition 3 A counting process $\{N(t),\ t \geq 0\}$ is an HPP if

1. The probability of an event occurring in $[t, t + \delta t)$ is $\lambda \delta t + o(\delta t)$.
2. The probability of two or more events occurring in $[t, t + \delta t)$ is $o(\delta t)$.
3. The occurrence of an event in $[t, t + \delta t)$ is independent of the number of events in $[0, t)$.

λ is called the *intensity* of the process.

Comment The above discussion illustrates the point that there is more than one way of characterising a counting process. In the context of reliability modelling, a particular characterisation may be more appropriate than alternate, equivalent characterisations. For example in the case of non-repairable items, Definition 2 is more appropriate; in the case of repairable items with the item being subjected to minimal repair after each failure, Definition 3 is more appropriate.

Expected Number of Events in [0, *t*)

Let $M(t)$ denote the expected number of events in $[0,t)$. Since $N(t)$ is distributed according to Poisson distribution with parameter λt, we have

$$M(t) = E[N(t)] = \lambda t \tag{B.7}$$

B.4.2.2 Non-stationary Poisson Process

In an HPP, the probability of an event occurring in $[t, t+\delta t)$ is $\lambda \delta t + o(\delta t)$, with λ a constant. A *non-stationary* or *non-homogeneous* Poisson process (NHPP) is a natural extension in which λ changes with time.

A counting process $\{N(t),\ t \geq 0\}$ is an NHPP if

1. N(0) = 0.
2. $\{N(t),\ t \geq 0\}$ has independent increments.
3. $P\{N(t+\delta t) - N(t) = 1\} = \lambda(t)\delta t + o(\delta t)$.
4. $P\{N(t+\delta t) - N(t) \geq 2\} = o(\delta t).\lambda(t)$ is called the *intensity function*. Let

$$\Lambda(t) = \int_0^t \lambda(x)\, dx \tag{B.8}$$

Then, it can be shown [see Ross (1970)] that

$$P\{N(t+s) - N(t) = j\} = \frac{e^{-\{\Lambda(t+s)-\Lambda(t)\}}\{\Lambda(t+s) - \Lambda(t)\}^j}{j!} \tag{B.9}$$

for $j \geq 0$.

This result may be used to define an NHPP in a manner similar to Definition 1 for an HPP.

Expected Number of Events in [0, *t*)

Since the probability of j events ($j \geq 0$) in [0, *t*) is given by

$$P\{N(t) = j\} = \frac{e^{-\Lambda(t)}\{\Lambda(t)\}^j}{j!} \tag{B.10}$$

the expected number of events in [0, *t*), $M(t)$, is given by

$$M(t) = E[N(t)] = \Lambda(t)$$

B.4.3 Renewal Processes

We first consider the ordinary renewal process and then discuss some extensions.

B.4.3.1 Ordinary Renewal Processes

As indicated earlier, a counting process characterised in terms of inter-event times is a stationary Poisson process if these times are independent and identically distributed exponential random variables. A natural generalisation is one where the inter-event times are independent and identically distributed with an *arbitrary* distribution.

A counting process $\{N(t),\ t \geq 0\}$ is an ordinary renewal process if

1. $N(0) = 0$.
2. X_1, the time to occurrence of the first event (from $t = 0$) and $X_j, j \geq 2$, the time between the $(j-1)$st and jth events, are a sequence of independent and identically distributed random variables with distribution function $F(x)$.
3. $N(t) = \text{Sup}\{n : S_n \leq t\}$, where

$$S_0 = 0, \quad S_n = \sum_{i=1}^{n} X_i,\ n \geq 1 \tag{B.12}$$

[*Note*: The HPP is a special case of the ordinary renewal process with $F(x)$ an exponential distribution function.]

Distribution of $N(t)$

Note that S_n is the time instant for the nth renewal (or event) and is the sum of n independent and identically distributed random variables. Since the X_i's are distributed with distribution function $F(x)$, from a result in Appendix A, the distribution of S_n is given by the n-fold convolution of F with itself—i.e.

$$P(S_n \leq x\} = F^{(n)}(x) = F(x) * F(x) * \cdots * F(x) \tag{B.13}$$

It is easily seen that $N(t) \geq n$ if and only if $S_n \leq t$. As a result,

$$\begin{aligned} P\{N(t) = n\} &= P\{N(t) \geq n\} \\ &\quad - P\{N(t) \geq (n+1)\} = P\{S_n \leq t\} - P(S_{n+1} \leq t\} \end{aligned} \tag{B.14}$$

for $n = 0, 1, \ldots$, where $S_0 \equiv 0$. Since

$$P\{S_n \leq t\} = F^{(n)}(t) \tag{B.15}$$

where $F^{(0)} \equiv 1$, we have

$$P\{N(t) = n\} = F^{(n)}(t) - F^{(n+1)}(t) \tag{B.16}$$

From this, expressions for the moments of $N(t)$ can be obtained. Of particular interest in reliability analysis is the first moment, the expected number of renewals in $[0, t)$.

Expected Number of Renewals in $[0, t)$

The expected number of renewals $M(t)$ is given by the integral equation

$$M(t) = E[N(t)] = \sum_{n=0}^{\infty} n\, P\{N(t) = n\} \tag{B.17}$$

Using (B.16), this can be written as

$$M(t) = \sum_{n=0}^{\infty} n\left\{F^{(n)}(t) - F^{(n+1)}(t)\right\} = \sum_{n=1}^{\infty} F^{(n)}(t) \tag{B.18}$$

Using Laplace transforms, it can be shown that

$$M(t) = F(t) + \int_0^t M(t-x)f(x)\mathrm{d}x \tag{B.19}$$

This equation for $M(t)$ can also be derived, using conditional expectation, as follows. Conditioned on X_1, the time to first failure, $M(t)$ can be written as

$$M(t) = \int_0^{\infty} E[N(t)|X_1 = x]\ \mathrm{d}F(x) \tag{B.20}$$

But,

$$E[N(t)|X_1 = x] = \begin{cases} 0, & \text{if } x > t \\ 1 + M(t-x), & \text{if } x \le t \end{cases} \tag{B.21}$$

Using (B.21) in (B.20) yields (B.19).*Comment* One is using the "renewal property" in deriving the above expression. If the first failure occurs at $x \le t$, then the renewals over $(t - x)$ occur according to an identical renewal process and hence the expected number of renewals over this period is $M(t-x)$.

Equation (B.19) is called the *renewal integral equation,* and $M(t)$ is called the *renewal function* associated with the distribution function $F(t)$. $M(t)$ plays an important role in reliability analysis. In general, it is difficult to obtain $M(t)$ analytically.

The renewal density function, $m(t)$, is given by

$$m(t) = \frac{\mathrm{d}M(t)}{\mathrm{d}t} \tag{B.22}$$

and satisfies the integral equation

$$m(t) = f(t) + \int_0^t m(t-x)f(x)\mathrm{d}x \tag{B.23}$$

where $f(t)$ is the density function associated with $F(t)$.

B.4.3.2 Delayed Renewal Process

A counting process $\{N(t),\ t \geq 0\}$ is a delayed renewal process if

1. $N(0) = 0$.
2. X_1, the time to the first event, is a non-negative random variable with distribution function $F(x)$.
3. $X_j, j \geq 2$, the time intervals between the jth and $(j-1)$st events, are independent and identically distributed random variables with a distribution function $G(x)$ different from $F(x)$.
4. $N(t) = \mathrm{Sup}\{n: S_n \leq t\}$ where S_n is given by (B.13).

[Note that when $G(x)$ equals $F(x)$, then the delayed renewal process reduces to an ordinary renewal process.]

Expected Number of Renewals in [0, *t*)

$M_d(t)$, the expected number of renewals over $[0, t)$ for the delayed renewal process is given by

$$M_d(t) = F(t) + \int_0^t M_g(t-x)\, f(x)\, \mathrm{d}x \tag{B.24}$$

where $M_g(t)$ is the renewal function associated with the distribution function $G(t)$.

B.4.3.3 Alternating Renewal Process

In an ordinary renewal process, the inter-event times are independent and identically distributed. In an alternating renewal process, the inter-event times are all independent but not identically distributed. More specifically, the odd numbered inter-event times $X_1, X_3, X_5, \ldots$ have a common distribution function $F(x)$ and the even numbered ones $X_2, X_4, X_6, \ldots$ have a common distribution function $G(x)$.

B.4.4 Additional Topics from Renewal Theory

B.4.4.1 Renewal-Type Equation

A renewal-type equation is an integral equation of the form

$$g(t) = h(t) + \int_0^t g(t - x)\, \mathrm{d}F(x) \tag{B.25}$$

where $h(\cdot)$ and $F(\cdot)$ are known functions and $g(\cdot)$ is the unknown function to be obtained as a solution to the integral equation. Then, g(t) given by

$$g(t) = h(t) + \int_0^t h(t - x)\, \mathrm{d}M(x) \tag{B.26}$$

where $M(x)$ is the renewal function associated with $F(x)$ is a solution of (B.19).

B.4.4.2 Renewal Reward Theorem

Consider an ordinary renewal process with inter-arrival times $X_1, X_2, X_3, \ldots$. Suppose that a reward of Z_i is earned at the time of the ith renewal. Then, the total reward earned by time t is given by

$$Z(t) = \sum_{i=1}^{N} Z_i \tag{B.27}$$

where $N(t)$ is the number of renewals in $[0, t)$. $Z(t)$ is a cumulative process with $N(t)$ given by a renewal process. If $E[|Y_i|]$ and $E[X_i]$ are finite, then

1. with probability 1, $\lim\limits_{t\to\infty} \frac{Z(t)}{t} \to \frac{E[Z_i]}{E[X_i]}$, and
2. $\lim\limits_{t\to\infty} E[\frac{Z(t)}{t}] \to \frac{E[Z_i]}{E[X_i]}$

   For a proof, see Ross (1970).

B.4.5 Marked Point Process

A marked point process is a point process with an auxiliary variable, called a mark, associated with each event. Let $Y_i,\ i \geq 1$, denote the mark attached to the ith event.

For example in the case of a multicomponent item, failure of a component can cause induced failures of one or more of the remaining components. If the number of components that must be replaced at the *i*th failure of the item is a random variable, then it can be viewed as a mark attached to an underlying point process characterising item failures.

B.4.5.1 A Simple Marked Point Process

A **simple marked** point process is characterised by

1. $\{N(t),\ t \geq 0\}$, a stationary Poisson process with intensity λ.
2. A sequence of independent and identically distributed random variables $\{Y_i\}$, called marks, which are independent of the Poisson process.This point process is also called a *Compound Poisson* process. Various extensions (e.g. a non-stationary point process and marks constituting a dependent sequence, to name a few) yield more complex marked point processes.

References

Ross SM (1970) Applied probability models with optimization applications. Holden-Day, San Francisco

Ross SM (1983) Introduction to probability models. Academic Press, New York

Ross SM (1983) Stochastic processes. Wiley, New York

Appendix C
Deterministic Optimisation

In this Appendix, some basic results for deterministic optimisation problems are presented. We first consider static optimisation and present results for both the unconstrained and constrained cases. Following this, we look at dynamic optimisation, where both discrete-time (multistage) and continuous-time formulations are considered.

C.1 Static Optimisation with a Scalar Objective Function

C.1.1 Single Variable Optimisation

In the simplest case, a decision maker (DM) has a single decision variable x that has to be selected optimally in order to maximise a scalar objective function $J(x)$[5]. $J(x)$ is differentiable, and the possible values of x belong to the interval $X = [a, b]$, where a and b are non-negative real numbers. X is termed the feasible region. The objective function represents the pay-off or reward earned, and it will also contain parameters that are fixed and so cannot be controlled by the DM.

This DM's problem can be expressed as

$$\max J(x), \text{ subject to } x \in [a, b]. \tag{C.1}$$

A *local maximum* of $J(x)$ occurs at the point $x^* \in [a, b]$ if $J(x^*) \geq J(x)$ for all x sufficiently close to x^*. $J(x)$ has a *global maximum* at x^* if $J(x^*) \geq J(x)$ for all $x \in [a, b]$. The global maximum is the optimal solution to (C.1).

A necessary condition for a local maximum at an interior point $x^* \in [a, b]$is that

$$J'(x^*) = \left.\frac{\mathrm{d}J(x)}{\mathrm{d}x}\right|_{x=x^*} = 0. \tag{C.2}$$

[5] If the DM's problem is to minimise the function $J(x)$, then it is equivalent to maximising the function $K(x)$ where $K(x) = -J(x)$. Hence, without loss of generality, we will confine our attention to maximisation.

D. N. P. Murthy and N. Jack, *Extended Warranties, Maintenance Service and Lease Contracts*, Springer Series in Reliability Engineering,
DOI: 10.1007/978-1-4471-6440-1,

Sufficient conditions for a local interior maximum at $x = x^*$ are

$$J'(x^*) = 0 \text{ and } J''(x^*) = \left.\frac{d^2 J(x)}{dx^2}\right|_{x=x^*} < 0 \tag{C.3}$$

The end points a and b of the interval $[a, b]$ also need to be checked. If $J'(a) \leq 0$ then a is a local maximum and if $J'(b) \geq 0$ then b is a local maximum.

The global maximum of $J(x)$ is the particular local maximum that produces the largest value of the objective function. If $J(x)$ is a *concave* function over the interval $[a, b]$, then the analysis becomes much simpler. The two end points cannot be local maxima so any interior point x^* with $J'(x^*) = 0$ is automatically the global maximum. In most cases, the equation $J'(x) = 0$ has to be solved numerically using a one-dimensional search procedure such as Binary search or Golden Section search. These numerical methods are discussed in Rao (2009).

C.1.2 Multivariable Unconstrained Optimisation

The DM now has a vector of decision variables $\underline{x} = (x_1, x_2, \ldots, x_n)$ to select in order to maximise the scalar objective function $J(\underline{x})$. $J(\underline{x})$ is assumed to be differentiable, and there is no constraint on any of the variables, so the feasible region X is n-dimensional Euclidean space R^n. The problem can be expressed as

$$\max J(\underline{x}), \text{ subject to } \underline{x} \in R^n. \tag{C.4}$$

The optimal solution to (C.4) is the global maximum of $J(\underline{x})$. The necessary condition for a local maximum is that

$$\frac{\partial J(\underline{x})}{\partial x_i} = 0 \text{ at } \underline{x} = \underline{x}^*, \qquad \text{for } i = 1, 2, \ldots, n. \tag{C.5}$$

If we define the *gradient* of the function $J(\underline{x})$ to be

$$\nabla J(\underline{x}) = \left(\frac{\partial J}{\partial x_1}, \frac{\partial J}{\partial x_2}, \ldots, \frac{\partial J}{\partial x_n}\right), \tag{C.6}$$

then the above necessary condition can be stated more succinctly as

$$\nabla J(\underline{x}) = \underline{0} \text{ at } \underline{x} = \underline{x}^*. \tag{C.7}$$

The *Hessian* is the $n \times n$ matrix whose element in the ith row and jth column $(1 \leq i, j \leq n)$ is given by

$$[H(\underline{x})]_{ij} = \frac{\partial^2 J(\underline{x})}{\partial x_i \partial x_j} \tag{C.8}$$

The kth *principal minor* of $H(\underline{x})$ is the $k \times k$ submatrix $H_k(\underline{x})$ obtained by deleting the last $n - k$ rows and columns of the Hessian. If $\nabla J(\underline{x}^*) = \underline{0}$ and the determinant of $H_k(\underline{x}^*)$ for $k = 1, 2, \ldots, n$ has the same sign as $(-1)^k$, then $\underline{x}^*$ is a local maximum.

The optimal solution to (C.4) is the global maximum of $J(\underline{x})$ and this point $\underline{x}^*$ is the particular local maximum that produces the largest value of the objective function. If the function $J(\underline{x})$ is concave, then any local maximum is also a global maximum. To locate possible local maxima, the equation $\nabla J(\underline{x}) = \underline{0}$ has to be solved. This can be done numerically using, for example, the multivariable gradient search (steepest ascent) procedure, Newton's or Quasi-Newton methods. These techniques are discussed in Rao (2009).

C.1.3 Multivariable Constrained Optimisation

We now consider maximisation problems where there are two types of constraints on the decision variables $\underline{x} = (x_1, x_2, \ldots, x_n)$—equality and inequality constraints.

C.1.3.1 Equality Constraints

Lagrange multipliers can be used to solve problems with equality constraints which take the form

$$\max J(\underline{x}), \text{ s.t. } g_j(\underline{x}) = b_j \quad \text{for } j = 1, 2, \ldots, m \tag{C.9}$$

We associate a *multiplier* λ_j with the jth constraint in (C.9) and construct the *Lagrangian*

$$L(\underline{x}, \underline{\lambda}) = J(x_1, x_2, \ldots, x_n) + \sum_{j=1}^{m} \lambda_j \left[b_j - g_j(x_1, x_2, \ldots, x_n)\right] \tag{C.10}$$

The necessary conditions for a constrained local maximum are obtained by setting $\partial L/\partial x_i = 0$ for $i = 1, 2, \ldots, n$ and $\partial \lambda/\partial x_i = 0$ for $j = 1, 2, \ldots, m$, giving

$$\begin{aligned} \frac{\partial J}{\partial x_i} - \sum_{j=1}^{m} \left[\lambda_j \frac{\partial g_j}{\partial x_i}\right] &= 0 \quad \text{for } i = 1, 2, \ldots, n, \\ b_j - g_j(x_1, x_2, \ldots, x_n) &= 0 \quad \text{for } j = 1, 2, \ldots, m. \end{aligned} \tag{C.11}$$

The solution of this system of $(n + m)$ equations in the $(n + m)$ unknowns yields all the possible local maxima which satisfy the constraints and, in general, has to be obtained numerically. The optimal solution (constrained global maximum) will be among the local maxima. If the function $J(\underline{x})$ is concave and each $g_j(x_1, x_2, \ldots, x_n)$ is a linear function, then any local maximum will also be a global maximum.

C.1.3.2 Inequality Constraints

The general maximisation problem with inequality constraints, also called a *non-linear programming problem*, takes the form

$$\max J(\underline{x}),\ \text{s.t. } g_j(\underline{x}) \leq b_j \quad \text{for } j = 1, 2, \ldots, m, \tag{C.12}$$

where the functions $J(\underline{x})$ and $g_1(\underline{x}), g_2(\underline{x}), \ldots, g_m(\underline{x})$ must satisfy some regularity conditions (see Bazarra et al. 2006).

If $\underline{x}^* = (x_1^*, x_2^*, \ldots, x_n^*)$ is an optimal solution to (C.12), then it must satisfy the m inequality constraints and there must exist m multipliers $\lambda_1, \lambda_2, \ldots, \lambda_m$ which satisfy the *Karush-Kuhn-Tucker* (KKT) conditions

$$\begin{aligned} \frac{\partial J(\underline{x}^*)}{\partial x_i} - \sum_{j=1}^{m} \left[\lambda_j \frac{\partial g_j(\underline{x}^*)}{\partial x_i} \right] &= 0 \quad \text{for } i = 1, 2, \ldots, n, \\ \lambda_j \left[b_j - g_j(\underline{x}^*) \right] &= 0 \quad \text{for } j = 1, 2, \ldots, m, \ \lambda_j \geq 0 \quad \text{for } j = 1, 2, \ldots, m. \end{aligned} \tag{C.13}$$

If the function $J(\underline{x})$ is concave and each $g_j(x_1, x_2, \ldots, x_n)$ is a convex function, then any point satisfying the conditions in (C.13) is an optimal solution to the problem given in (C.12).

C.2 Static Optimisation with a Vector Objective Function

The DM may be involved in a multiobjective optimisation problem, wishing to find the value of the single decision variable x that maximises the k objective functions $J_1(x), J_2(x), \ldots, J_k(x)$. In this case, the notion of optimality is not obvious because of the possible presence of conflicting objectives. In general, there will be no single optimal solution x^* for the DM that maximises all the objective functions simultaneously. A value of the decision variable x' is called *Pareto optimal* for the DM if there is no other value x that *dominates* x', so there is no x such that $J_i(x) > J_i(x')$, for all $i = 1, 2, \ldots, k$ and $J_j(x) > J_j(x')$, for at least one $j = 1, 2, \ldots, k$. When all the Pareto optimal solutions have been found, the DM has to identify the one that achieves the best compromise between all the conflicting objectives.

The DM may have a vector of decision variables $\underline{x} = (x_1, x_2, \ldots, x_n)$ to select. In this case, the objective functions would be $J_1(\underline{x}), J_2(\underline{x}), \ldots, J_k(\underline{x})$ and the DM then has a more complex multivariable optimisation problem to solve.

Techniques for optimisation of vector objective functions are discussed in Steuer (1986) and Coello et al. (2007).

C.3 Dynamic Optimisation with a Scalar Objective Function

C.3.1 Multistage (Discrete Time) Dynamic Optimisation

We now look at a dynamic optimisation problem where the DM has to make multiple decisions over time in order to maximise a specified objective function (e.g. total reward earned). The decisions may be made either at discrete-time points or continuously. We begin by looking at the discrete case where the time points or *stages* are denoted by $t = 0, 1, 2, \ldots, N$, and N is the length of the *time horizon*.

The technique that we now describe to solve this type of problem is called dynamic programming (DP). There are two major concepts used in this approach at each stage of the process—*state variables* and *decision* (*or control*) *variables*. A state variable s_t provides all the information about the "current position" that the DM needs to know. The DM then makes a decision x_t in order to change the state, and this decision results in a pay-off/reward $L(S_t, x_t)$ to the DM at this particular stage. There may be constraints on both variables at each stage.

The DM wants to determine the *policy* (sequence of decisions) that will maximise total reward earned over the N time periods which is given by

$$\sum_{t=0}^{N-1} L_t(s_t, x_t) + L_N(s_N), \tag{C.14}$$

where $L_N(S_N)$ denotes the possible reward earned by the DM at the end of the time horizon (terminal reward) if the process in state S_N.

The state variables undergo a transformation represented by the equation

$$s_{t+1} = f_t(s_t, x_t) \quad \text{for } t = 0, 1, 2, \ldots, N-1. \tag{C.15}$$

Bellman (1957) showed the dynamic optimisation problem described in (C.14) and (C.15) could be divided into a sequence of smaller problems. When the smaller problems have been solved, they are then combined to produce the solution to the complete problem. Bellman's *principle of optimality* says that given the current state, the optimal decision for each of the remaining stages does not depend on the previously reached states or previously chosen decisions.

To solve the optimisation problem, we define the *optimal value function* $V_t(S_t)$ as the maximum total reward earned by the DM using an optimal sequence of decisions for the remainder of the time horizon starting from state s_t at time (stage) $t = 0, 1, \ldots, N-1, N$. The optimal value function is a solution of the *functional equations*

$$\begin{aligned} V_t(s_t) &= \max_{x_t \in C_t}\{L_t(s_t,\ x_t) + V_{t+1}[f_t(s_t,\ x_t)]\}, \quad t = 0,\ 1,\ \ldots,\ N-1, \\ V_N(s_N) &= L_N(s_N). \end{aligned} \tag{C.16}$$

C_t is the set of possible decisions (feasible or constraint set) at time t. The optimal policy for the DM consists of the optimising decisions x_t^* which produce the optimal value function for each $t = 0, 1, \ldots, N - 1, N$.

C.3.2 Continuous-Time Dynamic Optimisation

We now focus on the situation where the DM has to make decisions continuously over a fixed time horizon of length T. The state variables and decision variables are now $s(t)$ and $x(t)$ for $0 \leq t \leq T$, and the *reward rate* earned by the DM at time t is $L(s(t), x(t), t)$. The DM wants to determine the policy $x^*(t)$ that will maximise total reward earned over the time horizon which is given by

$$\int_0^T L(s(t), x(t), t)\mathrm{d}t. \tag{C.17}$$

The objective function in (C.17) can easily be modified to include a terminal reward at time T. The state of the process evolves according to the differential equation

$$\dot{s}(t) = \frac{\mathrm{d}s(t)}{\mathrm{d}t} = f(s(t), x(t), t), \tag{C.18}$$

with the initial value $s(0) = s_0$ specified. The decision variable will also usually be constrained, so $x(t) \in C(t)$ for $0 \leq t \leq T$, where $C(t)$ is the feasible set of decisions at time t.

This is a standard *optimal control* problem (see, for example Bryson and Ho 1975 or Sethi and Thompson 2000). To obtain its optimal solution, we introduce a *co-state* variable $\lambda(t)$ and define the *Hamiltonian* function

$$H(s, x, t, \lambda) = L(s, x, t) + \lambda f(s, x, t). \tag{C.19}$$

Note that, in the Hamiltonian, the t dependence in the functions $s(t), x(t)$ and $\lambda(t)$ has been suppressed. The first-order (necessary) conditions for an optimal solution to the DM's problem are due to Pontryagin et al. (1962) and are known as the *maximum principle*. They are

$$\max_x H(s, x, t, \lambda) \quad \text{for all } t \in [0, T] \tag{C.20}$$

$$\dot{s} = \frac{\partial H}{\partial \lambda} \quad [\text{equation of motion for } s] \tag{C.21}$$

$$\dot{\lambda} = -\frac{\partial H}{\partial s} \quad [\text{equation of motion for } \lambda] \tag{C.22}$$

$$\lambda(T) = 0 \quad [\text{transversality condition}] \tag{C.23}$$

Note that (C.20), the first-order condition with respect to the decision variable is not stated as a derivative. This is to allow for the possibility of "end-point" solutions. The condition may be stated alternatively as

$$H(s, x^{*}, t, \lambda) \geq H(s, x, t, \lambda) \qquad \text{for all } t \in [0, T] \tag{C.24}$$

The transversality condition (C.23) implies that, in this problem, there is no binding constraint on the terminal value of the state variable. Conditions (C.21) and (C.22) produce two first-order differential equations for $s(t)$ and $\lambda(t)$, respectively. The general procedure is to first solve (C.22) with (C.23) as the required final condition. Then, (C.21) is solved with initial condition $s(0) = s_0$. Finally, the maximisation in (C.20) is performed.

Details of the derivation of the necessary conditions (C.20)–(C.23) along with the corresponding sufficient condition for an optimal solution can be found in Bryson and Ho (1975) or Kamien and Schwartz (1991). Note that, in the dynamic optimisation problems described in this section, only one state variable and one decision variable at each time t have been specified. The results that have been given can be generalised to the case of many state variables and decision variables and also where the number of decision variables need not equal the number of state variables.

References

Bazarra MS et al (2006) Non-Linear programming theory and algorithms, 3rd edn. Wiley, New Jersey

Bellman RE (1957) Dynamic programming. Princeton University Press, Princeton

Bryson AE, Ho YC (1975) Applied optimal control 2nd edn. Hemisphere, Washington DC

Coello CAC et al (2007) Evolutionary algorithms for solving multi-objective problems, 2nd edn. Springer, New York

Kamien MI, Schwartz NL (1991) Dynamic optimization: the calculus of variations and optimal control in economics and management. North Holland, Amsterdam

Pontryagin LS et al (1962) The mathematical theory of optimal processes. Wiley Interscience, New York

Rao SS (2009) Engineering optimization theory and practice, 4th edn. Wiley, New York

Sethi SP, Thompson GL (2000) Optimal control theory 2nd edn. Kluwer Academic Publishers, Berlin

Steuer RE (1986) Multiple criteria optimization theory and applications. Wiley, New York

Appendix D
Illustrative EWs, MSCs and LCs

The material given in this Appendix was obtained from various websites on the Internet.

D.1 Consumer Products [EWs and MSCs]

D.1.1 Case 1 Manufacturer's EW for Electrical and Electronic Products [Sony]

1 **Extended Warranty Services**

1.1 The benefits given to you in Sony's Extended Warranty are in addition to other rights and remedies you have under a law in relation to the product. Sony products come with guarantees that cannot be excluded under the Australian Consumer Law. You are entitled to a replacement or refund for a major failure and for compensation for any other reasonably foreseeable loss or damage. You are also entitled to have the products repaired or replaced if the products fail to be of acceptable quality, and the failure does not amount to a major failure. A "major failure" to comply with a consumer guarantee applying to products (goods) has a defined meaning under the Australian Consumer Law. One example of a major failure is if a reasonable consumer fully acquainted with the nature and extent of the failure would not have acquired the product.

1.2 Subject always to section 1.1 and the rest of these Extended Warranty Terms, Sony will provide to you the following benefits during the Standard Warranty Term and Extended Warranty Term for the product:

(a) When the product or any Sony accessory supplied with it does not perform in accordance with the manufacturer's specifications, Sony will repair or replace at Sony's cost the product or accessory.
(b) For any claim under section 1.2(a), Sony will provide you with access to your own Sony Extended Warranty Liaison for specialist support and end-to-end management of your claim.

D. N. P. Murthy and N. Jack, *Extended Warranties, Maintenance Service and Lease Contracts*, Springer Series in Reliability Engineering,
DOI: 10.1007/978-1-4471-6440-1, © Springer-Verlag London 2014

(c) For any claim under section 1.2(a), Sony will provide onsite pick-up and delivery for the product or accessory, if the pick-up address is within 25 kms of the nearest Sony Authorised Service Centre.
(d) If the claim under section 1.2(a) is for BRAVIA product, you can choose between an in-home service or onsite pick-up and delivery for the product, if the pick-up address is within 25 kms of the nearest Sony Authorised Service Centre. This means greater choice and more convenience for you.
(e) Should you need to make a claim or require assistance with your Sony product, your claim will be handled and serviced only by Sony Authorised repairers and support staff.
(f) For any claim under section 1.2(a), you will have access to a range of service bookings through the extended hours of operation from our Authorised Service Centres (subject to availability).

2 **Making a claim**

2.1 To make a claim under section 1.2(a) under the Extended Warranty, you will need to:

(a) Contact Sony to notify Sony of your claim. Contact details are as follows:
Ph: 1300 782 657
Service and Support Hotline
1300 13 SONY (7669)
(Service Centre locations, Product Information, Spare Parts, Support)
www.sony.com.au
(b) When making your claim, provide the plan number issued on your Extended Warranty Certificate. If you do not have your plan number, you may provide proof of purchase (e.g. Bill of sale, invoice or purchase receipt) with your claim.
(c) For claims, complete the claim form Sony provides to you and send your claim form to Sony as set out on the form. Your claim will need to provide Sony with sufficient details so we understand the nature of the problem.
(d) For claims, unless onsite service applies make the product available to Sony for onsite pick-up and delivery or provide the product to a Sony Authorised Service Centre (as applicable under these terms or as otherwise agreed with Sony). If you are required to return the product to a Sony Authorised Service Centre, Sony will provide details of the centre to you. Alternatively, to find the nearest Sony Authorised Service Centre, contact Sony using the above contact details. If service is in-home pick-up service, Sony will contact you to make arrangements for on-site service or pick-up.

2.2 The product will be at your risk while in transit to and from the Sony Authorised Service Centre, unless transported by Sony or its authorised representatives.

2.3 Sony and its Authorised Service Centres may seek reimbursement from you of any costs incurred by them when the product is found to be in good working order. To check if your product requires any type of service, please feel free to give our Customer Support line a call on 1300 13 SONY (7669) prior to going to any service centre.

3. **Repairs**

3.1 Products presented for repair may be replaced by refurbished products of the same type rather than being repaired. Refurbished parts may be used to repair the products. Replacement of the product or a part does not extend or restart the Standard Warranty Term or Extended Warranty Term. If the product is replaced during the Extended Warranty Term, the Extended Warranty automatically terminates upon replacement and a Standard Warranty Term will apply to the replacement product. You can ask your Extended Warranty Liaison about any special offers on a new Extended Warranty plan for the replacement product at that time.

3.2 If the product presented for repair is capable of retaining user-generated data, you are advised that repair of the product may result in loss of the data. Sony's dedicated team of Sony technicians and support staff can assist you in backing up any data of this type prior to any servicing of your product.

4. **Extended Warranty Term duration**

4.1 The Extended Warranty Term commences when Sony receives from you payment of the Extended Warranty fee, or when your Standard Warranty Term ends, whichever occurs last. The term of your cover will be listed in your Extended Warranty Certificate.

4.2 An Extended Warranty Certificate with a plan number will be issued to you within 28 days of the Extended Warranty fee being paid.

4.3 Sony is entitled to terminate the Extended Warranty by written notice to you if in Sony's opinion, the product is used contrary to its specifications, in which case Sony will pay a refund for the unexpired period of the Extended Warranty less administration expenses.

4.4 The Extended Warranty is transferable to a new owner, in the event of sale of the product, provided Sony is informed of the transfer in writing at the following address: Sony Warranty Support, Reply Paid 73765, NORTH RYDE BC, NSW 1670, AUSTRALIA.

5. **Limitations and exclusions to Extended Warranty coverage**

5.1 To the full extent permitted by law, but subject always to section 1.1, you do not have a right to make a claim under section 1.2(a):

(a) If the product has not been installed, operated, maintained or used in accordance with the manufacturer's instructions or specifications provided with the product.

(b) If the factory-applied serial number has been altered or removed from the product.
(c) For damage, malfunction or failure resulting from alterations, accident, misuse, abuse, fire, liquid spillage, mis-adjustment of customer controls, use on an incorrect voltage, power surges and dips, thunderstorm activity, acts of God, voltage supply problems, tampering or unauthorised repairs by any persons, use of defective or incompatible accessories, the operation of a computer virus of any kind, exposure to abnormally corrosive conditions or entry by any insect, vermin or foreign object in the product.
(d) For damage arising during transportation, installation or while moving the product, or to any transportation costs of the product or any parts thereof to and from the owner, unless otherwise specified in these warranty conditions.
(e) In relation to any third-party software or hardware not contained in the product as originally configured by the manufacturer.
(f) For any failure, to the extent that the failure is not a failure of the product to perform in accordance with its manufacturer's specifications.
(g) For replacement or repair of any (1) consumables (including batteries and cables), or (2) lost parts or accessories.
(h) While the product is outside Australia or New Zealand.
(i) For any wear and tear including to projector lamp or optical block assembly if the product is used in commercial, industrial, educational or rental applications.
(j) For the normal incidence of off-coloured or dark pixels in LCD screens as described in the User Manual for the product. Sony will only repair or replace the product if there are (1) 8 or more dark pixels in the screen (unless 2 or more are adjoining dark pixels) or (2) 2 or more bright pixels in the screen.
(k) If the product is a projector, to a claim for a replacement projector lamp or optical block assembly.

5.2 To the full extent permitted by law, but subject always to section 1.1:

(a) Sony will not be liable for any loss, damage or alterations to (1) third-party hardware or software; or (2) programs, data or information stored on any media or any part of the product, no matter how occurring; or for any loss or damage arising from loss of use, loss of profits or revenue, or for any resulting indirect or consequential loss or damage.
(b) Sony's aggregate liability in respect of all claims under the Extended Warranty shall not exceed the original purchase price of the product or, at Sony's option, the cost of replacing the product.
(c) Sony excludes all other warranties, conditions, terms, representations and undertakings relating to the product other than those expressly identified in these Extended Warranty Terms.

D.1.2 *Case 2 Retailer's EW for Electrical and IT Products [Harvey Norman]*

Take advantage of our comprehensive extended warranty and purchase those electrical and IT products you have always wanted! Prolong the life of your new purchase and enjoy the satisfaction and peace of mind of up to 5 year parts and labour coverage

No More Worries about Parts and Labour Costs

If a part is going to break down, undoubtedly it will be just after the warranty period has expired. No longer will this be a problem for you as all parts and labour costs are covered by the extended warranty

No More Worries about Beyond Repair Products

Harvey Norman is dedicated to providing a quality service. If for some reason there is a problem with your product that cannot be fixed, a replacement product will be provided to you

Extended Warranty Period

You can choose to cover your product for 3, 4 or 5 years. That will be 2, 3 or 4 years on top of the manufacturer's warranty

Can't Decide?

At Harvey Norman, we realise that major purchase requires some thought, so we are glad to extend up to 14 days after your purchase to decide whether you would like to take advantage of our Extended Warranty

Note:

Limit of liability (the sum of all repairs and/or replacement) shall not exceed the original purchase price of the product.
Harvey Norman Extended Warranty is for residential use only.
Harvey Norman Extended Warranty is not transferable.
Terms and Conditions apply.

D.1.3 *Case 3 Manufacturer's EW for Cars [Chrysler]*

Powertrain Care	Powertrain Care Plus	Added Care Plus	Maximum Care
■ Engine	■ Engine	■ Engine	■ Engine
■ Transmission	■ Transmission	■ Transmission	■ Transmission
■ Driveline	■ Driveline	■ Driveline	■ Driveline
COVERS 300+ COMPONENTS Our "Basic" Protection	PLUS	PLUS	PLUS
	■ Steering	■ Steering	■ Steering
	■ Air Conditioning	■ Air Conditioning	■ Air Conditioning
	COVERS 400+ COMPONENTS Our "Good" Protection. Available on 3/36 warranties only.	PLUS	PLUS
		■ Engine Cooling and Fuel	■ Engine Cooling and Fuel
		■ Front Suspension	■ Front Suspension
		■ Rear Suspension	■ Rear Suspension
		■ Electrical	■ Electrical
		■ Expanded Electrical	■ Expanded Electrical
		■ Instrumentation	■ Instrumentation
		■ Brakes	■ Brakes
		■ Anti-Lock Brakes	■ Anti-Lock Brakes
		■ Power Group	■ Power Group
		■ Luxury Group	■ Luxury Group
		COVERS 800+ COMPONENTS Our "Better" Protection.	PLUS
			■ Complete Mechanical Coverage for up to 5007+ components. "If it's Mechanical... It's Covered!"
			COVERS 5007+ COMPONENTS Our "Best" Protection.

D.2 Commercial and Industrial Products [EWs and MSCs]

D.2.1 *Case 4 Computer Servers [Hewlett Packard]*

The HP service contracts in the USA contain the following elements

1. Support Services[6]

---

[6] Support Services include the following:

- Constant monitoring and alerting on network components
- Full remote control and diagnostics of server equipment
- Immediate alert and response to all events
- Remote diagnostic and repair for all incidents and failures

2. Customer
3. Charges
4. Eligible Products
5. Limitations of Liability and Remedies
6. Timeliness of Action
7. Limitations of Service
8. Supported Software Versions
9. Non-HP Products
10. Customer Responsibilities
11. Off-Site Support and Exchange Services
12. On-Site Support for HP Network Connectivity Products
13. Maximum Use Limitations
14. Transfer of Service
15. Post Warranty Agreement Services
16. Term
17. Termination
18. Governing Laws
19. Entire Agreement

D.2.2 Case 5 Diesel Engines [Wartsila]

MSC-I: Supply Agreement
With supply agreement status, you get access to our global parts distribution network and are able to order and receive spare parts 24/7, including reconditioned components, wherever your facility is located and with the shortest possible lead time. We can also guarantee the availability of a global network of trained and skilled service professionals with the right tools and onboard/ on-site manpower to assist them.

Parts

24/7 global logistics of spare parts

Shortening of lead time

Correct spare parts

Information

Online services

Manpower

Availability to a global network of trained and skilled service professionals with right tools

---

(Footnote 6 continued)
- On-site hot swap exchange of failure devices
- Remote repair and fix, even of 'hung servers'
- Full incident logging and reporting
- 100 % cover for network server, clients and users.

On board/Onsite manpower supply

Workshop services

Global component drops for reconditioning

MSC-II: Technical Maintenance Agreement

A long-term service agreement covering maintenance planning and service crews wherever and whenever needed through the local and global presence of Wärtsilä's networks. We provide fixed prices for inspection, technical support, spare parts, training and maintenance work. Our dynamic maintenance concept leads to a better prediction of maintenance needs and the system's overall function.

Performance guarantee available.

Inspection

Regular inspections expert assistance and monthly reporting

Spare parts

Exchange programme

Reconditioning

MSC-III: Maintenance Agreement

A long-term service agreement covering maintenance planning and service crews wherever and whenever needed through the local and global presence of Wärtsilä's networks. We provide fixed prices for inspection, technical support, spare parts, training and maintenance work. Our dynamic maintenance concept leads to a better prediction of maintenance needs and the system's overall function.

Performance guarantee available.

Power plant agreement

Long-term service agreements with fixed fees for the duration of the agreements

Spare parts and/or labour supply for maintenance work

There can be performance guarantees

Inspection, technical support, spare parts, training/competence

Marine agreement

Long-term service agreements with fixed fees for the duration of the agreements

Spare parts and/or labour supply for maintenance work

There can be performance guarantees

Global agreements

Risk management

MSC-IV: Asset Management Agreement

More comprehensive than a maintenance agreement, an asset management agreement gives us full responsibility for performance and equipment life, so that you can concentrate on your core business. Our asset management agreements typically include full operation, management and maintenance services (O&M) as well as performance guarantees.

Power plant agreement
Full Operation Management and Maintenance services
Supply of required manpower, parts and knowledge to be able to take full responsibility for the operation of a plant
Performance guarantee
Responsibility for unscheduled maintenance and breakdowns
Risk management
Marine agreement
Wärtsilä provides manpower as a part of the crew
Supply of required manpower, parts and knowledge to be able to take full responsibility for the operation of the engine room
Performance guarantee
Responsibility for unscheduled maintenance and breakdowns
Risk management

D.3 Infrastructures [MSCs]

D.3.1 Case 5: New Zealand Transport Authority

Contract Agreement [*Contract for* NZTA1234, Highway Maintenance Example]
THIS AGREEMENT is made on ("date") **BETWEEN** ("the Contractor")

AND

The NZ Transport Agency, a Crown entity, established on 1 August 2008 by Section 93 of the Land Transport Management Act 2003 ("the Principal")
IT IS AGREED as follows:

1. **THE** Contractor shall carry out the obligations imposed on the Contractor by the Contract Documents.
2. **THE** Principal shall pay the Contractor the sum of $ _______ or such greater or lesser sum as shall become payable under the Contract Documents together with Goods and Services Tax at the times and in the manner provided in the Contract Documents.
3. **EACH** party shall carry out and fulfil all other obligations imposed on that party by the Contract Documents.
4. **THE** Contract Documents are this Contract Agreement and the following which form part of this agreement:

   (a) The Conditions of Tendering
   (b) Notices to Tenderers (give details with dates):
   (c) The Contractor's tender;
   (d) The notification of acceptance of tender;

(e) The General Conditions of Contract, NZS 3910:2003
(f) The Special Conditions of Contract;
(g) Specifications issued prior to the Date of Acceptance of Tender;
(h) Drawings issued prior to the Date of Acceptance of Tender;
(i) The Schedule of Prices
(j) The following additional documents: (*Identify any additional documents to be included for example agreed correspondence*)

D.4 Lease Contracts

D.4.1 Case 6: Automobile [British Columbia Transit]

THIS VEHICLE LEASE AGREEMENT dated the_____day of__________, 19__BETWEEN:
BRITISH COLUMBIA TRANSIT, a corporation incorporated pursuant to the British Columbia

Transit Act,
(hereinafter called "BC Transit")

OF THE FIRST PART

AND:

OPERATING COMPANY
(hereinafter called "Operating Company")

OF THE SECOND PART

A. **WHEREAS** the Operating Company is desirous of leasing and/or subleasing from BC Transit the vehicle (s) and equipment described in the list attached hereto as Schedule "A" (hereinafter collectively called the "equipment").
B. **AND WHEREAS** BC Transit either owns the equipment or is itself leasing the equipment (or part thereof as the case may be) from a third-party pursuant to the Head Lease Agreement (hereinafter called the "Head Lease Agreement").

THIS AGREEMENT WITNESSETH that for good and valuable consideration, the parties hereto agree and covenant is as follows:

1. **DEFINITIONS** In this indenture, the words "Lease" and "Lease Agreement" shall be deemed to mean, refer to and include the words "Sub-Lease" and "Sub-Lease Agreement", if applicable and as the context of this Lease Agreement so requires, as between BC Transit (Sublessor) and the Operating Company (Sublessee).

2. **LEASE** BC Transit hereby agrees to lease to the Operating Company and the Operating Company hereby agrees to lease from BC Transit the equipment, together with all accessories, additions, repairs and replacement parts affixed thereto, now or in the future.
3. **RENT** The Operating Company agrees to pay to BC Transit the sum of One Dollar ($1.00) forthwith and such payment shall be the rental charges payable by the Operating Company to BC Transit in respect of the equipment.
4. (A) **TERM** The term of this Lease Agreement shall commence on the date hereof and shall be terminated on that date of the following events first to occur:

   (a) The termination date provided for in the Annual Operating Agreement made pursuant to the BC Transit Act to which BC Transit and the Operating Company are party thereto which is to effect at the time this Lease is enacted, and/or any successor Annual Operating Agreement (hereinafter collectively called the Annual Operating Agreement); or
   (b) That date being 2 weeks after BC Transit has delivered to the Operating Company written notice of its intention to terminate this Lease Agreement; or
   (c) That date which BC Transit and the Operating Company mutually agree shall be an effective date of termination of this Lease Agreement.

4. (B) **TERMINATION OF HEAD LEASE** Notwithstanding the provisions of Section 4A herein, if any item of equipment is the subject of a Head Lease and if for any reason such Head Lease is terminated, then at the option of BC Transit, this Lease Agreement shall terminate with respect to such item of equipment.
5. **ACCEPTANCE** The Operating Company acknowledges that it has inspected the equipment and accepts the equipment as being in a good state of repair, except to the extent that the Operating Company notifies BC Transit in writing within 10 days of delivery (manufacturer's latent defects included).
6. **USE** The Operating Company shall use the equipment only for those purposes set out in the Annual Operating Agreement. The Operating Company shall not use the equipment for pleasure or any other business not contemplated in the Annual Operating Agreement. The Operating Company shall observe and adhere to the operating procedures and guidelines as issued by BC Transit and which relate to the use of the equipment.
7. **HEAD LEASE** BC Transit covenants with the Operating Company to perform and observe the covenants on its part contained in the Head Lease Agreement, if any. The Operating Company covenants with BC Transit to perform and observe the covenants on the part of BC Transit to be performed or observed under the provisions of the Head Lease, if any, other than the covenant to pay rent.
8. **LOCATION** The Operating Company shall cause the equipment to remain situate in the transit service area as designated in the Annual Operating Agreement, and the Operating Company shall not remove the

equipment from the said transit service area without the prior written consent of BC Transit.

9. **OWNERSHIP** Title to and ownership of the equipment, subject to the provisions of any Head Lease Agreement, shall at all times be and remain in the name of BC Transit and the Operating Company shall have no right of property therein, except the right to use the equipment in accordance with the terms of this Lease Agreement.
10. **OPERATING COSTS** The Operating Company shall pay all operating costs whatsoever of the equipment, including without limiting the generality of the foregoing, the cost of fuel, oil, insurance as prescribed in the Annual Operating Agreement, licences pursuant to the Motor Carrier Act, licence and registration fees pursuant to the Motor Vehicle Act, municipal licences and motor vehicle inspections fees (where applicable).
11. **REPAIRS** The Operating Company shall maintain and keep the equipment in good condition and repair to the satisfaction of BC Transit, adhering to the BC Transit Preventive Maintenance Program. The Operating Company further covenants that as component parts of the equipment either wear out or become otherwise inoperative, to replace the same with either parts which are approved by the manufacturer of the equipment or such substitute parts as BC Transit may from to time permit.
12. **INSPECTION** BC Transit shall have the right to inspect the equipment, without prior notice, at all reasonable times during the term of this Lease Agreement.
13. **ALTERATION** The Operating Company shall not alter or add or allow any other party to alter or add to the equipment in any way without the prior written approval of BC Transit. Any alterations, or additions to the equipment which are approved by BC Transit shall become and remain the property of BC Transit. The Operating Company shall affix on the equipment, any labels or insignias supplied by BC Transit. The Operating Company shall not permit any advertising to be posted on the exterior or the interior of the equipment, save and except as provided for in the Annual Operating Agreement.
14. **RECORDS** The Operating Company shall keep for each item of equipment and deliver to BC Transit as specified or upon request the following records:

    (a) Any record as required by the Annual Operating Agreement to be provided by the Operating Company.
    (b) Vehicle Daily Report Card (Form M098).
    (c) Preventative Maintenance Inspection Guide (Form M299).
    (d) Record of Preventive Maintenance Bus Inspections (Form M300).
    (e) Road Call Analysis (Form M301).
    (f) Monthly Bus Fuel Consumption Report (Form M307).
    (g) Unit Change Record (Form M310).
    (h) Accident/Incident Report (Form M318).

15. **MOTOR VEHICLE INSPECTION** The Operating Company shall be responsible for ensuring that the equipment is maintained in compliance with the British Columbia Motor Vehicle Act and Regulations, including the Commercial Vehicle Inspection Program. The Operating Company shall be responsible for ensuring the equipment is submitted for inspections pursuant to the provisions of Motor Vehicle Act, if so required by said provisions.
16. **LOSS OR DAMAGE** The Operating Company assumes and shall bear the entire risk of loss or damage to the equipment. No loss or damage to the equipment or any part thereof shall affect or impair any of the obligations of the Operating Company hereunder, and this Lease Agreement shall continue in full force and effect notwithstanding such loss or damage to the equipment. The Operating Company shall insure the equipment according to the laws in force and effect in the Province of British Columbia and in accordance with the provisions of the Annual Operating Agreement, and such provisions shall be incorporated into the terms and conditions of this Lease Agreement. The Operating Company shall punctually pay all insurance premiums when due in respect of any policies of insurance required to be purchased by it pursuant to the Annual Operating Agreement and the Operating Company shall provide BC Transit with copies of certificates of such insurance policies. In the event of loss or damage of any kind whatsoever to the equipment, the Operating Company shall forthwith comply with the reporting procedures in respect of such loss or damage as established by BC Transit. BC Transit at its sole discretion may either replace the lost or damaged equipment or alternatively direct the Operating Company to repair the damaged equipment, and the Operating Company shall comply with such direction.
17. **SURRENDER** Upon the termination of this Lease Agreement, the Operating Company shall forthwith return the equipment to BC Transit in good condition and repair, ordinary wear and tear resulting from the proper use of the equipment excepted, and the Operating Company shall, at its cost, return the equipment to BC Transit at a destination designated by BC Transit in the transit service areas as defined in the Annual Operating Agreement, and if the Operating Company fails to so deliver the equipment within 1 week from the termination of this Lease Agreement, BC Transit shall have the right to enter upon the premises where the equipment may be, and take possession of and remove it at the Operating Company s expense, all without legal process. The Operating Company covenants that, upon termination of this lease or upon surrender of the equipment for any other reason:

    (a) The equipment shall be in good condition and repair, in compliance with the BC Transit Maintenance Program;
    (b) The records for mechanical repairs listed in Section 14 of this agreement shall accompany each vehicle;
    (c) Average tire tread depth for all tires shall not be less than 8 mm (10/32″).

(d) A vehicle transfer form shall be executed by the Operating Company where applicable, and shall accompany each vehicle, and
(e) The Operating Company shall maintain insurance coverage in accordance with the provisions of Section 22 herein during the period of time that the equipment is being transferred to BC Transit, notwithstanding that this Lease Agreement may be terminated.

18. **LIENS AND CHARGES** The Operating Company shall, at all times, keep the equipment free from all levies, liens and encumbrances whatsoever and shall pay all licence fees, registration fees and assessments, charges and taxes, in accordance with the Annual Operating Agreement, which may be now or hereafter imposed directly upon the ownership, leasing, rent, possession or use of the equipment. If the Operating Company fails to pay any such levies, liens, encumbrances, assessments, charges or taxes, BC Transit may pay the same and in such event the costs thereof, together with interest calculated monthly at a rate equivalent to the prime rate established by The Royal Bank of Canada on the first day of each month, plus 2 % per annum, shall forthwith be due and payable by the Operating Company to BC Transit. Non-payment of such costs by the Operating Company to BC Transit forthwith upon demand by BC Transit shall be deemed to be a default under this Lease Agreement.
19. **WARRANTIES** The Operating Company acknowledges that BC Transit makes no warranties, either express or implied, as to any matter whatsoever, including without limiting the generality of the foregoing, the condition of the equipment nor its merchantability nor its fitness for any particular purpose.
20. **ASSIGNMENT, SUBLEASE** The Operating Company shall not transfer, deliver up possession of, or sublet the equipment, and the Operating Company's interest in this Lease Agreement shall not be assignable by the Operating Company without prior written consent of BC Transit; but nothing herein contained shall prevent BC Transit from assigning, pledging, mortgaging, transferring or otherwise disposing, either in whole or in part, of BC Transit s right hereunder. If the Operating Company is a corporation, then any sale or transfer of shares in the capital of the Operating Company shall be deemed to be an assignment under this Lease Agreement, and the written consent of BC Transit to such a sale or transfer shall be first had and obtained.
21. **INDEMNIFICATION** The Operating Company shall indemnify BC Transit against and hold BC Transit harmless from any and all claims, actions, suits, proceedings, costs, expenses, damages and liabilities including the costs arising out of, connected with or resulting from the equipment including without limitation the installation, possession, use, operation or return of the equipment or otherwise on account of any personal injury or death or damage to property occasioned by the operation of the said equipment during the term hereby granted.

22. **ANNUAL OPERATING AGREEMENT** The Operating Company covenants and agrees with BC Transit to perform each and every one of the conditions, terms, covenants and provisos contained in the Annual Operating Agreement, which on the part of the Operating Company are to be observed and performed.
23. **DEFAULT** Notwithstanding Section 4, the Operating Company covenants and agrees with BC Transit that BC Transit shall have the right to cancel and terminate this Lease Agreement forthwith by reason of any one or more of the following events:

    (a) If the Operating Company fails to observe and perform any of the terms, conditions, covenants and provisos contained in the Annual Operating Agreement, which on its part are to be observed and performed.
    (b) If the Operating Company fails to perform any of the terms, conditions, covenants and provisos contained in this Lease Agreement which on its part are to be observed and performed.
    (c) If a petition under any bankruptcy law shall be filed by or against the Operating Company or the Operating Company shall make any assignment for the benefit of its creditors or the Operating Company shall suffer or permit the appointment of any trustee or receiver or receiver–manager for the Operating Company's business or assets or any part thereof or otherwise becomes financially insolvent or if the Operating Company shall make or suffer any assignment, voluntary or involuntary, of the Operating Company's interest in any of the equipment included in this Lease Agreement or suffer any lien, attachment or levy of execution to become attached thereto.
    (d) If the Operating Company uses any equipment included in this Lease Agreement unreasonably or abusively resulting in damage to such equipment or an abnormal reduction in the life of the equipment or any part thereof.

24. **TERMINATION** Upon the termination of this Lease Agreement, the Operating Company shall forthwith return to BC Transit all items of equipment as referred to herein and the Operating Company shall be liable to BC Transit for damages and costs which BC Transit may sustain by reason of the Operating Company's default of this Lease Agreement, including, without limiting the generality of the foregoing, all legal fees and other expenses incurred by BC Transit in attempting to enforce the provisions of this Lease Agreement or to recover damages for default under this Lease Agreement, or to recover any equipment not forthwith returned by the Operating Company to BC Transit.
25. **WAIVER** No covenant or proviso contained in this Lease Agreement to be performed by the Operating Company may be waived by BC Transit, except by prior written consent of BC Transit, and any forbearance or indulgence by

BC Transit in this regard shall not constitute its waiver of such covenant or proviso to be performed by the Operating Company.

26. **REGULATIONS** Nothing in this Lease Agreement shall preclude BC Transit from setting "lease fees" chargeable in connection with the public transportation system operated by the Operating Company pursuant to the aforementioned Annual Operating Agreement. (Note: Lease fees are based upon the capital cost of vehicles and are covered by the Annual Operating Agreement budget).
27. **TIME OF THE ESSENCE** Time is to be of the essence of this Lease Agreement and each and all of its provisions.
28. **INTERPRETATION** It is hereby agreed by and between the parties hereto that wherever the singular or masculine is used throughout this Lease Agreement, the same shall be construed as meaning the plural or the feminine or body corporate or politic, respectively, and vice versa, where the context or the parties hereto so require and in the case where more than one Operating Company is a party hereto, the liability of each Operating Company shall be joint and several.
29. **GOVERNING, LAW** This Lease Agreement shall be interpreted and enforced in accordance with the laws of the Province of British Columbia.
30. **EXECUTORS, ADMINISTRATORS AND ASSIGNS** This Lease Agreement shall ensure to the benefit of and be binding upon the parties hereto, and their respective heirs, executors, administrators, successors and permitted assigns.

IN WITNESS WHEREOF the parties hereto set their hands and seals and where a party is a corporate entity, the seal of such party has been affixed hereto in the presence of its duly authorised officers, the day, month and year first above written.

BRITISH COLUMBIA TRANSIT

per:

The Corporate Seal of THE OPERATOR
was hereunto affixed

Authorised Signatory (ies)

D.4.2 Case 7: Automobile [Ford Company in USA]

1-800-727-7000	**Motor Vehicle Lease Agreement**	Lease Date:
	Lessee (and Co-Lessee) - Name and Address (including County):	
	Lessor - Name and Address:	

"Ford Credit" is Ford Motor Credit Company. The **"Holder"** is and its assigns.
By signing **"You"** (Lessee and Co-Lessee) agree to lease this Vehicle according to the terms on the front and back of this lease.

New/Used/Demo	Mileage at Delivery	Year/Make/Model	GVW # Truck (lbs.)	Vehicle ID #	Vehicle Use

1. Amount Due At Lease Signing or Delivery Below) $	**2. Monthly Payments** Your first monthly payment of $ is due on ,followed by payments of $ due on the .day of each month. The total of Your monthly payments is $	**3. Other Charges** (not part of Your monthly payment) Disposition fee (if You do not purchase the Vehicle) Total $	**4. Total of Payments** (The amount You will have paid by the end of the lease) $

Itemization of Amount Due at Lease Signing or Delivery

5. Amounts Due At Lease Signing or Delivery:	**6. How the Amount Due At Lease Signing or Delivery will be paid:**
a. Capitalized cost reduction $	**a.** Net trade-in allowance $
b. First monthly payment	**b.** Rebates and noncash credits $
c. Refundable security deposit	**c.** Amount to be paid in cash $
d. Title fees	**d.**
e. Registration fees	
f.	
g.	
h.	
i.	
Total $	Total $$

7. Your monthly payment is determined as shown below.

a. Gross capitalized cost. The agreed upon value of the Vehicle ($) and any items You pay over the lease term (such as service contracts, insurance, and any outstanding prior credit or lease balance) (Itemized below)** $
b. Capitalized cost reduction. The amount of any net trade-in allowance, rebate, noncash credit, or cash that You pay that reduces the gross capitalized cost -
c. Adjusted capitalized cost. The amount used in calculating Your base monthly payment =
d. Residual value. The value of the Vehicle at the end of the lease used in calculating Your base monthly payment -
e. Depreciation and any amortized amounts. The amounts charged for the Vehicle's decline in value through normal use and for other items paid over the lease term =
f. Rent charge. The amount charged in addition to the depreciation and any amortized amounts +
g. Total of base monthly payments. The depreciation and any amortized amounts plus the rent charge =
h. Lease payments. The number of payments in Your lease +
i. Base monthly payment =
j. Monthly sales / use tax +
k.
l.
m. Total monthly payment
n. Lease term in months.

$$	**Early Termination. You may have to pay a substantial charge if You end this lease early. The charge may be up to several thousand dollars. The actual charge will depend on when the lease is terminated. The earlier You end the lease, the greater this charge is likely to be**

8. Excess Wear and Use. You may be charged for excessive wear based on our standards for normal use. At the scheduled end of this lease, unless You purchase the Vehicle, **You must pay** to Lessor **cents** per mile for each mile in excess of miles shown on the odometer. See Items 19 and 23 on back for additional excess wear and use terms.

9. Extra Mileage Option Credit. At the scheduled end of this lease, You will receive credit of cents per unused mile for the number of unused miles between and miles, less any amounts You owe under this lease. You will not receive any credit if the Vehicle is destroyed, if You terminate Your lease early, exercise any purchase option, are in default or the credit is less than **$1.00.**

10. Purchase Option at End of Lease Term
$ plus official fees and taxes is Your lease end purchase option price. You have the option to purchase the Vehicle from Lessor in cash for the purchase option price at the end of this lease term if You are not not default.

Other Important Terms. See Your lease documents for additional information on early termination, purchase option and maintenance responsibilities, warranties, late and default charges, insurance, and any security interests, if applicable.

11. WARRANTY The Vehicle is covered by any warranty, extended warranty or service contract indicated below:

□ Standard new Vehicle warranty provided by the manufacturer or distributor of the Vehicle.

□

If the Vehicle is of a type normally used for personal use and the Lessor, or the Vehicle's manufacturer, extends a written warranty or service contract covering the Vehicle within 90 days from the date of this lease, You get implied warranties of merchantability and fitness for a particular purpose covering the Vehicle. Otherwise, You understand and agree that there are no such implied warranties, except as otherwise required by state law.

12. OFFICIAL FEES AND TAXES $
The estimated total amount You will pay for official and license fees, registration, title and taxes over the term of Your lease, whether included with Your monthly payments or assessed otherwise. The actual total of fees and taxes may be higher or lower depending on the tax rates in effect or the value of the leased property at the time a fee or tax is assessed.

13. LESSOR SERVICES
(See Item 18 on back)

14. LATE PAYMENTS You will pay a late charge on each payment that is not received within **10 days** after it is due. The charge is **7.5%** of the full amount of the scheduled payment or **$50.00** whichever is less.

15. LIFE, DISABILITY AND OTHER INSURANCE These coverages are **not required** to enter into this lease and will not be provided unless You sign below. If insurance is to be obtained by Lessor, the coverages are shown in a notice given to You this date and are for the term of this lease.

Life Insurance
Insurer **$ Initial Coverage Amount**
Insured **$ Premium**
Insured's Signature

Disability Insurance
Insurer **$ Monthly Coverage**
Insured **$ Premium**
Insured's Signature

Other Insurance
Insurer **$ Monthly Coverage**
Insured **$ Premium**
Insured's Signature
Total Premiums $

16. Itemization of Gross Capitalized Cost

Agreed Upon Value of the Vehicle $	Sales/Use Tax & Other Applicable Taxes $	Title Fees $	License & Registration Fees $	Extended Warranty & Service Contract $	Lessor Services $	Acquisition Fee $
Documentation Fee $	Life Insurance Premium $	Disability Insurance Premium $				**Total Gross Capitalized Cost** $

SIGNATURES AND IMPORTANT NOTICES

Modification: This lease sets forth all of the agreements of Lessor and You for the lease of the Vehicle. There is no other agreement. Any change in this lease must be in writing and signed by You and Ford Credit.

Lessee By Title

Co-Lessee By Title

NOTICE: (1) Do not sign this lease before You read it or if it has any blank space to be filled in. (2) You have the right to get a filled-in copy of this lease. You state that You have been given a filled-in copy of this lease at the time You sign it and notice of an assignment of this lease by the Lessor to Holder.

Lessee By Title

Co-Lessee By Title

YOU ACKNOWLEDGE THAT YOU HAVE READ AND AGREE TO BE BOUND BY THE ARBITRATION PROVISION ON THE REVERSE SIDE OF THIS CONTRACT

Lessor is hereby notified that Holder has assigned to "Intermediary," as defined in the Red Carpet Lease Assignment, its rights (but not its obligations) with respect to the purchase of this Vehicle and the sale of this Vehicle at lease termination.

Lessor accepts this lease and assigns it to Holder under the terms of the Red Carpet Lease - WOR Plan Agreement between Lessor and Holder unless otherwise indicated here: ☐ **LEV GUARRANTY**

Lessor By Title

VEHICLE MAINTENANCE, INSURANCE AND USE

17. VEHICLE USE AND SUBLEASING You will not use, or permit others to use the Vehicle (a) in violation of any law, (b) contrary to the provisions of any insurance policies covering the Vehicle, (c) outside the state where first titled or registered for more than **30** days without Ford Credit's written consent, (d) outside the United States, except for less than **30** days in Canada or (e) as a private or public carrier. You will keep this lease and Vehicle free of all liens and encumbrances. **You will not assign or sublease any interest in the Vehicle or lease without Ford Credit's written consent.**

18. VEHICLE MAINTENANCE AND OPERATING COSTS Proper Vehicle maintenance is Your responsibility. You must maintain and service the Vehicle at Your own expense, using materials that meet the manufacturer's specifications. This includes following the owner's manual and maintenance schedule, documenting maintenance performed, and making all needed repairs. You are also responsible for all operating costs such as gas and oil. Lessor will provide the service(s), if any, identified in the Lessor Services section under the terms of a separate agreement. The manufacturer will invalidate warranty coverage on parts affected by a failure to maintain the Vehicle as required by the manufacturer. (See Lessor Services on the front of lease.)

19. DAMAGE REPAIR You are responsible for repairs of **All Damage** which are not a result of normal wear and use. These repairs include, but are not limited to, those necessary to return the Vehicle to its pre-accident condition, including repairs to **Exterior Sheet Metal and Plastic Components,** and to **Vehicle Safety Systems,** including air bag, seat belt and bumper system components. Replacement of Sheet Metal must be made with Original Equipment Manufacturer Sheet Metal. All other repairs must be made with Original Equipment Manufacturer parts or those of equal quality. Discuss this requirement with Your insurance company prior to signing a collision repair estimate or before authorizing any collision repair work.

If You have not had the repairs made before the Vehicle is returned at the scheduled end of this lease, You will pay the estimated costs of such repairs, even if the repairs are not made prior to Holder's sale of the Vehicle.

20. VEHICLE INSURANCE You must insure the Vehicle during this lease. This insurance must be acceptable to Ford Credit and protect You and Holder with **(a)** comprehensive fire and theft insurance with a maximum deductible amount of **$1,000; and (b)** collision and upset insurance with a maximum deductible of **$1,000; and (c)** automobile liability insurance with minimum limits for bodily injury or death of **$25,000** for any one person and **$50,000** for any one accident, and **$10,000** for property damage. If the state in which You title/register the Vehicle establishes or changes the minimum automobile liability insurance limits greater than those listed above for bodily injury or death and property damage insurance, You must insure the Vehicle and the Holder at the higher minimum limits established by the state. **These amounts may not be sufficient to cover all Your liabilities. You may wish to consult Your insurance advisor about obtaining additional coverage.** You will list the loss payee and additional insured as requested by Lessor. You must give Ford Credit evidence of this insurance.

You authorize Ford Credit, on Your behalf, to receive and endorse checks or drafts, and settle or release any claim under the insurance related to Holder's ownership of the Vehicle. You also assign to Holder any other insurance proceeds related to this lease or Holder's i anterest in the Vehicle.

If You or Ford Credit obtain a refund for amounts paid to third parties for insurance, service contracts, or any other amount paid to a third party included in the Gross Capitalized Cost of this lease, You must pay to the Holder the entire amount of the refund and You authorize the Holder to subtract the refund from the amount You owe under this lease.

LESSOR IS NOT PROVIDING VEHICLE INSURANCE OR LIABILITY INSURANCE

If you title/register the Vehicle in, or change the garage location of the Vehicle to a state where Ford Credit has established minimum automobile liability insurance limits greater than those listed above for bodily injury or death and property damage insurance, You must insure the Vehicle and the Holder at the higher minimum limits established by Ford Credit

ENDING YOUR LEASE

21. TERMINATION This lease will terminate (end) upon (a) the end of the term of this lease, (b) the return of the Vehicle to Lessor, and (c) the payment by You of all amounts owed under this lease. Ford Credit may cancel this lease if You default.

22. RETURN OF VEHICLE If You do not buy the Vehicle, at lease end You must return it to Lessor unless Ford Credit specifies another place. If You fail to return the Vehicle, You must continue to pay the monthly payments **plus** other damages to Ford Credit, including amounts payable under default. Payment of these amounts will not allow You to keep the Vehicle.

23. STANDARDS FOR EXCESS WEAR AND USE You are responsible for all repairs to the Vehicle that are not the result of normal wear and use. These repairs include, but are not limited to those necessary to repair or replace: (a) **Tires** which are unmatched, unsafe or have less than **1/8** inch of remaining tread in any place; (b) **Electrical or Mechanical** defects or malfunctions; (c) **Glass, Paint, Body Panels, Trim and Grill Work** that are broken, mismatched, chipped, scratched, pitted, cracked, or if applicable, dented or rusted; (d) **Interior** rips, stains, burns or worn areas; and (e) **All Damage** which would be covered by collision or comprehensive insurance whether or not such insurance is actually in force. Replacement of Sheet Metal must be made with Original Equipment Manufacturer Sheet Metal. All other repairs must be made with Original Equipment Manufacturer parts or those of equal quality. Your use or repair of the Vehicle must not invalidate any warranty.

If You have not had the repairs made before the Vehicle is returned at the scheduled end of this lease, You will pay the estimated costs of

25. VOLUNTARY EARLY TERMINATION AND RETURN THE VEHICLE You may terminate this lease early, if You are not in default, by returning the Vehicle to Lessor and paying the following: (a) an early termination fee of **$200, plus** (b) the difference, if any, between the Unpaid Adjusted Capitalized Cost and the Vehicle's Fair Market Wholesale Value, **plus** (c) all other amounts then due under this lease. You will never pay more than the sum of the remaining unpaid lease payments, plus any excess wear and use and mileage charges, and all other amounts then due under this lease.

VOLUNTARY EARLY TERMINATION AND PURCHASE THE VEHICLE You may purchase the Vehicle from Lessor at any time for the sum of the remaining payments, less any unearned Rent Charges, plus the purchase option price and all other amounts then due under this lease.

Unpaid Adjusted Capitalized Cost is reduced on each payment due date. It is calculated by reducing the Adjusted Capitalized Cost each month by the difference between the Base Monthly Payment and the part of the Rent Charges earned in that month on an **actuarial basis.** Rent Charges are earned when due. Lessor or Ford Credit will provide You with a written explanation of the actuarial method upon Your request.

Fair Market Wholesale Value, at Your option, will be: (a) an amount agreed to by You and the Lessor, or (b) the value which could be realized at the wholesale sale of the Vehicle, as determined by a professional appraisal obtained by You at Your expense within **10** days from termination from an independent third party agreeable to Ford Credit, or (c) if not established by agreement or appraisal, the net amount received by Ford Credit upon the sale of the Vehicle at wholesale.

Please contact Ford Credit at 1-800-727-7000 or

such repairs, even if the repairs are not made prior to Holder's sale of the Vehicle.

24. ODOMETER STATEMENT Federal law requires You to complete a statement of the Vehicle's mileage at the end of this lease.

www.fordcredit.com if You have any questions regarding terminating Your Red Carpet Lease.

DEFAULT AND LOSS OF VEHICLE

26. DEFAULT You will be in default if (a) You fail to make any payment when due, or (b) a bankruptcy petition is filed by or against You, or (c) any governmental authority seizes the Vehicle and does not promptly and unconditionally release the Vehicle to You, or (d) You have provided false or misleading material information when applying for this lease, or (e) You fail to keep any other agreement in this lease.

If You are in default, Ford Credit may cancel this lease, take back the Vehicle and sell it at a public or private sale. You also give Ford Credit the right to go on Your property to peacefully retake the Vehicle. Even if Ford Credit retakes the Vehicle, You must still pay at once: (a) the difference, if any, between the Unpaid Adjusted Capitalized Cost and the value which could be realized at the sale of the Vehicle, **plus** (b) all other amounts then due under this lease. The value which could be realized at the sale of the Vehicle at Your option will be: (a) the net amount received by Ford Credit upon the sale of the vehicle at wholesale, or (b) as determined by a professional appraisal obtained by You at Your expense within **10** days from default, from an independent third party agreeable to Ford Credit. You must also pay all expenses, including reasonable attorney's fees, payable by Ford Credit to obtain, hold and sell the Vehicle, collect amounts due and enforce Holder's rights under this lease. You authorize Ford Credit to cancel Your insurance and apply any proceeds to Your obligation.

27. LOSS OR DESTRUCTION OF VEHICLE If the Vehicle is stolen or destroyed, You will pay to Ford Credit: (a) the Unpaid Adjusted Capitalized Cost, **plus** (b) all other amounts then due under this lease, **minus** (c) any insurance proceeds received by Ford Credit. Gap Waiver: If You had in effect the insurance required under this lease and Ford Credit receives the full proceeds, You will pay to Ford Credit: (a) any past due monthly lease payments, **plus** (b) the amount of the applicable insurance deductible, **plus** (c) all other amounts then due under this lease. Even if the Vehicle is insured, until Ford Credit receives the appropriate amount above, You are responsible for the scheduled monthly payments.

ADDITIONAL INFORMATION

28. ASSIGNMENT AND ADMINISTRATION When You and Lessor sign this lease, Lessor will assign it to Holder. Ford Credit or a substitute will administer this lease. You must then pay all amounts due under this lease to Ford Credit.

If Ford Credit is not the Holder of this lease, Holder has appointed Ford Credit as its agent. As agent for Holder, Ford Credit has the power to act on Holder's behalf to administer, enforce, and defend this lease. If Lessor has agreed to repair or maintain the Vehicle, obtain any insurance or perform any other service, You will look only to the Lessor for these services.

29. TAXES You will promptly pay all fees, charges, and taxes relating to the lease or Vehicle (except for Lessor's or Holder's income taxes). You will pay these amounts even if they are assessed after lease end.

30. TITLING The Vehicle will be titled in the name of Holder. You will register the Vehicle as directed by Ford Credit. You will pay all license, title and registration costs.

31. LIFE INSURANCE If Ford Credit receives the benefits paid under any life insurance described on the reverse side, this lease will continue if there is a Co-Lessee. Any Co-Lessee will pay when due all amounts not paid by the insurance. If there is no Co-Lessee, Ford Credit will accept a reasonable replacement designated by Your estate who agrees in writing to perform Your obligations not covered by the insurance.

32. INDEMNITY You will indemnify and hold harmless Lessor, Ford Credit and Holder and their assigns from any loss or damage to the Vehicle and its contents and from all claims, losses, injuries, expenses and costs related to the use, maintenance, or condition of the Vehicle. You will promptly pay all fines and tickets imposed on the Vehicle or its driver. If You do not pay, You will reimburse Ford Credit and pay a **$20** administration fee, unless prohibited by law, for every such fine, ticket, or penalty that must be paid on Your behalf.

33. SECURITY DEPOSIT Your security deposit may be used by Ford Credit to pay all amounts that You fail to pay under this Lease. You will not receive any interest, profits or other earnings on Your security dt posit(s).

34. CONSUMER REPORTS: You authorize Ford Credit to obtain consumer credit reports from consumer reporting agencies (credit bureaus) for any reason and at any time in connection with this lease.

35. GENERAL Except as otherwise provided by the law of the state where You reside, the law that will apply to this lease is the law of the state where the Lessor's place of business is, as set forth on the front of the lease. If that law does not allow any of the agreements in this lease, the ones that are not allowed will be void. The rest of this lease will still be good.

READ THIS ARBITRATION PROVISION CAREFULLY AND IN ITS ENTIRETY

ARBITRATION

Arbitration is a method of resolving any claim, dispute, or controversy (collectively, a "Claim") without filing a lawsuit incourt. Either you or Lessor ("us" or "we") (each, a "Party") may choose at anytime, including after a lawsuit is filed, to have any Claim related to this contract decided by arbitration. Such Claims include but are not limited to the following: 1) Claims in contract, tort, regulatory or otherwise; 2) Claims regarding the interpretation, scope, or validity of this clause, or arbitrability of any issue; 3) Claims between you and us, our employees, agents, successors, assigns, subsidiaries, or affiliates; 4) Claims arising out of or relating to your application for credit, this contract, or any resulting transaction or relationship, including that with the dealer, or any such relationship with third parties who do not sign this contract.

RIGHTS YOU AND WE AGREE TO GIVE UP

If either you or we choose to arbitrate a Claim, then you and we agree to waive the following rights:

- RIGHT TO A TRIAL, WHETHER BY A JUDGE OR JURY
- RIGHT TO PARTICIPATE AS A CLASS REPRESENTATIVE OR A CLASS MEMBER IN ANY CLASS CLAIM YOU MAY HAVE
- AGAINST US WHETHER IN COURT OR IN ARBITRATION
- BROAD RIGHTS TO DISCOVERY AS ARE AVAILABLE IN A LAWSUIT
- RIGHT TO APPEAL THE DECISION OF AN ARBITRATOR
- OTHER RIGHTS THAT ARE AVAILABLE IN A LAWSUIT

Rights You And We Do Not Give Up: If a Claim is arbitrated, you and we will continue to have the following rights, without waiving this arbitration provision as to any Claim: 1) Right to file bankruptcy in court; 2) Right to enforce the ownership interest in the vehicle, whether by repossession or through a court of law; 3) Right to take legal action to enforce the arbitrator's decision; and 4) Right to request that a court of law review whether the arbitrator exceeded its authority.

Either Party must contact any association below and the other Party to start arbitration. The applicable rules (the "Rules") may be obtained from the association.

- American Arbitration Association ("AAA"), at 1-800-778-7879, or www.adr.org;
- J.A.M.S./Endispute, at 1-800-440-1000, or www.jamsadr.com;
- National Arbitration Forum, at 1-800-474-2371, or www.arb-forum.com.

If there is a conflict between the Rules and this contract, this contract shall govern. This contract is subject to the Federal Arbitration Act (9 U.S.C. § 1 et seq.) and the Federal Rules of Evidence. The arbitration decision shall be in writing with a supporting opinion. We will pay your total reasonable arbitration fees and expenses (not including attorney fees, except where applicable law otherwise provides) in excess of $125. We will pay the whole filing fee if we demand arbitration first. Any portion of this arbitration clause that is unenforceable shall be severed, and the remaining provisions shall be enforced

D.4.3 Case 8: WENDT EQUIPMENT LEASING TERMS AND CONDITIONS

ARTICLE 1. THE PARTIES. Wendt, LLP, ("Lessor") agrees to lease to the customer (the "Lessee") identified on the front page or order form of this lease agreement (the "Lease") the equipment as described on the front page or order form of this Lease (the "Equipment").

ARTICLE 2. THE RENTAL PERIOD. The rental period extends from the time the Equipment leaves the Lessor's yard until it is returned to the Lessor's yard in satisfactory working condition. On out-of-town shipments of Equipment, the date of the bill of lading is the beginning of the rental period and it ends on the date the Equipment is returned to the Lessor's yard or siding, or on the date of return bill of lading, if stipulated by the Lessor.

ARTICLE 3. RENT. The rental rates are set forth on the face of this Lease. Rental rates are based on 8 h per day, 5 days per week and 22 eight-hour days in any 30 consecutive day period. Should the Equipment be used longer, the overtime rates, set forth in Article 3, shall apply. The Lessee shall pay rent for the entire period on each piece of Equipment. Rent is not subject to any deductions on account of non-working time. The monthly rates are not subject to deductions on account of non-working time. Fractions of the month at the beginning or the end of the rental period shall be at the monthly rental rate, pro-rated, but only after one full month of rental. If Lessee fails to take possession of the Equipment reserved for it or cancels this Lease, the Lessee agrees to pay a cancellation fee to the Lessor in the amount of 2 % of the value of the Equipment as noted on the face of this Lease and 4 % of the value of the Equipment if it has been loaded for transit to the Lessee.

ARTICLE 4. OVERTIME RATE BASIS. One of the following schedules of overtime charges should be agreed upon, and noted on the front page or order form of this Lease.

Schedule A: On the daily rate, add 1/8th of the daily rate for each hour worked in excess of 8 h in any one day; 1/40th of the weekly rate for each hour worked in excess of 40 h in any one week; and 1/176th of the monthly rate for each hour in excess of 176 h worked in any 30 consecutive day period.

Schedule B: On the daily rate, for each hour over 8 h, 1/16th of the daily rate shall be charged. On the weekly or monthly rate, two shifts are charged at 11/2 times the single shift, and three shifts are charged at 2 times the single shift rate. If no overtime rate schedule is referenced on the front page or order form of this Lease. Schedule "A" shall apply. Lessee agrees to state in writing the number of excess hours the Equipment is used and to pay the Lessor the appropriate rent amount.

ARTICLE 5. TERMS OF PAYMENT. Rentals shall be paid on the 15th of the month following the first use of the Equipment unless otherwise stipulated on the first page or order form of this Lease. Lessor shall be entitled to reimbursement of all costs and expenses, including court costs and attorneys fees, incurred in collecting payment from Lessee. Any past due accounts shall have interest accruing at a rate of 2 % per month. Any payments made on past due accounts shall first be applied to collection costs and expenses, then late payment fees, then to interest, then to rent. Payment of late payment fees and interest shall not waive the Lessor's right to terminate this Lease as hereinafter provided.

In addition to any other rights available to Lessor under this Lease, if any rent is not paid within 30 days of due date, the Lessee shall be in breach of the terms of this Lease. If the Lessee is in breach of this Lease or becomes subject to any of bankruptcy, receivership or insolvency proceeding, the Lessor may, without notice, declare the entire amount of rent under this Lease due and payable, terminate this Lease without court order and take possession of the Equipment without being in breach of this Lease or liable to Lessee for trespass. Lessee will be responsible for any and all legal and transportation costs incurred by Lessor in any such repossession.

ARTICLE 6. LOADING AND FREIGHT CHARGES. The Equipment is rented F.O.B. to the Lessor's yard or siding. Any additional charges incurred in loading, unloading, erection, dismantling, are the responsibility of the Lessee. If the Lessee does not furnish shipping instructions, the Lessor will select the means of conveyance for Lessor.

ARTICLE 7. NOTICE OF RETURN OR RECALL. The Lessor may recall any or all Equipment upon 30 days written notice to the Lessee. The Lessee may return any or all Equipment to the Lessor upon 30 days written notice to Lessor.

ARTICLE 8. SUBLEASING. No Equipment listed herein may be subleased by the Lessee. The Lessee further agrees not to assign or transfer any interest in this Lease without written consent of the Lessor.

ARTICLE 9. RELOCATION EQUIPMENT. Lessee agrees not to move the Equipment to another location without the express written consent of the Lessor.

ARTICLE 10. REPAIRS AND MAINTENANCE. The Lessor is required to supply the Equipment in good operating condition. The Lessee acknowledges by signing this Lease that it has carefully examined the Equipment and accepts the Equipment as being in good operating condition. The Lessee agrees that it will pay all cost of repairs during the rental period, including labour, material, parts and other items, except for normal wear and tear. Rent continues until the Equipment is returned to Lessor with all necessary repairs made to the Equipment and with it in normal operating condition. "Normal wear and tear" is defined as use of the Equipment under normal work conditions, with qualified personnel providing proper operation, maintenance and service. If repairs exceeding the normal wear and tear are necessary upon return of the Equipment, Lessor is authorised to make such repairs and Lessee agrees to pay Lessor the reasonable costs of such repairs to the Equipment and rent while such repairs are being made. Lessee agrees not to cover, alter, substitute, or remove any identifying insignia displayed on the Equipment. Lessee will not permit the Equipment to be abused, overloaded, and used beyond its capacity. Lessee will not alter the Equipment in any fashion and shall use and operate the Equipment in accordance with all applicable laws and the manufacturer's operating manual. The Equipment furnished is standard from manufacturer only. Any modification or additions or optional equipment to be at added to the Equipment shall be at an additional cost to Lessee. Equipment to be used by Lessee under normal working conditions as designed and specified by manufacturer. Unusual or abnormal working conditions, requiring work in rock, excessive mud, abrasives, etc., or tying down, towing, demolition, adding additional or excessive weight will be billed to the Lessee as additional wear and tear and/or cost of repairs as provided herein.

ARTICLE 11. INSPECTION. Before shipment is made, the Lessee may require inspection of the Equipment. If it is not in substantially the condition required by this Lease, the cost of inspection will be paid by the Lessor, and Lessee may cancel the Lease at its option or require the Lessor to supply Equipment in normal operating condition. The Lessor will have the right at any time to inspect Equipment and will be given free access by Lessee to it and the necessary facilities to accomplish the inspection.

ARTICLE 12. INSURANCE & INDEMNIFICATION. Lessee, at its own expense, shall carry and maintain in force at all times during the term of this Lease insurance of the type and minimum coverage limits as follows:

(1) Worker's Compensation—Statutory amount under the laws of the state where the Lessee is operating the Equipment.
(2) Commercial General Liability—$1,000,000 per occurrence.
(3) Property/Casualty insurance—with coverage limits sufficient to cover the full replacement cost of the Equipment.

All such insurance shall be in form and with companies reasonably satisfactory to the Lessor. Evidence of adequate insurance shall be delivered to Lessor within 10 days after execution of this Lease, and thereafter certificates of renewal policies shall be delivered to Lessor within 10 days prior to the expiration of the term of such policy. Any policies of insurance carried by the Lessee shall provide that as against Lessor, the Lessee and insurers shall waive any rights of subrogation, set off, counterclaim or any other deduction, whether by attachment or otherwise.

Notwithstanding Lessee's responsibility for insurance hereunder, Lessee shall defend, indemnify and hold the Lessor harmless from and against any and all loss or damage to the Equipment or liability through use of the Equipment during the term of this Lease. If the Equipment is destroyed through fire, flood, explosion, or any other cause, the Lessee will repay the Lessor the full replacement cost of the Equipment. Rent shall continue to accrue through the date Lessor receives payment in full of the replacement cost of the Equipment.

ARTICLE 13. TITLE. Title to the Equipment shall at all times remain vested in the Lessor. The Lessee agrees to keep the Equipment free and clear of any claims, liens or encumbrances. Lessee further agrees to use the Equipment in accordance with all applicable government regulations, ordinances or laws. The Lessee shall give the Lessor immediate notice in case any Equipment is levied upon or becomes subject to seizure.

ARTICLE 14. TAXES. Lessee agrees to pay all government taxes or other assessments against this Equipment except as provided on the first page or order form of this Lease.

ARTICLE 15. WAIVERS. No waivers of any part or article of this Lease shall be construed to be a waiver of any other part or article or be recognised unless it is in writing and signed by both parties.

ARTICLE 16. LIMITED LIABILITY: LESSEE AGREES LESSOR DOES NOT AND CANNOT GUARANTEE OR WARRANT THE SUCCESS OR FAILURE OF THE USE OF ANY EQUIPMENT LEASED HEREUNDER. BECAUSE LESSOR CANNOT GUARANTEE OR WARRANT THE OUTCOME FROM ANY USE OF EQUIPMENT LEASED HEREUNDER, LESSEE AGREES IT SHALL RENT THE EQUIPMENT FROM LESSOR ON THE BASIS THAT SUCH EQUIPMENT MAY BE WHOLLY INEFFECTIVE AT THE INTENDED PURPOSE FOR WHICH IT HAS BEEN LEASED. BASED ON THE FOREGOING, LESSOR DISCLAIMS ALL WARRANTIES, EXPRESS OR IMPLIED, INCLUDING ANY WARRANTIES OF MERCHANTABILITY OR FITNESS FOR A PARTICULAR PURPOSE. LESSEE FURTHER AGREES THAT LESSOR SHALL NOT BE LIABLE TO LESSEE, OR ANY OF ITS AGENTS, EMPLOYEES, CUSTOMERS OR CONTRACTORS FOR ANY LOSS OR INJURY ARISING OUT OF, IN WHOLE OR IN PART, THE EQUIPMENT LEASED HEREUNDER. NOTWITHSTANDING THE FOREGOING AND BASED UPON THE NEGOTIATED RENT FOR THE EQUIPMENT LEASED HEREUNDER, LESSOR'S MAXIMUM LIABILITY FOR ANY CLAIM BROUGHT AGAINST IT HEREUNDER SHALL BE THE LESSER OF: (I) THE AMOUNT OF RENT PAID BY LESSEE TO LESSOR FOR THE EQUIPMENT AT ISSUE, OR (II) ONE MONTH'S RENT FOR THE EQUIPMENT AT ISSUE. UNDER NO CIRCUMSTANCES SHALL LESSOR BE RESPONSIBLE FOR ANY BUSINESS INTERRUPTION DAMAGES INCURRED BY LESSEE OR ANY OTHER THIRD PARTY RELATING IN ANY MANNER TO THIS LEASE OR THE EQUIPMENT THAT IS THE SUBJECT OF THIS LEASE.

ARTICLE 17. INDEMNITY. Lessee agrees to indemnify, defend and hold harmless Lessor, its affiliates, employees, successors and assigns (all referred to as "Lessor") from and against any losses, damages, claims, fines, penalties and expenses (including reasonable attorneys fees) that arise out of or result from injuries or death to persons or damage to property in any way arising out of or caused or alleged to have been caused by services or Equipment provided by Lessor.

Index

D. N. P. Murthy and N. Jack, *Extended Warranties, Maintenance Service and Lease Contracts*, Springer Series in Reliability Engineering,
DOI: 10.1007/978-1-4471-6440-1, © Springer-Verlag London 2014

MIX
Papier aus verantwortungsvollen Quellen
Paper from responsible sources
FSC® C105338

If you have any concerns about our products,
you can contact us on
ProductSafety@springernature.com

In case Publisher is established outside the EU,
the EU authorized representative is:
Springer Nature Customer Service Center GmbH
Europaplatz 3, 69115 Heidelberg, Germany

Printed by Libri Plureos GmbH
in Hamburg, Germany